# ELEMENTARY MATRIX THEORY

# ELEMENTARY MATRIX THEORY

## Howard Eves

Professor Emeritus of Mathematics
University of Maine at Orono

Dover Publications, Inc.
New York

Published in Canada by General Publishing Company, Ltd.,
30 Lesmill Road, Don Mills, Toronto, Ontario.
Published in the United Kingdom by Constable and Company, Ltd., 10 Orange Street, London WC2H 7EG.

This Dover edition, first published in 1980, is an unabridged
and corrected republication of the work originally published in
1966 by Allyn and Bacon, Inc., Boston.

*International Standard Book Number: 0-486-63946-0*
*Library of Congress Catalog Card Number: 79-56333*

Manufactured in the United States of America
Dover Publications, Inc.
180 Varick Street
New York, N.Y. 10014

TO
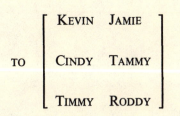

$$\begin{bmatrix} \text{KEVIN} & \text{JAMIE} \\ \text{CINDY} & \text{TAMMY} \\ \text{TIMMY} & \text{RODDY} \end{bmatrix}$$

# PREFACE

The present text is designed to serve as a one-semester introduction to the theory of matrices for the general student at the undergraduate level. Thus, in addition to mathematics majors, the author has had in mind the growing number of students from other fields of study (such as physics, engineering, the social sciences, and business) who find they must know something of matrix theory. It is also hoped that the text may be of use, in both summer institutes and in-service courses, to teachers of mathematics. Accordingly, the elementary, or concrete, approach is employed and no previous course in abstract algebra is required.

Matrices are studied over the field of complex numbers, with occasional reference to the field of real numbers, and then, in a final optional chapter of the book, the extension to arbitrary number rings and number fields and to abstract rings and abstract fields is indicated. In the final chapter also will be found an indication of the important study of matrices via linear algebra and vector spaces.

The theory of matrices is far more extensive than the uninitiated can possibly realize, and it has extended greatly beyond those areas of application that originally motivated its study. Therefore it becomes necessary, in a one-semester undergraduate text, to select material with some care. The aim of the present text has been to choose fundamental material that either can be used by students of various disciplines or is required as preliminary to any deeper study or application of the theory.

The first five chapters of the text can ordinarily be covered in a three-hour, one-semester college course, and the first three or four chapters can be used for shorter, more elementary, or more slow-moving courses. The text is arranged so that the material between any two successive problem lists can be dealt with in one class period, and a selection of the problems immediately following such a segment of material can then be assigned for homework. It should be noted that these problems are not all just routine calculation exercises, but that most of them involve proof and the extension of some of the concepts and material of the text. If it is desired, in a three-hour one-semester course, to cover all or part of the material of Chapter 6, then it will be necessary to omit a number of the later sections of Chapters 4 and 5.

Following each chapter (except the final Chapter 6) are addenda containing brief statements and/or suggestions of additional allied material. These are not part of the course proper. They are included as extra material for the interested, the superior, or the advanced student. They can be used for special assignments in the form of short papers or to introduce a student to "junior" research. Some institutions permit beginning graduate students to take, for graduate credit, certain essentially undergraduate courses. Should the matrix course be one of these, the addenda can be used to differentiate the work of the graduate from that of the undergraduate. Chapter 6 can also be used for this purpose.

In any case, the Prolegomenon should be discussed first (in a couple of class sessions). It serves as motivation for, and introduction to, the text proper; it sets the stage for the concrete approach to matrix theory and the consideration of matrices as hypercomplex numbers. The Epilegomenon is designed to direct the interested student to some possible lines of further work in the theory of matrices. Here also will be found a first Bibliography to other sources of material.

*Howard Eves*

*Stillwater, Maine*

# CONTENTS

PREFACE                                                                    vii

## 0. PROLEGOMENON

0.1  SIGNIFICANT ORDERED RECTANGULAR ARRAYS OF
     NUMBERS                                                                 1

PROBLEMS                                                                     5

0.2  ALGEBRAS OF ORDERED ARRAYS OF NUMBERS                                   7

PROBLEMS                                                                    12

## 1. FUNDAMENTAL CONCEPTS AND OPERATIONS

1.1  $S$ MATRICES, $\mathscr{R}$ MATRICES, AND $\mathscr{C}$ MATRICES      14

1.2  ADDITION AND SCALAR MULTIPLICATION OF
     MATRICES                                                               16

PROBLEMS                                                                    18

1.3  CAYLEY MULTIPLICATION OF MATRICES                                      19

PROBLEMS                                                                    23

1.4   SOME SPECIAL MATRICES                                                 25

PROBLEMS                                                                    29

1.5   TRANSPOSITION                                                         30

1.6   SYMMETRIC AND SKEW-SYMMETRIC MATRICES                                 31

PROBLEMS                                                                    33

1.7   CONJUGATION AND TRANJUGATION                                         33

1.8   HERMITIAN AND SKEW-HERMITIAN MATRICES                                35

PROBLEMS                                                                    37

1.9   PARTITIONED MATRICES                                                  37

PROBLEMS                                                                    41

# ADDENDA

1.1A   LINEAR TRANSFORMATIONS                                               43

1.2A   BILINEAR, QUADRATIC, AND HERMITIAN FORMS                             44

1.3A   A MATRIX APPROACH TO COMPLEX NUMBERS                                 46

1.4A   A BUSINESS APPLICATION                                               48

1.5A   ENUMERATION OF $k$-STAGE ROUTES                                      49

1.6A   APPLICATION TO MATHEMATICAL SYSTEMS                                  51

1.7A   JORDAN AND LIE PRODUCTS OF MATRICES                                  53

1.8A   SQUARE ROOTS OF MATRICES                                            56

1.9A   PRIMITIVE FACTORIZATION OF MATRICES                                  57

# 2. EQUIVALENCE

2.1   ROW EQUIVALENCE OF MATRICES                                          59

PROBLEMS                                                                    64

2.2   NONSINGULAR MATRICES                                                  65

PROBLEMS                                                                                        68

2.3   COLUMN EQUIVALENCE OF MATRICES                               70

2.4   EQUIVALENCE OF MATRICES                                                 72

PROBLEMS                                                                                        74

2.5   LINEAR DEPENDENCE AND INDEPENDENCE OF A
       SET OF VECTORS                                                                   75

2.6   ROW RANK AND COLUMN RANK OF A MATRIX               77

PROBLEMS                                                                                        79

2.7   RANK OF A MATRIX                                                                81

PROBLEMS                                                                                        84

2.8   APPLICATION TO THE SOLUTION OF SYSTEMS OF
       LINEAR EQUATIONS                                                               86

PROBLEMS                                                                                        90

2.9   LINEARLY INDEPENDENT SOLUTIONS OF SYSTEMS
       OF LINEAR EQUATIONS                                                        92

PROBLEMS                                                                                        97

# ADDENDA

2.1A   LEFT AND RIGHT INVERSES                                               98

2.2A   SOME FURTHER METHODS OF MATRIX INVERSION      99

2.3A   LINEAR DEPENDENCE AND INDEPENDENCE OF A
         SET OF MATRICES                                                             102

2.4A   LINES IN A PLANE AND PLANES IN SPACE                   103

2.5A   AN AFFINE CLASSIFICATION OF CONICS AND
         CONICOIDS ACCORDING TO THE RANKS OF
         THEIR ASSOCIATED MATRICES                                      104

2.6A   KRONECKER PRODUCT OF MATRICES                        107

2.7A   DIRECT SUM OF MATRICES                                            107

# 3. DETERMINANTS

3.1 PERMUTATIONS 109

3.2 THE NOTION OF DETERMINANT 111

PROBLEMS 112

3.3 SOME ELEMENTARY PROPERTIES OF DETERMINANTS 114

PROBLEMS 117

3.4 COFACTORS 119

PROBLEMS 123

3.5 CYCLIC DETERMINANTS AND VANDERMONDE DETERMINANTS 125

PROBLEMS 127

3.6 CHIO'S EXPANSION 129

APPENDIX 133

PROBLEMS 134

3.7 LAPLACE'S EXPANSION 136

PROBLEMS 139

3.8 THE PRODUCT THEOREM 140

PROBLEMS 145

3.9 DETERMINANT RANK OF A MATRIX 148

PROBLEMS 152

3.10 ADJOINT OF A SQUARE MATRIX 153

PROBLEMS 157

# ADDENDA

3.1A A GEOMETRIC STUDY OF PERMUTATIONS 160

3.2A PERMUTATION MATRICES 163

3.3A   PERMANENTS                                              163

3.4A   POSTULATIONAL DEFINITIONS OF DETERMINANT   165

3.5A   THE SWEEP-OUT PROCESS FOR EVALUATING
       DETERMINANTS                                      165

3.6A   PFAFFIANS                                               167

3.7A   SOLUTION OF SYSTEMS OF LINEAR EQUATIONS     169

3.8A   CONTINUANTS                                          172

3.9A   AN APPLICATION OF DETERMINANTS TO
       TRIANGLES AND TETRAHEDRA                       173

3.10A  QUANTITATIVE ASPECT OF LINEAR
       INDEPENDENCE OF VECTORS                        174

3.11A  SYLVESTER'S DIALYTIC METHOD OF
       ELIMINATION                                         178

## 4. MATRICES WITH POLYNOMIAL ELEMENTS

4.1   REVIEW OF SOME POLYNOMIAL THEORY            179

PROBLEMS                                                      183

4.2   LAMBDA MATRICES                                     184

PROBLEMS                                                      189

4.3   THE SMITH NORMAL FORM                           190

PROBLEMS                                                      194

4.4   INVARIANT FACTORS AND ELEMENTARY DIVISORS   195

PROBLEMS                                                      198

4.5   THE CHARACTERISTIC FUNCTION OF A SQUARE
      MATRIX                                                 199

PROBLEMS                                                      202

4.6   SOME RESULTS RELATED TO THE CHARACTERISTIC
      FUNCTION OF A SQUARE MATRIX                      204

PROBLEMS                                                          206

4.7   CHARACTERISTIC VECTORS OF A SQUARE MATRIX           207

PROBLEMS                                                          209

4.8   THE MINIMUM FUNCTION OF A SQUARE MATRIX            210

PROBLEMS                                                          213

4.9   FINDING THE MINIMUM FUNCTION OF A SQUARE
      MATRIX                                                      213

PROBLEMS                                                          216

## ADDENDA

4.1A   ELEMENTARY $\lambda$ MATRICES                             217

4.2A   SYSTEMS OF LINEAR DIFFERENTIAL EQUATIONS
       WITH CONSTANT COEFFICIENTS                                218

4.3A   EQUIVALENCE OF PAIRS OF MATRICES                          218

4.4A   $k$TH ROOTS OF NONSINGULAR MATRICES                       219

4.5A   THE COEFFICIENTS IN THE CHARACTERISTIC
       FUNCTION                                                  220

4.6A   COMPUTATION OF $A^{-1}$ BY THE HAMILTON-
       CAYLEY EQUATION                                           220

4.7A   FRAME'S RECURSION FORMULA FOR INVERTING
       A MATRIX                                                  221

4.8A   CHARACTERISTIC ROOTS OF A POLYNOMIAL
       FUNCTION OF A MATRIX $A$                                  221

## 5. SIMILARITY AND CONGRUENCE

5.1   SIMILAR MATRICES                                           222

PROBLEMS                                                          225

5.2   SIMILAR MATRICES (CONTINUED)                               226

PROBLEMS                                                            230

5.3   CONGRUENT MATRICES                                            230

PROBLEMS                                                            233

5.4   CANONICAL FORMS UNDER CONGRUENCY FOR
      SKEW-SYMMETRIC 𝒞 MATRICES                                     235

PROBLEMS                                                            237

5.5   CANONICAL FORMS UNDER CONGRUENCY FOR
      SYMMETRIC ℛ MATRICES                                          237

PROBLEMS                                                            241

5.6   CONJUNCTIVITY, OR HERMITIAN CONGRUENCE                        242

PROBLEMS                                                            245

5.7   ORTHOGONAL MATRICES AND ORTHOGONAL
      SIMILARITY                                                    246

PROBLEMS                                                            252

5.8   UNITARY MATRICES AND UNITARY SIMILARITY                       253

PROBLEMS                                                            258

5.9   NORMAL MATRICES                                               259

PROBLEMS                                                            262

# ADDENDA

5.1A   COMPANION MATRICES                                          263

5.2A   REGULAR SYMMETRIC MATRICES                                   264

5.3A   ROTATIONS IN 3-SPACE                                         264

5.4A   CAYLEY'S CONSTRUCTION OF REAL
       ORTHOGONAL MATRICES                                          265

5.5A   THE CHARACTERISTIC ROOTS OF AN
       ORTHOGONAL MATRIX                                            267

5.6A   DEFINITE, SEMIDEFINITE, AND INDEFINITE REAL
       SYMMETRIC MATRICES                                           269

5.7A   GRAM MATRICES                                          270

5.8A   SOME THEOREMS OF AUTONNE                               271

5.9A   SIMULTANEOUS REDUCTION OF A PAIR OF
       QUADRATIC FORMS                                        272

5.10A  HADAMARD MATRICES                                      273

5.11A  EQUITABLE MATRICES                                     274

# 6. TOWARD ABSTRACTION

6.1    NUMBER RINGS AND NUMBER FIELDS                         276

PROBLEMS                                                      279

6.2    GENERAL RINGS AND GENERAL FIELDS                       281

PROBLEMS                                                      284

6.3    MATRIX REALIZATION                                     286

PROBLEMS                                                      289

6.4    $k$-VECTOR SPACES OVER A FIELD $F$                     291

PROBLEMS                                                      293

6.5    GENERAL VECTOR SPACES                                  294

PROBLEMS                                                      297

6.6    LINEAR TRANSFORMATIONS OF VECTOR SPACES                297

PROBLEMS                                                      303

6.7    JORDAN AND LIE ALGEBRAS                                304

PROBLEMS                                                      305

# 7. EPILEGOMENON

                                                              307

# BIBLIOGRAPHY

                                                              310

# INDEX

                                                              315

# ELEMENTARY
# MATRIX THEORY

# 0. PROLEGOMENON

## 0.1 Significant ordered rectangular arrays of numbers

In mathematics and its applications, numbers frequently appear, or can be made to appear, in significant ordered rectangular arrays. Following are a few instances with which the reader is perhaps already familiar.

*Example* 1.   The complex number $a + ib$, where $a$ and $b$ are real and $i$ is the imaginary unit, is completely determined by the $1 \times 2$ array, or ordered pair, $(a, b)$ of real numbers. It follows that, associated with the algebra of complex numbers, there is a certain algebra of ordered pairs of real numbers. We shall comment further on this in Illustration 1, Section 0.2.

*Example* 2.   The Cartesian coordinates of a point in three-dimensional space constitute an ordered triple $(x, y, z)$ of real numbers $x$, $y$, and $z$.

*Example* 3.   Cartesian direction numbers of a line in three-dimensional space are given as an ordered triple $(l, m, n)$ of real numbers $l$, $m$, $n$, where

$$l^2 + m^2 + n^2 \neq 0.$$

It follows from Examples 2 and 3 that much of Cartesian analysis can perhaps be neatly studied by means of a suitable algebra of ordered triples of real numbers. Part of this algebra is briefly considered in Illustration 2, Section 0.2.

*Example* 4.   The conic section with Cartesian equation

$$ax^2 + 2hxy + by^2 + 2gx + 2fy + c = 0$$

is completely determined by the symmetric $3 \times 3$ array of numbers

$$\begin{bmatrix} a & h & g \\ h & b & f \\ g & f & c \end{bmatrix}.$$

It follows that properties of the conic section can be expressed in terms of properties of its associated array. For example, if we let $D$ denote the determinant of the array and let $A$ denote the subdeterminant

$$\begin{vmatrix} a & h \\ h & b \end{vmatrix},$$

it is shown in analytic geometry texts that the conic is: (1) an ellipse if and only if $D \neq 0$ and $A > 0$, (2) a hyperbola if and only if $D \neq 0$ and $A < 0$, (3) a parabola if and only if $D \neq 0$ and $A = 0$, (4) a pair of intersecting straight lines if and only if $D = 0$ and $A \neq 0$, (5) a pair of parallel or coincident straight lines if and only if $D = 0$ and $A = 0$.

*Example* 5.   The linear transformation

$$u = ax + by + cz,$$
$$v = dx + ey + fz,$$

which maps points $(x, y, z)$ of three-space onto points $(u, v)$ of the plane, is completely determined by the $2 \times 3$ array of numbers

$$\begin{bmatrix} a & b & c \\ d & e & f \end{bmatrix}.$$

It follows that linear transformations can be studied by means of an appropriate algebra of rectangular arrays of numbers. Illustration 3, Section 0.2, indicates the beginnings of such an algebra.

*Example* 6.   The system of linear equations

$$a_{11}x_1 + a_{12}x_2 + a_{13}x_3 = c_1$$
$$a_{21}x_1 + a_{22}x_2 + a_{23}x_3 = c_2$$
$$a_{31}x_1 + a_{32}x_2 + a_{33}x_3 = c_3$$

is completely determined by the 3 × 4 array of coefficient numbers

$$\begin{bmatrix} a_{11} & a_{12} & a_{13} & c_1 \\ a_{21} & a_{22} & a_{23} & c_2 \\ a_{31} & a_{32} & a_{33} & c_3 \end{bmatrix}.$$

It follows that information concerning the number of solutions of the system of linear equations must be expressible in terms of properties of the associated array. We shall actually look into this matter later in the book.

*Example* 7.   The bilinear form

$$a_{11}x_1y_1 + a_{12}x_1y_2 + a_{13}x_1y_3$$

$$+ a_{21}x_2y_1 + a_{22}x_2y_2 + a_{23}x_2y_3$$

$$+ a_{31}x_3y_1 + a_{32}x_3y_2 + a_{33}x_3y_3$$

is completely determined by the 3 × 3 array of numbers

$$\begin{bmatrix} a_{11} & a_{12} & a_{13} \\ a_{21} & a_{22} & a_{23} \\ a_{31} & a_{32} & a_{33} \end{bmatrix}.$$

*Example* 8.   The quadratic form

$$ax^2 + by^2 + cz^2 + 2fyz + 2gzx + 2hxy$$

is completely determined by the symmetric 3 × 3 array of numbers

$$\begin{bmatrix} a & h & g \\ h & b & f \\ g & f & c \end{bmatrix}.$$

*Example* 9.   Statistical information can often be neatly organized in the form of a rectangular array of numbers. Suppose, for example, that a building contractor constructs ranch style, Cape Cod style, and Colonial style homes. Suppose, further, that 50, 70, and 60 units of steel are needed in the construction of the three styles of homes, respectively; similarly, 200, 180, and 250 units of wood; 160, 120, and 80 units of glass; 70, 90, and 50 units of paint; and 170, 210, and 130 units of labor, respectively. This information can be presented by the following 3 × 5 array of numbers:

$$\begin{array}{r} \\ \text{ranch} \\ \text{Cape Cod} \\ \text{Colonial} \end{array} \begin{array}{ccccc} \text{steel} & \text{wood} & \text{glass} & \text{paint} & \text{labor} \\ \begin{bmatrix} 50 & 200 & 160 & 70 & 170 \\ 70 & 180 & 120 & 90 & 210 \\ 60 & 250 & 80 & 50 & 130 \end{bmatrix} \end{array}.$$

*Example* 10.    Consider the graphical network pictured in Figure 1. The incidence relations of the vertices and edges of this network are described by the 4 × 6 array

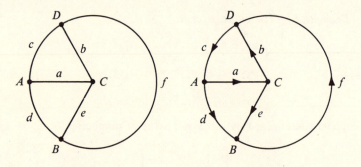

FIG. 1                        FIG. 2

$$\begin{array}{c} \\ A \\ B \\ C \\ D \end{array} \begin{array}{cccccc} a & b & c & d & e & f \\ \begin{bmatrix} 1 & 0 & 1 & 1 & 0 & 0 \\ 0 & 0 & 0 & 1 & 1 & 1 \\ 1 & 1 & 0 & 0 & 1 & 0 \\ 0 & 1 & 1 & 0 & 0 & 1 \end{bmatrix} \end{array},$$

wherein a 1 in the intersection of a given row and a given column indicates that the vertex heading the row is an end point of the edge heading the column.

If the edges of the graph are directed as pictured in Figure 2, the network can be described by the 4 × 6 array

$$\begin{array}{c} \\ A \\ B \\ C \\ D \end{array} \begin{array}{cccccc} a & b & c & d & e & f \\ \begin{bmatrix} 1 & 0 & -1 & 1 & 0 & 0 \\ 0 & 0 & 0 & -1 & -1 & 1 \\ -1 & 1 & 0 & 0 & 1 & 0 \\ 0 & -1 & 1 & 0 & 0 & -1 \end{bmatrix} \end{array},$$

wherein a plus 1 or a minus 1 indicates that the concerned vertex is the initial or terminal point, respectively, of the concerned directed edge.

The first array in this example is called the *undirected incidence matrix* of its associated network, and the second array is called the *directed incidence matrix* of its associated network. If $m$ is a loop of a directed network, so that $m$ originates and terminates at a vertex $V$, then the symbol $\pm 1$ is placed in the directed incidence matrix at the intersection of the row headed by $V$ and the column headed by $m$.

## PROBLEMS

**0.1–1** Construct both the undirected and the directed incidence matrices for the network pictured in Figure 3.

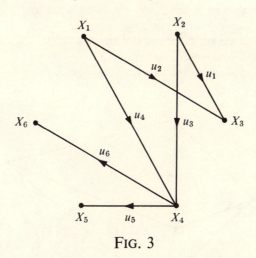

FIG. 3

**0.1–2** **(a)** Suppose six football teams $A$, $B$, $C$, $D$, $E$, $F$ belong to a league and suppose that by midseason: $A$ has played $C$ and $D$; $B$ has played $C$ and $D$; $C$ has played $A$ and $B$; $D$ has played $A$, $B$, $E$, and $F$; $E$ has played $D$; $F$ has played $D$. Design a square array of 1's and 0's to describe this information.

**(b)** How might you alter the array (by changing some of the 1's to $-1$'s) to include also the following information? $A$ won game $A$-$C$; $A$ won game $A$-$D$; $B$ won game $B$-$C$; $B$ won game $B$-$D$; $D$ won game $D$-$E$; $D$ won game $D$-$F$.

(c) Suppose the games were played in the order *B-C*, *A-C*, *B-D*, *A-D*, *D-E*, *D-F*. Construct the incident matrices of the network whose vertices are *A*, *B*, *C*, *D*, *E*, *F* and whose directed edges are *BC*, *AC*, *BD*, *AD*, *DE*, *DF*, and compare these arrays with the incidence matrices of Problem 0.1–1.

**0.1–3**  Suppose a manufacturer produces four distinct models, *A*, *B*, *C*, *D*, of a certain item. Each model contains a certain number of three possible different subassemblies *a*, *b*, *c*, and each subassembly is made up of a certain number of five possible different parts $\alpha$, $\beta$, $\gamma$, $\delta$, $\epsilon$. Suppose model *A* contains one *a*, one *b*, one *c*; model *B* contains one *a*, one *b*, two *c*'s; model *C* contains two *a*'s, one *b*, two *c*'s; model *D* contains two *a*'s, three *b*'s, four *c*'s. Further, suppose that each subassembly *a* contains eight $\alpha$'s, three $\beta$'s, four $\gamma$'s, zero $\delta$'s, one $\epsilon$; each subassembly *b* contains four $\alpha$'s, six $\beta$'s, three $\gamma$'s, zero $\delta$'s, zero $\epsilon$'s; each subassembly *c* contains zero $\alpha$'s, one $\beta$, one $\gamma$, two $\delta$'s, two $\epsilon$'s. Set up

(a) a 4 × 3 *models-subassemblies array*,
(b) a 3 × 5 *subassemblies-parts array*,
(c) a 4 × 5 *models-parts array*.

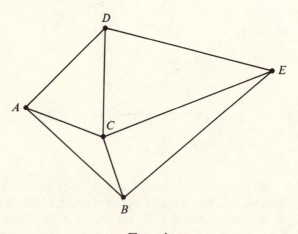

FIG. 4

**0.1–4**  Consider five cities, *A*, *B*, *C*, *D*, *E*, certain pairs of which are connected by two-way bus routes, as indicated in Figure 4. Construct symmetric 5 × 5 arrays to describe: (a) the number of one-stage routes from each city to each city, (**b**) the number of two-stage routes from each city to each city, (**c**) the number of three-stage routes from each city to each city.

## 0.2   Algebras of ordered arrays of numbers

We have seen that many examples can be furnished in which mathematical entities or situations can be described by suitable ordered rectangular arrays of numbers. In a discussion about a given set of such entities or situations, the arrays representing a pair of entities or situations frequently combine in some specific manner to yield the array of a certain related entity or situation. Let us clarify this last remark with three illustrations.

*Illustration* 1.   Consider the representation of complex numbers as ordered pairs of real numbers, as described above in Example 1 of Section 0.1. Associated with the two complex numbers $a + ib$ and $c + id$ are the complex numbers

$$(a + c) + i(b + d) \qquad \text{and} \qquad (ac - bd) + i(bc + ad)$$

representing their sum and product, respectively. It is natural, then, to define the ordered pairs

$$(a + c, b + d) \qquad \text{and} \qquad (ac - bd, bc + ad)$$

to be the *sum* and the *product*, respectively, of the two ordered pairs $(a, b)$ and $(c, d)$. One can now easily show, directly from the preceding definitions, that addition of ordered pairs of real numbers is commutative and associative, and that multiplication of ordered pairs of real numbers is commutative, associative, and distributive over addition. It is convenient further to define $k(a, b)$, where $k$ is a real number, to be the ordered pair $(ka, kb)$. In this way we begin to build up a certain algebra of ordered pairs of real numbers, and, if these ordered pairs of real numbers are treated as single entities represented by single letters, one has the following algebraic laws: If $u$, $v$, $w$ represent any three of our ordered pairs of real numbers, and if $m$ and $n$ denote any two real numbers, then

(1) $u + v = v + u$,
(2) $u + (v + w) = (u + v) + w$,
(3) $m(u + v) = (mu) + (mv)$,
(4) $(m + n)u = (mu) + (nu)$,

(5) $u \times v = v \times u$,

(6) $u \times (v \times w) = (u \times v) \times w$,

(7) $u \times (v + w) = (u \times v) + (u \times w)$,

(8) $m(u \times v) = (mu) \times v$.

*Illustration* 2.   Consider the geometrical study of directed line segments radiating from a given point in three-dimensional space. Superimposing a three-dimensional rectangular Cartesian frame of reference with origin at the given point, any such directed line segment can be characterized by the ordered triple of real numbers constituting the Cartesian coordinates of the terminal point of the segment. If $a = (a_1, a_2, a_3)$ and $b = (b_1, b_2, b_3)$ are such number triples for two given noncollinear segments $\alpha$ and $\beta$ radiating from the origin, it can be shown that the number triple

$$c = (a_2 b_3 - a_3 b_2, a_3 b_1 - a_1 b_3, a_1 b_2 - a_2 b_1)$$

represents a directed line segment $\gamma$, which is perpendicular to both $\alpha$ and $\beta$. The number triple $c$ is called the *vector product* of the number triples $a$ and $b$, taken in this order, and we write $c = a \times b$. Defining $a + b$ to be the triple

$$(a_1 + b_1, a_2 + b_2, a_3 + b_3),$$

and defining $ka$, where $k$ is a real number, to be the triple

$$(ka_1, ka_2, ka_3),$$

one can easily show that the following algebraic laws hold: If $u$, $v$, $w$ represent any three of our ordered triples of real numbers, and if $m$ and $n$ denote any two real numbers, then

(1) $u + v = v + u$,

(2) $u + (v + w) = (u + v) + w$,

(3) $m(u + v) = (mu) + (mv)$,

(4) $(m + n)u = (mu) + (nu)$,

(5) $u \times v = -(v \times u)$,

(6) in general $u \times (v \times w) \neq (u \times v) \times w$,

(7) $u \times (v + w) = (u \times v) + (u \times w)$,

(8) $m(u \times v) = (mu) \times v$.

We have here the beginning of a significant algebra of ordered triples of real numbers, which plays an important role in physics and in geometry. It is to be noted that this algebra differs, in some of its laws, from the algebra of Illustration 1.

*Illustration* 3.    Consider a study of linear transformations of the type

$$x' = ax + by,$$
$$y' = cx + dy,$$

where $a$, $b$, $c$, $d$ are real numbers. The foregoing transformation may be thought of as mapping the point $(x, y)$ of the Cartesian plane onto the point $(x', y')$ of the same plane. Clearly, the transformation is completely determined by the four coefficients $a$, $b$, $c$, $d$, and may therefore be represented by the square array

$$\begin{bmatrix} a & b \\ c & d \end{bmatrix}.$$

If the linear transformation given above is followed by the linear transformation

$$x'' = ex' + fy',$$
$$y'' = gx' + hy',$$

the result can be shown by elementary algebra to be the linear transformation

$$x'' = (ea + fc)x + (eb + fd)y,$$
$$y'' = (ga + hc)x + (gb + hd)y.$$

Since the transformation resulting from one transformation followed by another is called the *product* of the two original transformations, one is motivated, in our considered study of linear transformations, to formulate the following definition of the product of the 2 × 2 arrays of numbers:

$$\begin{bmatrix} e & f \\ g & h \end{bmatrix} \begin{bmatrix} a & b \\ c & d \end{bmatrix} = \begin{bmatrix} ea + fc & eb + fd \\ ga + hc & gb + hd \end{bmatrix}.$$

This multiplication of 2 × 2 arrays of numbers can be shown to be associative, but simple numerical examples will show that it is not, in general, commutative.

An important point brought out by our three illustrations of algebras of certain ordered rectangular arrays of numbers is that there is no single way to define, say, the "product" of two such arrays. Any adopted definition is motivated by the application of the arrays that one has in mind. Thus there exist in the literature a number of different ways of "multiplying" two rectangular arrays of numbers to secure a third rectangular array of numbers, and we shall encounter some of these ways in later parts of the book. Interestingly enough, however, there is one "product" of pairs of rectangular arrays that stands out more prominently than the others, simply because it happens to have a surprising number of important applications and because some of the other "products" can be neatly defined in terms of this fundamental one. The "product" we are referring to is an extension of that given in Illustration 3 above, and is historically one of the oldest "products" of rectangular arrays of numbers to have been considered.

Algebras of ordered arrays of numbers can perhaps be said to have originated with the Irish mathematician and physicist Sir William Rowan Hamilton (1805–1865) when, in 1837 and 1843, he devised his treatment of complex numbers as ordered pairs of real numbers (see Illustration 1) and his real quaternion algebra as an algebra of ordered quadruples of real numbers. More general algebras of ordered $n$-tuples of real numbers were considered by the German mathematician Hermann Günther Grassmann (1809–1877) in 1844 and 1862, when he published the first and second editions of his remarkable treatise on space analysis, entitled *Die Ausdehnungslehre* (The Calculus of Extension). A highly useful algebra of ordered triples of real numbers (hinted at in Illustration 2) arises from a coordinate treatment of the vector analysis of the American physicist Josiah Willard Gibbs (1838–1903), first described by Gibbs in a small pamphlet privately distributed among his students in 1881–1884. The consideration of $n \times n$ square arrays of numbers, where $n > 1$, originated in the theory of determinants, and the latter had its origin in the theory of systems of linear equations.

The name *matrix* was first assigned to a rectangular array of numbers by James Joseph Sylvester (1814–1897) in 1850. The Scotch-American matrix scholar, J. H. M. Wedderburn (1882–1948), considered Sir William Rowan Hamilton's paper, "Linear and vector functions," of 1853, to contain the beginnings of a *theory* of matrices. But it was Arthur Cayley (1821–1895) who, in his paper, "A memoir on the theory of matrices," of 1858, first considered general $m \times n$ matrices as single entities subject to certain laws of combination. Cayley's approach was motivated by a study of linear transformations similar to that considered in our Illustration 3 above.

Since Cayley's time, the theory of matrices has expanded prodigiously and has found applications in many, many areas. For example, it was in 1925 that Heisenberg recognized matrix theory as exactly the tool he needed to develop quantum mechanics. Some of the most striking applications of matrix theory are found in modern atomic physics, which has given rise to such terms as the *scattering matrix, spin matrices, annihilation matrices*, and *creation matrices*. Relativity theory employs matrix concepts and theory. The applications to mechanics, which involve such matters as angular velocity and acceleration, moving axes, principal axes of inertia, kinetic energy, and oscillation theory, have led to the subject of *matrix mechanics*. Large parts of electromagnetic theory are most neatly and compactly handled by matrix methods, and matrix theory is a prerequisite for many modern treatments of circuit analysis and synthesis. In engineering appear *stress* and *strain matrices*.

Matrix theory has become indispensable in modern statistical studies, and many important papers developing matrix theory have been written by statisticians. In statistics we find *data matrices, correlation matrices, covariance matrices*, and *stochastic matrices*. Matrix algebra is the chief mathematical tool used in the multiple factor analysis of psychometrics. In fact, the social sciences have recently found matrix concepts and procedures of enormous value, and are responsible for many terms in matrix work, such as *communication matrices* and *dominance matrices*. Modern aerodynamics bristles with matrix theory and matrix computation. Indeed, matrices play a major role in the attack of many numerical problems using high-speed computing machinery, and such matters as the computation of *eigenvectors, eigenvalues*, and *inverses* of matrices are standard procedure at computing centers. The study and solution of systems of linear differential equations has been greatly compactified by the employment of matrix theory. Matrix methods in geometry have been extensively cultivated; indeed, much of matrix theory originated in geometrical applications, such as linear transformation theory or vector space theory and the analytical study of conics and conicoids.

Students of function theory become familiar with the so-called *Jacobian matrix* and *Hessian matrix*. In advanced function theory one encounters *infinite matrices* and such important special matrices as *Mittag-Leffler matrices, Borel matrices, Lindelöf matrices, Argand matrices, Bessel function matrices, Nörlund matrices, Euler-Knopp matrices, Hausdorff matrices, Kojima matrices, Toeplitz matrices, Le Roy matrices, Raff matrices*, $\alpha$, $\beta$, and $\gamma$ *matrices*. Many parts of classical algebra, such as the study of quadratic, bilinear, and Hermitian forms, not only are more neatly carried out via matrix theory, but also have actually contributed much to the development of the theory. Modern algebraists have found that nearly every abstract algebraic system can be given a concrete *matrix representation*. Matrices have invaded the business world, and such subjects as linear programming utilize matrix notation and procedures; in economics one encounters, among others, the *input-output matrix*. In game theory there is the *payoff matrix*. In graph, or network, theory there are the *incidence matrices* and the *cyclomatic matrix*. Matrix theory, as a useful tool, certainly needs no justification.

## PROBLEMS

**0.2–1**    Let $X$ and $Y$ be vectors radiating from a fixed point in three-dimensional space and interpret these vectors as forces acting on a particle located at the fixed point. If $X$ and $Y$ are represented by the coordinates $(x_1, x_2, x_3)$ and $(y_1, y_2, y_3)$ of their end points, referred to some superimposed rectangular Cartesian frame having its origin at the fixed point, how should one define $(x_1, x_2, x_3) + (y_1, y_2, y_3)$ and $k(x_1, x_2, x_3)$, where $k$ is a real number?

**0.2.–2**    It is shown in analytic geometry that any conic passing through the four points of intersection of two conics given by the Cartesian equations

$$f_1(x, y) \equiv a_1 x^2 + 2h_1 xy + b_1 y^2 + 2g_1 x + 2f_1 y + c_1 = 0$$

and

$$f_2(x, y) \equiv a_2 x^2 + 2h_2 xy + b_2 y^2 + 2g_2 x + 2f_2 y + c_2 = 0$$

is given by an equation of the form $mf_1(x, y) + nf_2(x, y) = 0$, where $m$ and $n$ are appropriate real numbers. If we represent a conic

$$ax^2 + 2hxy + by^2 + 2gx + 2fy + c = 0$$

by the symmetric $3 \times 3$ array

$$\begin{bmatrix} a & h & g \\ h & b & f \\ g & f & c \end{bmatrix},$$

how should one define: (**a**) the product of such an array and a real number, (**b**) the sum of two such arrays?

**0.2–3**    The product of a real number and a real bilinear form in $x$ and $y$ is a real bilinear form in $x$ and $y$; also, the sum of two real bilinear forms in $x$ and $y$ is a real bilinear form in $x$ and $y$. If a real bilinear form in $x$ and $y$ is represented by a $3 \times 3$ array of real numbers, as shown in Example 6 of Section 0.1, how should one define: (**a**) the product of such an array and a real number, (**b**) the sum of two such arrays?

**0.2–4**  Defining the sum and product of ordered pairs of real numbers by

$$(a, b) + (c, d) = (a + c, b + d),$$
$$(a, b)(c, d) = (ac - bd, bc + ad),$$

and further defining

$$k(a, b) = (ka, kb),$$

where $k$ is a real number, establish the eight algebraic laws listed in Illustration 1.

**0.2–5**  Defining the sum and product of ordered triples of real numbers by

$$(a_1, a_2, a_3) + (b_1, b_2, b_3) = (a_1 + b_1, a_2 + b_2, a_3 + b_3),$$
$$(a_1, a_2, a_3)(b_1, b_2, b_3) = (a_2 b_3 - a_3 b_2, a_3 b_1 - a_1 b_3, a_1 b_2 - a_2 b_1),$$

and further defining

$$k(a_1, a_2, a_3) = (ka_1, ka_2, ka_3),$$

where $k$ is a real number, establish the eight algebraic laws listed in Illustration 2.

**0.2–6**  Verify the details of Illustration 3, showing in particular that the product there defined of $2 \times 2$ arrays of numbers is associative but not, in general, commutative.

# 1. FUNDAMENTAL CONCEPTS AND OPERATIONS

*1.1. S matrices, $\mathscr{R}$ matrices, and $\mathscr{C}$ matrices. 1.2 Addition and scalar multiplication of matrices. Problems. 1.3 Cayley multiplication of matrices. Problems. 1.4 Some special matrices. Problems. 1.5 Transposition. 1.6 Symmetric and skew-symmetric matrices. Problems. 1.7 Conjugation and tranjugation. 1.8 Hermitian and skew-Hermitian matrices. Problems. 1.9 Partitioned matrices. Problems. ADDENDA. 1.1A. Linear transformations. 1.2A. Bilinear, quadratic, and Hermitian forms. 1.3A. A matrix approach to complex numbers. 1.4A. A business application. 1.5A. Enumeration of k-stage routes. 1.6A. Application to mathematical systems. 1.7A. Jordan and Lie products of matrices. 1.8A. Square roots of matrices. 1.9A. Primitive factorization of matrices.*

## 1.1 $S$ matrices, $\mathscr{R}$ matrices, and $\mathscr{C}$ matrices

In this brief initial section we introduce some basic terminology and notation. Of the various notations that have been adopted for displaying a matrix, we choose square brackets. Like much notation in mathematics, this choice is made to accommodate the compositor.

**1.1.1** DEFINITIONS. A rectangular array

$$A = \begin{bmatrix} a_{11} & a_{12} & \cdots & a_{1n} \\ a_{21} & a_{22} & \cdots & a_{2n} \\ \cdots & \cdots & \cdots & \cdots \\ a_{m1} & a_{m2} & \cdots & a_{mn} \end{bmatrix}$$

of $mn$ elements, chosen from a set $S$ of elements and arranged in $m$ rows and $n$ columns, as illustrated, is called an $m \times n$ (read "*m by n*") *S matrix*, or an *$m \times n$ matrix over the set $S$*, or simply, if the underlying set $S$ need not be stressed, an *$m \times n$ matrix*. The elements of the set $S$ are called *scalars*, and the scalars that make up an $S$ matrix $A$ are called the *elements* of $A$.

An $m \times n$ matrix is said to be of *order* $(m, n)$. A $1 \times n$ (that is, one row) matrix is called a *row vector of order $n$*; an $m \times 1$ (that is, one column) matrix is called a *column vector of order $m$*. If $m = n$, the matrix is called a *square matrix of order $n$*. If matrix $A$ is square of order $n$, the elements $a_{11}, a_{22}, \cdots, a_{nn}$ are said to constitute the (*principal*) *diagonal* of $A$.

**1.1.2** NOTATIONS. Matrices will commonly be denoted by capital letters, as $A$, in which case the element of the matrix in the intersection of the $i$th row, numbered from the top, and the $j$th column, numbered from the left, will be denoted by the corresponding small letter with the subscript $ij$ attached, as $a_{ij}$. In case of possible confusion, as when $i = 12$ and $j = 24$, we write $a_{12,24}$, with a comma between the values of $i$ and $j$. Sometimes the matrix $A$ of order $(m, n)$ will be denoted by $[a_{ij}]_{(m,n)}$, or, if the order is clear and need not be specified, simply by $[a_{ij}]$. If the matrix $A$ is square of order $n$, the former notation will be simplified to $[a_{ij}]_{(n)}$. Two alternative notations used by other authors for an $m \times n$ matrix $A$ are $(a_{ij})_{(m,n)}$ and $\|a_{ij}\|_{(m,n)}$. A column vector

$$\begin{bmatrix} u_1 \\ u_2 \\ \cdot \\ \cdot \\ \cdot \\ u_m \end{bmatrix}$$

will frequently, for convenience and space-saving purposes, be written horizontally as

$$\{u_1, u_2, \cdots, u_m\}.$$

Finally, we shall on occasion denote the $i$th row of matrix $A$ by $A_i$ and the $j$th column by $A'_j$.


**1.1.3** REMARKS. (1) Some authors, with good logic, make a distinction (which in our work we ignore) between an $m \times n$ rectangular array of $mn$ elements and an $m \times n$ matrix of $mn$ elements. The distinction is that a matrix is a rectangular array, and more; it is a rectangular array that is a member of a system of rectangular arrays in which operations of addition and multiplication are defined in certain definite ways. These definitions are formulated in the next two sections, and there it will be seen that addition and multiplication of matrices require an ability to add and multiply the elements of the matrices. It follows, then, that the elements of the set $S$ over which the matrices are taken must themselves be members of some system in which operations of addition and multiplication are defined.

(2) Very important in applications of matrices are matrices taken over the set $\mathscr{C}$ of complex numbers or over the set $\mathscr{R}$ of real numbers. Such matrices will be called *complex matrices* and *real matrices*, respectively, and will, when necessary, be denoted as $\mathscr{C}$ *matrices* and $\mathscr{R}$ *matrices*. From this point on, when the word "matrix" is used with no qualifying adjective or letter, it will be understood that we mean "complex matrix" or "$\mathscr{C}$ matrix." Clearly, an $\mathscr{R}$ matrix is a special kind of $\mathscr{C}$ matrix.

(3) Matrices over the sets of systems more general than the system of complex numbers or the system of real numbers will be considered in a later chapter, where it will be found that most of our theorems (with their proofs) and most of our formulated definitions continue to hold for these more general situations.


## 1.2  Addition and scalar multiplication of matrices

In this section we introduce the fundamental operations on matrices, called *addition* and *scalar multiplication*. We commence with a definition of equality of two matrices.


**1.2.1** DEFINITION. Two matrices $A = [a_{ij}]$ and $B = [b_{ij}]$ are said to be *equal*, and we write $A = B$, if and only if $A$ and $B$ have the same order and corresponding elements of $A$ and $B$ are equal.

Since equality of matrices is equality in the sense of identity, the following theorem is obviously true.

**1.2.2** THEOREM. *Equality of matrices is determinative, reflexive, symmetric, and transitive. That is,*

(1) *if* A *and* B *are any two matrices, either* A = B *or* A ≠ B (the *determinative* property);

(2) *if* A *is any matrix, then* A = A (the *reflexive* property);

(3) *if* A = B, *then* B = A (the *symmetric* property);

(4) *if* A = B *and* B = C, *then* A = C (the *transitive* property).

**1.2.3** DEFINITION. By the *sum*, $A + B$, of two $m \times n$ matrices $A = [a_{ij}]$ and $B = [b_{ij}]$ is meant the $m \times n$ matrix $C = [c_{ij}]$, where $c_{ij} = a_{ij} + b_{ij}$.

*Example*
$$\begin{bmatrix} 1 & -1 & 2 \\ 3 & 0 & 1 \end{bmatrix} + \begin{bmatrix} 2 & 2 & 2 \\ 1 & 0 & -1 \end{bmatrix} = \begin{bmatrix} 3 & 1 & 4 \\ 4 & 0 & 0 \end{bmatrix}.$$

**1.2.4** DEFINITION AND NOTATION. A matrix each of whose elements is 0 is called a *zero matrix*, and will be denoted by $O$, or by $O_{(n)}$ or $O_{(m,n)}$ if the order needs to be emphasized.

*Example*
$$O_{(2,3)} = \begin{bmatrix} 0 & 0 & 0 \\ 0 & 0 & 0 \end{bmatrix}.$$

**1.2.5** DEFINITION. By the *scalar product*, $kA$, of an $m \times n$ matrix $A = [a_{ij}]$ and a scalar $k$ is meant the $m \times n$ matrix $C = [c_{ij}]$, where $c_{ij} = ka_{ij}$.

*Example*
$$2\begin{bmatrix} 1 & -1 & 2 \\ 3 & 0 & 1 \end{bmatrix} = \begin{bmatrix} 2 & -2 & 4 \\ 6 & 0 & 2 \end{bmatrix}.$$

**1.2.6** NOTATION. We agree to write $(-1)A$ simply as $-A$, and $A + (-B)$, if $A$ and $B$ are of the same order, simply as $A - B$.

On the basis of Definitions 1.2.1, 1.2.3, 1.2.4, and 1.2.5, the reader will find no trouble in establishing all parts of the following theorem.

**1.2.7**   THEOREM.   *If* A, B, C *are* m × n *matrices and if* u *and* v *are scalars, then*

(1)  A + B = B + A (the *commutative law* of addition);
(2)  A + (B + C) = (A + B) + C (the *associative law* of addition)
(3)  u(A + B) = uA + uB   )(the *distributive laws* of scalar
(4)  (u + v)A = uA + vA   ∫  multiplication);
(5)  u(vA) = (uv)A (the *associative law* of scalar multiplication);
(6)  −A = $[-a_{ij}]$;
(7)  A − A = $O_{(m,n)}$;
(8)  A + $O_{(m,n)}$ = A;
(9)  A + C = B *if and only if* C = B − A;
(10) uA = O *if and only if* u = 0 *or* A = $O_{(m,n)}$,
(11) uA = uB *and* u ≠ 0 *imply* A = B.

## PROBLEMS

**1.2.1**   Establish all parts of Theorem 1.2.7.

**1.2–2**   If $A$, $B$, $C$ are $m \times n$ matrices, show that $A + C = B + C$ if and only if $A = B$. (This is called the *cancellation law for addition* of matrices.)

**1.2–3**   If $A$ and $B$ are $m \times n$ matrices, show that the matrix equation $X + A = B$ has the unique solution $X = B - A$.

**1.2–4**   Solve the matrix equation

$$\begin{bmatrix} i & 0 \\ 1 & -i \end{bmatrix} + X = \begin{bmatrix} i & 2 \\ 3 & 4 + i \end{bmatrix} - X.$$

**1.2–5**   If $A$, $B$, $C$ are $m \times n$ matrices, show that $(A + B) - C = A + (B - C)$.

**1.2–6**   The sum of the diagonal elements of a square matrix $A$ is called the *trace* (or *spur*) of $A$, and is denoted by tr $A$. If $A$ and $B$ are $n$th-order matrices and $k$ is a scalar, show that **(a)** tr $(A + B) = $ tr $A + $ tr $B$. **(b)** tr $(kA) = k$ tr $A$.

**1.2–7**   Determine whether the "addition" operations ⊕, defined for $m \times n$ matrices as follows, obey the commutative and associative laws:

**(a)**   $A \oplus B = C$, where $c_{ij} = a_{ij} + 2b_{ij}$;
**(b)**   $A \oplus B = C$, where $c_{ij} = a_{ij} + (b_{ij})^2$;
**(c)**   $A \oplus B = C$, where $c_{ij} = (a_{ij})^2 + (b_{ij})^2$;
**(d)**   $A \oplus B = C$, where $c_{ij} = a_{ij}b_{ij}$;
**(e)**   $A \oplus B = C$, where $c_{ij} = 2(a_{ij} + b_{ij})$.

**1.2–8** Let $S$ be a set of elements for which an addition of the elements is defined which is both commutative and associative. Defining the sum of two $m \times n$ $S$ matrices as in Definition 1.2.3, show that this addition is both commutative and associative.

**1.2–9** Let $S$ be a set of elements for which a multiplication of the elements is defined which is both commutative and associative. Defining the scalar product $kA$ of an $S$ matrix $A$ and a scalar $k$ of $S$ as in Definition 1.2.5, establish the associative law of scalar multiplication.

**1.2–10** Let $S$ be a set of elements for which both an addition and a multiplication of the elements is defined which are such that: (1) addition is commutative and associative, (2) multiplication is commutative and associative, (3) multiplication is distributive over addition. Defining addition of $S$ matrices and scalar multiplication of $S$ matrices as in Definitions 1.2.3 and 1.2.5, show that the first five parts of Theorem 1.2.7 continue to hold.

### 1.3  Cayley multiplication of matrices

Preparatory to defining the very important *Cayley product* of two matrices of suitable orders, we define the so-called *inner product* of two vectors of the same order.

**1.3.1**  DEFINITION.  Let $X$ be a row or column vector with elements $x_1, \cdots, x_n$ and let $Y$ be a row or column vector of the same order with elements $y_1, \cdots, y_n$. By the *inner product*,† $X \cdot Y$, of $X$ and $Y$ is meant the scalar

$$x_1 y_1 + x_2 y_2 + \cdots + x_n y_n.$$

*Example.*  If $X = [2, -1, 3]$ and $Y = \{2, 0, -2\}$, then $X \cdot Y = (2)(2) + (-1)(0) + (3)(-2) = -2$.

**1.3.2**  DEFINITION.  Let $A = [a_{ij}]$ be an $m \times n$ matrix and $B = [b_{ij}]$ be an $n \times p$ matrix. Since the order of any row $A_i$ of $A$ is equal to

---

† In vector analysis, the inner product of two vectors $X$ and $Y$ of the same order is sometimes referred to as their *dot product* (from the notation for the product) and sometimes as their *scalar product* (since the result is a scalar). This latter term is avoided here, since its use would lead to confusion with the scalar product of Definition 1.2.5. There are other notations prevalent in the literature for the inner product of the two vectors $X$ and $Y$, one of the most common being $(X|Y)$.

the order of any column $B'_j$ of $B$, it is possible to form the inner product $A_i \cdot B'_j$. By the *Cayley product AB* of the matrices $A$ and $B$, taken in this order, is meant the $m \times p$ matrix $C = [c_{ij}]$, where $c_{ij} = A_i \cdot B'_j$. Hereafter, when we speak of the "product" of two matrices $A$ and $B$ without any qualifying adjective, we shall mean the "Cayley product" of $A$ and $B$.

Alternatively we may say, by the *Cayley product AB* of an $m \times n$ matrix $A$ and an $n \times p$ matrix $B$, taken in this order, is meant the $m \times p$ matrix $C = [c_{ij}]$, where

$$c_{ij} = a_{i1}b_{1j} + a_{i2}b_{2j} + \cdots + a_{in}b_{nj} = \sum_{t=1}^{n} a_{it}b_{tj}.$$

*Examples*

(1) $\begin{bmatrix} 1 & -1 & 2 \\ 3 & 0 & 1 \end{bmatrix} \begin{bmatrix} 1 & 2 & 0 \\ 0 & -1 & 1 \\ 1 & 2 & -1 \end{bmatrix} = \begin{bmatrix} 3 & 7 & -3 \\ 4 & 8 & -1 \end{bmatrix}.$

(2) $[x_1, x_2] \begin{bmatrix} 1 & 2 & 4 \\ -1 & 3 & -2 \end{bmatrix} = [x_1 - x_2, 2x_1 + 3x_2, 4x_1 - 2x_2].$

(3) $[x_1, x_2, x_3] \begin{bmatrix} y_1 \\ y_2 \\ y_3 \end{bmatrix} = [x_1y_1 + x_2y_2 + x_3y_3]$. Note that if $X =$

$[x_1, x_2, x_3]$ and $Y = \{y_1, y_2, y_3\}$, then $XY$ is the $1 \times 1$ matrix $[X \cdot Y]$.

**1.3.3** DEFINITIONS. When the number of columns of a matrix $A$ is equal to the number of rows of a matrix $B$, we say that $A$ *is conformable to B* (in this order) *for multiplication*. In the product $AB$ we say that $B$ is *premultiplied* by $A$ and that $A$ is *postmultiplied* by $B$.

**1.3.4** THEOREM. *Multiplication of matrices is not always commutative, not even for square matrices of the same order.*

If $A$ has order $(m, n)$ and $B$ has order $(n, p)$, and $p \neq m$, then $AB$ is defined but $BA$ is not. If $p = m \neq n$, then $AB$ and $BA$ are both defined, but $AB \neq BA$, since $AB$ is square of order $m$ and $BA$ is square of order $n$. Even if $p = m = n$, we do not necessarily have $AB = BA$, as the following example testifies:

$$\begin{bmatrix} 0 & 1 \\ 1 & 0 \end{bmatrix}\begin{bmatrix} 0 & 0 \\ 1 & 0 \end{bmatrix} = \begin{bmatrix} 1 & 0 \\ 0 & 0 \end{bmatrix}, \qquad \begin{bmatrix} 0 & 0 \\ 1 & 0 \end{bmatrix}\begin{bmatrix} 0 & 1 \\ 1 & 0 \end{bmatrix} = \begin{bmatrix} 0 & 0 \\ 0 & 1 \end{bmatrix}.$$

As a consequence of the above proof we have

**1.3.5**  THEOREM.  *If* A *and* B *are matrices such that* AB $=$ BA, *then* A *and* B *must be square and of the same order.*

The noncommutative character of Cayley multiplication is an important feature of matrix algebra, which causes that algebra to differ from the familiar algebra of real numbers studied in high school. In fact, matrix algebra was historically one of the earliest so-called *noncommutative algebras*—that is, algebras possessing a noncommutative multiplication—to be devised. Because of this feature of matrix algebra, one must pay careful attention to the order of the factors in any product of matrices. But this noncommutative nature of matrix multiplication is not the only difference between matrix multiplication and the ordinary multiplication of real numbers. The following theorem points up a second major difference.

**1.3.6**  THEOREM.  *If* A *and* B *are matrices such that* AB $=$ O, *it does not necessarily follow that either* A $=$ O *or* B $=$ O.

For we have, for example,

$$\begin{bmatrix} 0 & 1 \\ 0 & 0 \end{bmatrix}\begin{bmatrix} 1 & 0 \\ 0 & 0 \end{bmatrix} = \begin{bmatrix} 0 & 0 \\ 0 & 0 \end{bmatrix}.$$

Though there are some radical differences between matrix multiplication and the familiar multiplication of real numbers, we now show that these two products do have some important properties in common, in particular distributivity over addition and associativity.

**1.3.7**  THEOREM.  *If* A *is an* m $\times$ n *matrix and* B *and* C *are* n $\times$ p *matrices, then* A(B $+$ C) $=$ AB $+$ AC.

The order of each side of the desired equation is $(m, p)$. Moreover, the $(ij)$th element of the left side is

$$\sum_{t=1}^{n} a_{it}(b_{tj} + c_{tj}) = \sum_{t=1}^{n}(a_{it}b_{tj} + a_{it}c_{tj}) = \sum_{t=1}^{n} a_{it}b_{tj} + \sum_{t=1}^{n} a_{it}c_{tj},$$

which is the $(ij)$th element of the right side.

The reader may similarly prove

**1.3.8**  THEOREM.   *If* B *and* C *are* m × n *matrices and* A *is an* n × p *matrix, then* (B + C)A = BA + CA.

**1.3.9**  THEOREM.   *If* A, B, C *are* m × n, n × p, p × q *matrices, respectively, then* A(BC) = (AB)C.

The order of each side of the desired equation is $(m, q)$. Set $BC = D$ and $AB = E$. Then, by Definition 1.3.2,

$$d_{sj} = \sum_{t=1}^{p} b_{st}c_{tj}, \qquad e_{it} = \sum_{s=1}^{n} a_{is}b_{st}.$$

Now the $(ij)$th element of $A(BC) = AD$ is, by Definition 1.3.2,

$$\sum_{s=1}^{n} a_{is}d_{sj} = \sum_{s=1}^{n} a_{is}\left(\sum_{t=1}^{p} b_{st}c_{tj}\right)$$

$$= \sum_{s=1}^{n} \left(\sum_{t=1}^{p} a_{is}b_{st}c_{tj}\right)$$

(since a factor independent of the index of summation may be introduced under the summation sign)

$$= \sum_{t=1}^{p} \left(\sum_{s=1}^{n} a_{is}b_{st}c_{tj}\right)$$

(since the order of summation in a finite double sum can be commuted)

$$= \sum_{t=1}^{p} \left(\sum_{s=1}^{n} a_{is}b_{st}\right)c_{tj}$$

(since a factor independent of the index of summation may be lifted from under the summation sign)

$$= \sum_{t=1}^{p} e_{it}c_{tj},$$

which, by Definition 1.3.2, is the $(ij)$th element of $EC = (AB)C$.

Finally, we let the reader establish the following easy but important theorem.

**1.3.10** THEOREM. *If* A *and* B *are two matrices conformable for multiplication, and if* u *and* v *are scalars, then*:

(1) $(uA)(vB) = (uv)(AB)$,

(2) $(-A)(-B) = AB$,

(3) $A(uB) = (uA)B = u(AB)$,

(4) $A(-B) = (-A)B = -(AB)$.

## PROBLEMS

**1.3–1** If $X$, $Y$, $Z$ are three row vectors of order $n$, show that

(a) $X \cdot Y = Y \cdot X$,

(b) $X \cdot (Y + Z) = (X \cdot Y) + (X \cdot Z)$,

(c) $(Y + Z) \cdot X = (Y \cdot X) + (Z \cdot X)$.

**1.3–2** Prove Theorem 1.3.8.

**1.3–3** Establish all parts of Theorem 1.3.10.

**1.3–4** Let $A$ and $B$ be square matrices of order $n$ and let

$$C = kA + lB, \qquad D = rA + sB,$$

where $k, l, r, s$ are scalars such that $ks - lr \neq 0$. Show that $CD = DC$ if and only if $AB = BA$.

**1.3–5** Show that for any scalars $a, b, c, d$, the two matrices

$$A = \begin{bmatrix} a & b \\ -b & a \end{bmatrix} \quad \text{and} \quad B = \begin{bmatrix} c & d \\ -d & c \end{bmatrix}$$

commute under multiplication.

**1.3–6** If

$$A(\theta_i) = \begin{bmatrix} \cos \theta_i & -\sin \theta_i \\ \sin \theta_i & \cos \theta_i \end{bmatrix}, \qquad i = 1, 2,$$

show that $A(\theta_1)$ and $A(\theta_2)$ commute under multiplication.

**1.3–7** Two matrices $A$ and $B$ are said to *anticommute* with each other if $AB = -BA$. Show that the *Pauli spin matrices* (used in the study of electron spin in quantum mechanics)

$$\sigma_x = \begin{bmatrix} 0 & 1 \\ 1 & 0 \end{bmatrix}, \qquad \sigma_y = \begin{bmatrix} 0 & -i \\ i & 0 \end{bmatrix}, \qquad \sigma_z = \begin{bmatrix} 1 & 0 \\ 0 & -1 \end{bmatrix},$$

where $i^2 = -1$, anticommute with each other.

**1.3–8** If $C$ is a matrix of order $(m, n)$ and $D$ a matrix of order $(n, m)$, show that $\text{tr}\,(CD) = \text{tr}\,(DC)$. (See Problem 1.2–6 for the definition of $\text{tr}\,A$.)

**1.3–9**   Let $A$ and $B$ be matrices such that $A$ is conformable to $B$ for multiplication. Prove that:

(a) If $A$ has a pair of identical rows, then $AB$ has the same pair of rows identical.

(b) If $B$ has a pair of identical columns, then $AB$ has the same pair of columns identical.

**1.3–10**   Let $A$, $B$, $C$ be three matrices such that $AB = C$. Let $X$ be a column vector whose elements are the sums of the corresponding rows of $B$; let $Y$ be a column vector whose elements are the sums of the corresponding rows of $C$. Show that $AX = Y$. (This fact may be used to check a matrix multiplication and is useful in machine computation.)

**1.3–11**   If $A$, $B$, $C$ are matrices such that $AC = CB$, prove that $A$ and $B$ must be square, though not necessarily of the same order.

**1.3–12**   If $A$ and $B$ are $n$th order matrices, explain why, in general, $(A + B)^2 \neq A^2 + 2AB + B^2$ and $A^2 - B^2 \neq (A + B)(A - B)$.

**1.3–13**   (a) Show that

$$\begin{bmatrix} 0 & 1 \\ 0 & 0 \end{bmatrix}\begin{bmatrix} 3 & 0 \\ 0 & 0 \end{bmatrix} = \begin{bmatrix} 0 & 1 \\ 0 & 0 \end{bmatrix}\begin{bmatrix} 2 & 0 \\ 0 & 0 \end{bmatrix}.$$

(b) Let

$$A = \begin{bmatrix} 1 & -2 & 1 \\ 2 & 1 & -3 \\ -5 & 2 & 3 \end{bmatrix}, \qquad X = \begin{bmatrix} 2 & 5 & -1 & -7 \\ -2 & 1 & 3 & 4 \\ 3 & 2 & 1 & 2 \end{bmatrix},$$

$$Y = \begin{bmatrix} 3 & 6 & 0 & -6 \\ -1 & 2 & 4 & 5 \\ 4 & 3 & 2 & 3 \end{bmatrix}.$$

Show that $AX = AY$, although $A \neq O$ and $X \neq Y$. (It follows that the cancellation law that holds for multiplication of real numbers does not hold for Cayley multiplication of matrices.)

**1.3–14**   Show that the set of eight matrices

$$\begin{bmatrix} 1 & 0 \\ 0 & 1 \end{bmatrix}, \quad \begin{bmatrix} 0 & 1 \\ -1 & 0 \end{bmatrix}, \quad \begin{bmatrix} 0 & -1 \\ 1 & 0 \end{bmatrix}, \quad \begin{bmatrix} -1 & 0 \\ 0 & -1 \end{bmatrix},$$

$$\begin{bmatrix} i & 0 \\ 0 & -i \end{bmatrix}, \quad \begin{bmatrix} -i & 0 \\ 0 & i \end{bmatrix}, \quad \begin{bmatrix} 0 & -i \\ -i & 0 \end{bmatrix}, \quad \begin{bmatrix} 0 & i \\ i & 0 \end{bmatrix},$$

where $i^2 = -1$, is closed under Cayley multiplication.

**1.3–15**   Show that conformability of matrices for Cayley multiplication is not an equivalence relation.

**1.3–16**  Let $A$, $B$, $C$ be matrices such that $AC = CA$ and $BC = CB$. Show that $(AB + BA)C = C(AB + BA)$.

**1.3–17**  Show that if $A$, $X$, $Y$ are $n \times n$ matrices such that $AX = XA = AY = YA$, it does not necessarily follow that $X = Y$.

## 1.4   Some special matrices

There are many special matrices, like the zero matrices of Definition 1.2.4, that play important roles in various parts of matrix theory. We consider some of these special matrices in this section; other special matrices will appear in subsequent parts of the chapter and of the book.

**1.4.1**  DEFINITION AND NOTATION.  A square matrix $[a_{ij}]_{(n)}$ such that $a_{ij} = 0$ if $i \neq j$ and $a_{ij} = 1$ if $i = j$ is called an *identity matrix*, and will be denoted by $I_{(n)}$.

*Example*

$$I_{(3)} = \begin{bmatrix} 1 & 0 & 0 \\ 0 & 1 & 0 \\ 0 & 0 & 1 \end{bmatrix}.$$

The reader should find no trouble in establishing the following theorem.

**1.4.2**  THEOREM.  *If* A *is an* m $\times$ n *matrix, then* $I_{(m)}A = AI_{(n)} = A$, *and* $I_{(m)}$ *and* $I_{(n)}$ *are the only matrices such that this is true for all* m $\times$ n *matrices* A.

On the other hand, one must not jump to the conclusion that if $BA = A$, then $B$ is an identity matrix, for consider

$$\begin{bmatrix} 0 & 0 \\ 0 & 1 \end{bmatrix} \begin{bmatrix} 0 & 0 \\ 2 & 3 \end{bmatrix} = \begin{bmatrix} 0 & 0 \\ 2 & 3 \end{bmatrix}.$$

**1.4.3**  NOTATION.  If $A$ is a square matrix of order $n$, we set $A^0 = I_{(n)}$ and write $A^{k+1}$ for $A^k A$, where $k$ is any nonnegative integer.

**1.4.4**    THEOREM.    *If* A *is a square matrix of order* n *and if* p *and* q *are nonnegative integers, then*: (1) $A^1 = A$, (2) $A^p A^q = A^{p+q}$, (3) $(A^p)^q = A^{pq}$.

(1) $A^1 = A^{0+1} = A^0 A = I_{(n)} A = A$.

(2) Clearly, $A^p A^0 = A^p I_{(n)} = A^p = A^{p+0}$. Now suppose $k$ is a nonnegative integer such that $A^p A^k = A^{p+k}$. Then

$$A^p A^{k+1} = A^p (A^k A) = (A^p A^k) A = A^{p+k} A = A^{(p+k)+1} = A^{p+(k+1)},$$

and the desired result follows by the principle of mathematical induction.

(3) Clearly, $(A^p)^0 = I_{(n)} = A^0 = A^{p \cdot 0}$. Now suppose $k$ is a nonnegative integer such that $(A^p)^k = A^{pk}$. Then, using part (2),

$$(A^p)^{k+1} = (A^p)^k A^p = A^{pk} A^p = A^{pk+p} = A^{p(k+1)},$$

and the desired result follows by the principle of mathematical induction.

**1.4.5**    DEFINITION.    A square matrix $[a_{ij}]$ such that $a_{ij} = 0$ if $i \neq j$ and $a_{ij} = a$, where $a$ is a scalar, if $i = j$, is called a *scalar matrix*. An $n \times n$ scalar matrix with $a$'s along the principal diagonal can be written in the form $aI_{(n)}$.

*Example*

$$\begin{bmatrix} 2 & 0 & 0 \\ 0 & 2 & 0 \\ 0 & 0 & 2 \end{bmatrix} = 2I_{(3)}.$$

The easy proofs of the following two theorems are left to the reader.

**1.4.6**    THEOREM.    *If* A *is an* mth-*order scalar matrix with* a's *along the diagonal and* B *is an* m $\times$ n *matrix, then* AB = aB.

**1.4.7**    THEOREM.    *The sum and product of two* nth-*order scalar matrices are* nth-*order scalar matrices.*

**1.4.8**    DEFINITION AND NOTATION.    A square matrix $[a_{ij}]$ such that $a_{ij} = 0$ if $i \neq j$ is called a *quasi-scalar*, or *diagonal*, *matrix*. The $n \times n$ diagonal matrix $D$ having $d_1, d_2, \cdots, d_n$ along the diagonal is often expressed as

$$D = \text{diag}(d_1, d_2, \cdots, d_n).$$

*Example*

$$\begin{bmatrix} 2 & 0 & 0 \\ 0 & -1 & 0 \\ 0 & 0 & 0 \end{bmatrix} = \text{diag}(2, -1, 0).$$

It is to be noted that scalar matrices are special examples of quasi-scalar matrices.

The easy proofs of the following three theorems about quasi-scalar matrices are left to the reader.

**1.4.9** THEOREM. *Premultiplication of an* m × n *matrix* A *by an* mth-*order quasi-scalar matrix* M *yields a matrix* B *whose* ith *row* (i = 1, ···, m) *is the* ith *row of* A *multiplied by the* (i, i)th *element of* M.

**1.4.10** THEOREM. *Postmultiplication of an* m × n *matrix* A *by an* nth-*order quasi-scalar matrix* N *yields a matrix* C *whose* jth *column* (j = 1, ···, n) *is the* jth *column of* A *multiplied by the* (j, j)th *element of* N.

**1.4.11** THEOREM. *If* A *and* B *are two* nth-*order quasi-scalar matrices, then* A + B *and* AB *are* nth-*order quasi-scalar matrices and* AB = BA.

**1.4.12** DEFINITION. A square matrix $A$ such that $A^2 = A$ is called an *idempotent matrix*.

*Example*

$$\begin{bmatrix} 4 & -2 \\ 6 & -3 \end{bmatrix}^2 = \begin{bmatrix} 4 & -2 \\ 6 & -3 \end{bmatrix}.$$

**1.4.13** THEOREM. *If* A *is an idempotent matrix, then* $A^n$ = A *for all positive integers* n.

Clearly, $A^1 = A$. Suppose $k$ is a positive integer such that $A^k = A$. Then $A^{k+1} = A^k A = AA = A^2 = A$, and the theorem follows by the principle of mathematical induction.

**1.4.14** DEFINITIONS. A square matrix $A$ such that $A^r = O$ for some positive integer $r$ is called a *nilpotent matrix*, and the smallest positive integral exponent $r$ such that $A^r = O$ is called the *index* of $A$.

*Example*

$$\begin{bmatrix} 1 & 1 \\ -1 & -1 \end{bmatrix}$$ is nilpotent with index 2, since

$$\begin{bmatrix} 1 & 1 \\ -1 & -1 \end{bmatrix}^2 = \begin{bmatrix} 0 & 0 \\ 0 & 0 \end{bmatrix}.$$

**1.4.15**  DEFINITIONS.    A square matrix $A = [a_{ij}]$ such that $a_{ij} = 0$ if $i > j$ is called an *upper triangular matrix*. A square matrix $A = [a_{ij}]$ such that $a_{ij} = 0$ if $i \geqq j$ is called an *upper matrix*.

*Examples*

$$\begin{bmatrix} 1 & 2 & 3 & 4 \\ 0 & 0 & 2 & 0 \\ 0 & 0 & 1 & 1 \\ 0 & 0 & 0 & 3 \end{bmatrix}$$

is an upper triangular matrix.

$$\begin{bmatrix} 0 & a & b & c \\ 0 & 0 & d & e \\ 0 & 0 & 0 & f \\ 0 & 0 & 0 & 0 \end{bmatrix}$$

is an upper matrix.

As a very easy theorem we have

**1.4.16**  THEOREM.    *The sum and product of two nth-order upper (upper triangular) matrices are upper (upper triangular) matrices.*

A somewhat more difficult theorem, whose proof we leave to the reader, is

**1.4.17**  THEOREM.    *Every upper matrix is nilpotent.*

**1.4.18**  DEFINITION.    A square matrix $A$ such that $A^2 = I$ is called an *involutoric matrix*.

*Example*

$$\begin{bmatrix} 1 & 0 \\ 0 & -1 \end{bmatrix}^2 = \begin{bmatrix} 1 & 0 \\ 0 & 1 \end{bmatrix}.$$

# PROBLEMS

**1.4–1**  Prove Theorem 1.4.2.

**1.4–2**  Show that if

$$B = \begin{bmatrix} 2 & 2 & 3 \\ 2 & 5 & 6 \\ 3 & 6 & 10 \end{bmatrix} \quad \text{and} \quad A = \begin{bmatrix} 1 & 1 & 1 \\ 1 & 1 & 1 \\ -1 & -1 & -1 \end{bmatrix},$$

then $BA = A$.

**1.4–3**  Explain why, in general, $(AB)^p \neq A^p B^p$, where $A$ and $B$ are $n$th-order matrices and $p$ is a positive integer.

**1.4–4**  If $A = \begin{bmatrix} 0 & i \\ i & 0 \end{bmatrix}$, where $i^2 = -1$, compute $A^2$, $A^3$, $A^4$. Give a general rule for $A^n$, where $n$ is a positive integer.

**1.4–5**  Show that any two polynomial functions, with complex coefficients, of a square matrix $A$ commute.

**1.4–6**  Show that arbitrary polynomials $f(A)$ and $g(B)$, with complex coefficients, in fixed matrices $A$ and $B$ of order $n$ commute if and only if $A$ and $B$ commute.

**1.4–7**  Let $A$ and $B$ be square matrices of the same order and such that $A$ and $AB - BA$ commute. Show that for any positive integer $n$,

$$A^n B - BA^n = n(AB - BA)A^{n-1}.$$

**1.4–8**  Prove Theorems 1.4.6 and 1.4.7.

**1.4–9**  Show that the only $n$th-order matrices that commute with all $n$th-order matrices are the $n \times n$ scalar matrices.

**1.4–10**  Prove Theorems 1.4.9, 1.4.10, 1.4.11.

**1.4–11**  Show that

$$U = \begin{bmatrix} 2 & -2 & -4 \\ -1 & 3 & 4 \\ 1 & -2 & -3 \end{bmatrix} \quad \text{and} \quad V = \begin{bmatrix} -1 & 2 & 4 \\ 1 & -2 & -4 \\ -1 & 2 & 4 \end{bmatrix}$$

are idempotent and that $UV = O$.

**1.4–12**  Characterize all $n$th-order quasi-scalar idempotent $\mathscr{C}$ matrices.

**1.4–13**  If $AB = A$ and $BA = B$, show that $A$ and $B$ are idempotent.

**1.4–14**  Show that

$$\begin{bmatrix} 1 & 2 & 3 \\ 1 & 2 & 3 \\ -1 & -2 & -3 \end{bmatrix} \quad \text{and} \quad \begin{bmatrix} -4 & 4 & -4 \\ 1 & -1 & 1 \\ 5 & -5 & 5 \end{bmatrix}$$

are nilpotent with index 2.

**1.4–15** Show that

$$\begin{bmatrix} 0 & 1 & 0 & 0 \\ 0 & 0 & 1 & 0 \\ 0 & 0 & 0 & 1 \\ 0 & 0 & 0 & 0 \end{bmatrix}$$

is nilpotent with index 4. Generalize.

**1.4–16** If $B$ and $C$ are of order $n$, $A = B + C$, $C^2 = O$, and $BC = CB$, show that $A^{k+1} = B^k[B + (k + 1)C]$ for all positive integers $k$.

**1.4–17** If $A$ is nilpotent with index 2, show that $A(I + A)^n = A$ for all positive integral $n$.

**1.4–18** Show that every $2 \times 2$ nilpotent $\mathscr{C}$ matrix with index 2 has the form

$$\begin{bmatrix} ab & b^2 \\ -a^2 & -ab \end{bmatrix},$$

and that every nonzero $\mathscr{C}$ matrix of this form is nilpotent with index 2.

**1.4–19** If $A$ and $B$ are nonzero matrices such that $AB = O$, then each is called a *proper divisor of zero*. Show that every nilpotent matrix is a proper divisor of zero.

**1.4–20** Prove Theorem 1.4.16.

**1.4–21** Prove Theorem 1.4.17.

**1.4–22** Formulate a definition of a *lower matrix* and state two theorems concerning such matrices.

**1.4–23** (a) Show that matrix $A$ is involutoric if and only if

$$(I - A)(I + A) = O.$$

(b) If $A$ is involutoric, show that $A^n = A$ or $I$ according as $n$ is an odd or an even nonnegative integer.

**1.4–24** Show that

$$\begin{bmatrix} 3 - 4k & 2 - 4k & 2 - 4k \\ -1 + 2k & 2k & -1 + 2k \\ -3 + 2k & -3 + 2k & -2 + 2k \end{bmatrix}$$

is involutoric for all complex numbers $k$.

## 1.5 Transposition

Addition and multiplication of matrices are *binary* operations in that each acts on an *ordered pair* of matrices to produce a third matrix. In this very brief section we introduce an important *unary* operation on matrices, that is, an operation performed on a *single* matrix to produce a second matrix.

**1.5.1** DEFINITION AND NOTATION. If $A$ is an $m \times n$ matrix, then the $n \times m$ matrix whose successive rows are the successive columns of $A$ is called the *transpose* of $A$ and will be denoted by $A'$. In other words, the $(i, j)$th element of $A'$ is the $(j, i)$th element of $A$.

Some authors denote the transpose of $A$ by $A^T$.

*Example.* If $A = \begin{bmatrix} 1 & -2 & 0 \\ 1 & 1 & 2 \end{bmatrix}$, then $A' = \begin{bmatrix} 1 & 1 \\ -2 & 1 \\ 0 & 2 \end{bmatrix}$.

**1.5.2** THEOREM. *If* A *and* B *are matrices of suitable orders, and* k *is a scalar, then*:

(1) $(A')' = A$;

(2) $(A + B)' = A' + B'$;

(3) $(kA)' = kA'$;

(4) $(AB)' = B'A'$; *and, more generally,* $(ABC \cdots MN)' = N'M' \cdots C'B'A'$;

(5) $(A^n)' = (A')^n$ *for all nonnegative integers* n.

We shall establish part (4) of the theorem and leave the easy proofs of the remaining parts to the reader.

Set $A = [a_{ij}]_{(m,n)}$, $B = [b_{ij}]_{(n,p)}$, $A' = [a'_{ij}]_{(n,m)}$, $B' = [b'_{ij}]_{(p,n)}$, $AB = [c_{ij}]_{(m,p)}$, and $B'A' = [c'_{ij}]_{(p,m)}$, where, of course, $a'_{ij} = a_{ji}$ and $b'_{ij} = b_{ji}$. Now

$$c_{ji} = \sum_{t=1}^{n} a_{jt}b_{ti} = \sum_{t=1}^{n} b'_{it}a'_{tj} = c'_{ij}.$$

It follows that $B'A' = (AB)'$. The extension to any number of factors is easily accomplished by the process of mathematical induction.

## 1.6 Symmetric and skew-symmetric matrices

In applications one frequently encounters matrices that are equal to their transposes and other matrices that are equal to the negatives of their transposes. We are thus led to formulate the following definitions.

**1.6.1** DEFINITIONS. A matrix $A$ is said to be *symmetric* if $A = A'$, and is *skew-symmetric* if $A = -A'$.

*Examples*

$$\begin{bmatrix} 1 & 2 & 3 \\ 2 & 0 & -4 \\ 3 & -4 & -2 \end{bmatrix} \text{ is symmetric;}$$

$$\begin{bmatrix} 0 & 2 & 3 \\ -2 & 0 & -4 \\ -3 & 4 & 0 \end{bmatrix} \text{ is skew-symmetric.}$$

The proofs of the first three of the following theorems are left to the reader.

**1.6.2** THEOREM. *Symmetric and skew-symmetric matrices are square.*

**1.6.3** THEOREM. *In a symmetric matrix* A, $a_{ij} = a_{ji}$. *That is,* A *is symmetrical in its principal diagonal.*

**1.6.4** THEOREM. *In a skew-symmetric 𝒞 matrix* A, $a_{ij} = -a_{ji}$, *and the elements along the principal diagonal are all equal to zero.*

**1.6.5** THEOREM. *If* A *is a matrix, then* AA′ *is symmetric.*
For we have $(AA')' = (A')'A' = AA'$.

**1.6.6** COROLLARY. *If* X *is a column vector of order* n, *then* XX′ *is a symmetric matrix of order* n.

**1.6.7** THEOREM. *If* A *is a square matrix, then* A + A′ *is symmetric and* A − A′ *is skew-symmetric.*
For we have
$$(A + A')' = A' + (A')' = A' + A = A + A'$$
and
$$(A - A')' = [A + (-A')]' = A' + (-A')'$$
$$= A' - (A')' = A' - A = -(A - A').$$

**1.6.8** THEOREM. *Any square matrix* A *can be expressed as the sum of a symmetric and a skew-symmetric matrix in one and only one way.*

For, since $A$ is square, $A = (1/2)(A + A') + (1/2)(A - A')$. But, by Theorem 1.6.7, $(1/2)(A + A')$ is symmetric and $(1/2)(A - A')$ is skew-symmetric.

Now suppose $A = M + N$, where $M$ is symmetric and $N$ is skew-symmetric. Then $A' = M' + N' = M - N$. Solving, we find

$$M = (1/2)(A + A'), \quad N = (1/2)(A - A').$$

## PROBLEMS

**1.5–1**  Prove parts (1), (2), (3), (5) of Theorem 1.5.2.

**1.5–2**  If $A$, $B$, $C$, $\cdots$, $M$, $N$ are matrices of suitable orders, prove that $(ABC \cdots MN)' = N'M' \cdots C'B'A'$.

**1.6–1**  Establish Theorem 1.6.2.

**1.6–2**  Establish Theorems 1.6.3 and 1.6.4.

**1.6–3**  Prove that a symmetric matrix of order $n$ has at most $n(n + 1)/2$ distinct elements.

**1.6–4**  Prove that a skew-symmetric $\mathscr{C}$ matrix of order $n$ has at most $n(n - 1)/2$ elements having distinct nonzero numerical values.

**1.6–5**  If $A$ and $B$ are symmetric $n$th-order matrices such that $AB = BA$, show that $AB$ is symmetric.

**1.6–6**  If $A$ is a symmetric or a skew-symmetric matrix, show that $AA' = A'A$ and that $A^2$ is symmetric. What can you say about $A^k$, where $k$ is a positive integer?

**1.6–7**  If $A$ is a symmetric(skew-symmetric)$m$th-order matrix and $P$ is an $m \times n$ matrix, show that $P'AP$ is a symmetric (skew-symmetric) matrix.

## 1.7  Conjugation and tranjugation

In this and the next section we shall consider two important unary operations that can be performed on $\mathscr{C}$ matrices. We first recall that if $u = a + ib$ is a complex number, then $\bar{u} = a - ib$ is known as the *complex conjugate* of $u$.

**1.7.1**  DEFINITIONS AND NOTATION.  If $A = [a_{ij}]$ is a matrix with complex numbers as elements, then the matrix $\bar{A} = [\bar{a}_{ij}]$, where $\bar{a}_{ij}$ is the complex conjugate of $a_{ij}$, is called the *conjugate* of $A$. The matrix $A^* = (\overline{A})'$ is called the *tranjugate* (transposed conjugate) of $A$.

Some authors denote the conjugate of $A$ by $A^C$ and the tranjugate of $A$ by $A^{CT}$.

*Example.* If

$$A = \begin{bmatrix} 1-i & i & 3 \\ 0 & 1 & 2+i \\ -3i & -4 & 10-2i \end{bmatrix},$$

then

$$\overline{A} = \begin{bmatrix} 1+i & -i & 3 \\ 0 & 1 & 2-i \\ 3i & -4 & 10+2i \end{bmatrix}$$

and

$$A^* = \begin{bmatrix} 1+i & 0 & 3i \\ -i & 1 & -4 \\ 3 & 2-i & 10+2i \end{bmatrix}.$$

If $u$ and $v$ are complex numbers, then $\overline{\overline{u}} = u$, $\overline{u+v} = \overline{u} + \overline{v}$, and $\overline{uv} = \overline{u} \cdot \overline{v}$. Using these properties, one can easily establish all parts of the following theorem:

**1.7.2    THEOREM.**    *If* A *and* B *are matrices of suitable orders and* k *is a scalar, then:*

(1) $\overline{\overline{A}} = A$;

(2) $\overline{A+B} = \overline{A} + \overline{B}$;

(3) $\overline{kA} = \overline{k}\,\overline{A}$;

(4) $\overline{AB} = \overline{A}\,\overline{B}$;

(5) $(\overline{A})' = \overline{A'}$.

**1.7.3    THEOREM.**    *If* A *and* B *are matrices of suitable orders, and* k *is a scalar, then:*

(1) $(A^*)^* = A$:

(2) $(A+B)^* = A^* + B^*$;

(3) $(kA)^* = \bar{k}\, A^*$;

(4) $(AB)^* = B^*A^*$.

For we have:

(1) $(A^*)^* = [(\bar{A})']^* = (\overline{\bar{A}'})' = (\overline{\overline{A}'})' = (A')' = A$;

(2) $(A + B)^* = (\overline{A + B})' = (\bar{A} + \bar{B})' = (\bar{A})' + (\bar{B})' = A^* + B^*$;

(3) $(kA)^* = (\overline{kA})' = (\bar{k}\,\bar{A})' = \bar{k}(\bar{A})' = \bar{k}A^*$;

(4) $(AB)^* = (\overline{AB})' = (\bar{A}\,\bar{B})' = (\bar{B})'(\bar{A})' = B^*A^*$.

## 1.8    Hermitian and skew-Hermitian matrices

Hermitian and skew-Hermitian matrices are generalizations of real symmetric and real skew-symmetric matrices.

**1.8.1**   DEFINITIONS.   A matrix $A$ with complex elements is said to be *Hermitian*† if $A = A^*$, and *skew-Hermitian* if $A = -A^*$.

*Examples.*   If $a, b, c, d, e, f, g, h, k$ are real numbers and if $i = \sqrt{-1}$, then

$$\begin{bmatrix} a & b + ic & e + if \\ b - ic & d & h + ik \\ e - if & h - ik & g \end{bmatrix}$$

is Hermitian and

$$\begin{bmatrix} ia & b + ic & e + if \\ -b + ic & id & h + ik \\ -e + if & -h + ik & ig \end{bmatrix}$$

is skew-Hermitian.

We first note:

---

† Charles Hermite (1822–1901), in 1855, proved an important theorem concerning any $\mathscr{C}$ matrix $A$ for which $A = A^*$ (he proved that the characteristic roots of such a matrix are all real), resulting in such matrices being named after him.

**1.8.2** THEOREM. *A Hermitian (skew-Hermitian) matrix is square, and a real Hermitian (skew-Hermitian) matrix is symmetric (skew-symmetric).*

The proof that Hermitian and skew-Hermitian matrices are square is left to the reader. If $A$ is real Hermitian, then $A = A^* = (\bar{A})' = A'$, and $A$ is symmetric; if $A$ is real skew-Hermitian, then $A = -A^* = -(\bar{A})' = -A'$, and $A$ is skew-symmetric.

We now list some theorems whose straightforward proofs are left as exercises for the reader.

**1.8.3** THEOREM. *If* A *is Hermitian (skew-Hermitian), then* iA *and* $-$iA *are skew-Hermitian (Hermitian).*

**1.8.4** THEOREM. *If* A *is Hermitian, then* A $=$ M $+$ iN, *where* M *is real symmetric and* N *is real skew-symmetric.*

**1.8.5** THEOREM. *If* A *is a square* $\mathscr{C}$ *matrix, then* A $+$ A* *is Hermitian and* A $-$ A* *is skew-Hermitian.*

**1.8.6** THEOREM. *Any square* $\mathscr{C}$ *matrix can be expressed as the sum of a Hermitian and a skew-Hermitian matrix.*

**1.8.7** THEOREM. *If* A *is any* $\mathscr{C}$ *matrix* (not necessarily square), *then* A*A *and* AA* *are both Hermitian.*

**1.8.8** THEOREM. *If* A *is Hermitian, then so also is* B*AB *for every conformable* $\mathscr{C}$ *matrix* B.

**1.8.9** THEOREM. *If* A *is a Hermitian matrix of order* n *and if* S *is a* $\mathscr{C}$ *matrix of order* (n, 1), *then the scalar* $\bar{S}'AS$ *is real.*

For we have $\overline{\bar{S}'AS} = (\overline{\bar{S}'AS})' = (\bar{\bar{S}}'\bar{A}\bar{S})' = (S'\bar{A}\bar{S})' = \bar{S}'\bar{A}'S = \bar{S}'A^*S = \bar{S}'AS$, and the scalar is real (since a complex number that is equal to its conjugate complex number is real).

# PROBLEMS

**1.7–1**  Establish all parts of Theorem 1.7.2.

**1.7–2**  Give authorities for each step in the proofs in the text of the various parts of Theorem 1.7.3.

**1.8–1**  Prove that Hermitian and skew-Hermitian matrices are square.

**1.8–2**  Prove Theorem 1.8.3.

**1.8–3**  Prove Theorem 1.8.4.

**1.8–4**  Prove Theorem 1.8.5.

**1.8–5**  Prove Theorem 1.8.6.

**1.8–6**  Prove Theorem 1.8.7.

**1.8–7**  Prove Theorem 1.8.8.

**1.8–8**  If $A, B, C, \cdots, M, N$ are $n$th-order $\mathscr{C}$ matrices, prove that $(ABC \cdots MN)^* = N^*M^* \cdots C^*B^*A^*$.

**1.8–9**  Show that the diagonal elements of a Hermitian matrix are all real numbers and that the diagonal elements of a skew-Hermitian matrix are all pure imaginary numbers.

**1.8–10**  If $A$ is the Hermitian matrix of Theorem 1.8.4, prove that $A^*A$ is real if and only if $M$ and $N$ anticommute.

**1.8–11**  If $A$ is skew-Hermitian, show that $A = M + iN$, where $M$ is real skew-symmetric and $N$ is real symmetric.

## 1.9  Partitioned matrices

We shall find that it is sometimes convenient to subdivide a matrix into rectangular blocks of elements. This leads us to consider so-called *partitioned*, or *block*, *matrices*.

**1.9.1**  DEFINITION.  The array of elements belonging to (not necessarily consecutive) rows $i_1, i_2, \cdots, i_r$ and columns $j_1, j_2, \cdots, j_s$ of a matrix $A$ is called a *submatrix* of $A$ of order $(r, s)$.

**1.9.2**  THEOREM.  *If* A, B, C *are matrices such that* AB = C, *then the submatrix contained in rows* $i_1, \cdots, i_r$ *and columns* $j_1, \cdots, j_s$ *of* C *is equal to the product of the submatrix of* A *consisting of these rows of* A *and the submatrix of* B *consisting of these columns of* B.

This theorem is an immediate consequence of the definition of multiplication of matrices.

*Example.* We have

$$
\begin{matrix} \rightarrow \\ \rightarrow \\ \\ \rightarrow \end{matrix}
\begin{bmatrix} 0 & 1 & 2 \\ 1 & 2 & 3 \\ 2 & 3 & 0 \\ 3 & 0 & 1 \end{bmatrix}
\begin{matrix} \downarrow & \downarrow \\ \end{matrix}
\begin{bmatrix} 0 & 1 & 2 & 3 \\ 1 & 2 & 3 & 0 \\ 2 & 3 & 0 & 1 \end{bmatrix}
=
\begin{matrix} \downarrow & \downarrow \\ \end{matrix}
\begin{bmatrix} 5 & 8 & 3 & 2 \\ 8 & 14 & 8 & 6 \\ 3 & 8 & 13 & 6 \\ 2 & 6 & 6 & 10 \end{bmatrix}
\begin{matrix} \leftarrow \\ \leftarrow \\ \\ \leftarrow \end{matrix}.
$$

Now take $i_1 = 1$, $i_2 = 2$, $i_3 = 4$, and $j_1 = 2$, $j_2 = 3$. Note that

$$
\begin{bmatrix} 0 & 1 & 2 \\ 1 & 2 & 3 \\ 3 & 0 & 1 \end{bmatrix}
\begin{bmatrix} 1 & 2 \\ 2 & 3 \\ 3 & 0 \end{bmatrix}
=
\begin{bmatrix} 8 & 3 \\ 14 & 8 \\ 6 & 6 \end{bmatrix}.
$$

**1.9.3**  THEOREM.  *Let matrix* $A_{(m,n)}$ *be partitioned by groups of consecutive rows into submatrices*† $A_1, \cdots, A_i, \cdots, A_r$, *and let matrix* $B_{(n,p)}$ *be partitioned by groups of consecutive columns into submatrices* $B_1, \cdots, B_j$, $\cdots, B_s$. *Let* $C_{(m,p)} = AB$ *be partitioned into submatrices by row groups exactly as* A *and by column groups exactly as* B, *and denote by* $C_{ij}$ *the submatrix of* C *belonging to the ith row group and the jth column group. Then* $C_{ij} = A_i B_j$.

This theorem is an immediate consequence of Theorem 1.9.2.

*Example.* We have

$$
\begin{bmatrix} 0 & 1 & 2 \\ 1 & 2 & 3 \\ 2 & 3 & 0 \\ 3 & 0 & 1 \end{bmatrix}
\left[\begin{array}{ccc|c} 0 & 1 & 2 & 3 \\ 1 & 2 & 3 & 0 \\ 2 & 3 & 3 & 1 \end{array}\right]
=
\left[\begin{array}{c|ccc} 5 & 8 & 3 & 2 \\ \hline 8 & 14 & 8 & 6 \\ 3 & 8 & 13 & 6 \\ 2 & 6 & 6 & 10 \end{array}\right],
$$

or, schematically,

$$
\left[\frac{A_1}{A_2}\right][B_1 \mid B_2 \mid B_3] = \left[\begin{array}{c|c|c} C_{11} & C_{12} & C_{13} \\ \hline C_{21} & C_{22} & C_{23} \end{array}\right],
$$

where, by Theorem 1.9.2, $C_{ij} = A_i B_j$.

**1.9.4**  THEOREM.  *Let matrix* $A_{(m,n)}$ *be partitioned into column groups* $A_1, \cdots, A_r$, *containing* $n_1, \cdots, n_r$ *columns, respectively, where*

† In this section, $A_i$ does not necessarily mean the *i*th row of *A*.

$n_1 + \cdots + n_r = n$, *and let matrix* $B_{(n,p)}$ *be partitioned into row groups* $B_1, \cdots, B_r$ *in exactly the same way. Then*

$$C = AB = [A_1B_1 + \cdots + A_rB_r].$$

For we have

$$c_{ij} = \sum_{k=1}^{n} a_{ik}b_{kj}$$

$$= \sum_{k=1}^{n_1} a_{ik}b_{kj} + \sum_{k=n_1+1}^{n_1+n_2} a_{ik}b_{kj} + \cdots + \sum_{k=n_1+\cdots+n_{r-1}+1}^{n} a_{ik}b_{kj}$$

$$= [A_1B_1]_{ij} + [A_2B_2]_{ij} + \cdots + [A_rB_r]_{ij}$$

$$= [A_1B_1 + A_2B_2 + \cdots + A_rB_r]_{ij},$$

where, for convenience, we have temporarily introduced the notation $[M]_{ij}$ for the $(i,j)$th element of matrix $M$.

**1.9.5**  DEFINITION.    A partitioning as in Theorem 1.9.4 is called a *conformable partitioning* of $A$ and $B$.

Combining Theorems 1.9.2 and 1.9.4, we obtain the following remarkable theorem.

**1.9.6**  THEOREM.    *Let* $A_{(m,n)}$ *and* $B_{(n,p)}$ *be two conformably partitioned matrices. Further, let the rows of* A *be partitioned arbitrarily into groups, and the columns of* B *arbitrarily into groups. Denote the submatrices of* A *and* B *so arising by* $A_{ij}$, $B_{ij}$, *where* i *denotes the occurrence in row groups,* j *the occurrence in column groups. Finally, let the product* C = AB *be partitioned according to the row-partitioning of* A *and the column-partitioning of* B. *Then the* (i, j)*th submatrix of* C *is*

$$C_{ij} = \sum_{k} A_{ik}B_{kj}.$$

*Example.*    We have

$$\begin{bmatrix} 0 & 1 & 2 \\ 1 & 2 & 3 \\ 2 & 3 & 0 \\ 3 & 0 & 1 \end{bmatrix} \begin{bmatrix} 0 & 1 & 2 & 3 \\ 1 & 2 & 3 & 0 \\ 2 & 3 & 0 & 1 \end{bmatrix} = \begin{bmatrix} 5 & 8 & 3 & 2 \\ 8 & 14 & 8 & 6 \\ 3 & 8 & 13 & 6 \\ 2 & 6 & 6 & 10 \end{bmatrix},$$

or, schematically,

$$\begin{bmatrix} A_{11} & A_{12} \\ A_{21} & A_{22} \end{bmatrix} \begin{bmatrix} B_{11} & B_{12} & B_{13} \\ B_{21} & B_{22} & B_{23} \end{bmatrix}$$

$$= \begin{bmatrix} A_{11}B_{11} + A_{12}B_{21} & A_{11}B_{12} + A_{12}B_{22} & A_{11}B_{13} + A_{12}B_{23} \\ A_{21}B_{11} + A_{22}B_{21} & A_{21}B_{12} + A_{22}B_{22} & A_{21}B_{13} + A_{22}B_{23} \end{bmatrix}.$$

We note that Theorem 1.9.6 is an extension of the Cayley rule of multiplication of matrices, wherein elements of the matrices have been replaced by blocks of elements arising from a suitable partitioning of the matrices involved. When the blocks of elements are taken as the elements themselves, Theorem 1.9.6 is seen to reduce to the Cayley rule for multiplication of conformable matrices.

Block multiplication of matrices is of considerable theoretical use, and we shall employ the device on numerous occasions in proofs of certain important theorems; indeed, this is the reason we introduced the idea. But block multiplication of matrices also has practical use in large-scale computation. Consider, for example, the computation of the product $C$ of two $50 \times 50$ matrices $A$ and $B$ on a medium-size computer such as an IBM 650. Now there are 2500 elements in each matrix, and an IBM 650 has storage capacity for only the same number of numbers along with the operational instructions for the machine. One possible way of performing the multiplication on this computer is to partition $A$ and $B$ into $25 \times 25$ submatrices:

$$A = \begin{bmatrix} A_{11} & A_{12} \\ A_{21} & A_{22} \end{bmatrix}, \qquad B = \begin{bmatrix} B_{11} & B_{12} \\ B_{21} & B_{22} \end{bmatrix}.$$

Each submatrix possesses 625 elements, and so the machine can successively compute the products $A_{11}B_{11}, A_{11}B_{21}, A_{12}B_{21}, A_{12}B_{22}, \cdots$, recording each product on punched cards so as to clear the machine storage for the next product, and then finally compute $C_{11} = A_{11}B_{11} + A_{12}B_{22}, \cdots$.

Some of the following problems will show that block multiplication of matrices is even useful to human calculators where special patterns appear in the matrices. Before proceeding to the problems, we introduce a useful extension of Definition and Notation 1.4.8.

**1.9.7** Definition and notation. Let $A_1, A_2, \cdots, A_s$ denote square matrices of respective orders $m_1, m_2, \cdots, m_s$. Then the block matrix

$$\begin{bmatrix} A_1 & O & \cdots & O \\ O & A_2 & \cdots & O \\ \cdot & \cdot & \cdots & \cdot \\ O & O & \cdots & A_s \end{bmatrix}$$

is called the *direct sum* of the matrices $A_1, A_2, \cdots, A_s$, taken in this order, and is denoted by

$$\mathrm{diag}(A_1, A_2, \cdots, A_s).$$

## PROBLEMS

**1.9–1** (a) Show that in computing the Cayley product of an $m \times n$ matrix and an $n \times p$ matrix, there are altogether $mpn$ products and $mp(n - 1)$ sums to be performed.

(b) Show that in computing the Cayley product of two square matrices of order $n$, there are altogether $n^3$ products and $n^3 - n^2$ sums to be performed.

(c) Assuming that a multiplication of two numbers takes ten times as long to perform as an addition of two numbers, and that an addition takes ten seconds, how many hours would it take to compute the Cayley product of two tenth-order matrices?

**1.9–2** Compute very simply, by appropriate partitioning,

(a) $\begin{bmatrix} 0 & 0 & 1 & 2 \\ 0 & 0 & 1 & -1 \end{bmatrix} \begin{bmatrix} 1 & 2 \\ 3 & 4 \\ 1 & 0 \\ 0 & 1 \end{bmatrix}$,

(b) $\begin{bmatrix} 1 & 0 & 0 & 1 \\ 0 & 1 & 0 & 2 \\ 0 & 0 & 1 & 3 \end{bmatrix} \begin{bmatrix} 1 & 0 & 0 \\ 0 & 1 & 0 \\ 0 & 0 & 1 \\ 3 & 2 & 1 \end{bmatrix}$.

**1.9–3** If $A = \begin{bmatrix} 1 & 0 & 0 & 0 \\ 0 & 1 & 0 & 0 \\ a & b & -1 & 0 \\ c & d & 0 & -1 \end{bmatrix}$, find $A^{18}$.

**1.9–4**   Let

$$A = \begin{bmatrix} 2 & 0 & 0 & 0 & 0 \\ 0 & 2 & 0 & 0 & 0 \\ 1 & 0 & a & b & c \\ 0 & 1 & d & e & f \end{bmatrix}, \qquad B = [b_{ij}]_{(5,3)}.$$

Show, with appropriate partitioning, that the only nontrivial computation required to find the product $AB$ is that involved in obtaining the product of a $2 \times 3$ matrix and a $3 \times 3$ matrix.

**1.9–5**   **(a)** If

$$A = \begin{bmatrix} I_{(2)} & O \\ O & -I_{(2)} \end{bmatrix}, \qquad B = \begin{bmatrix} -I_{(2)} & G \\ O & I_{(2)} \end{bmatrix}, \qquad C = \begin{bmatrix} -I_{(2)} & O \\ H & I_{(2)} \end{bmatrix},$$

show that $A^2 = B^2 = C^2 = I_{(4)}$ and $AB + BA = AC + CA = -2I_{(4)}$.

**(b)** If, further, $GH = HG = O$, show that $BC + CD = 2I_{(4)}$.

**1.9–6**   Let $A$ and $B$ be $n \times n$ matrices having the block $O_{(k,n-k)}$ in their top right corners. Show, by block multiplication, that matrix $AB$ has the same form.

**1.9–7**   **(a)** If $A = \text{diag}(A_1, A_2, \cdots, A_s)$ and $B = \text{diag}(B_1, B_2, \cdots, B_s)$, where $A_i$ and $B_i$ have the same order for each $i = 1, 2, \cdots, s$, show that

$$AB = \text{diag}(A_1 B_1, A_2 B_2, \cdots, A_s B_s).$$

**(b)** Compute

$$\begin{bmatrix} 2 & 0 & 0 & 0 & 0 \\ 0 & 1 & 2 & 0 & 0 \\ 0 & 0 & -1 & 0 & 0 \\ 0 & 0 & 0 & 2 & 3 \\ 0 & 0 & 0 & 1 & -2 \end{bmatrix}^2.$$

**1.9–8**   **(a)** Show that an $m \times n$ matrix possesses $(2^m - 1)(2^n - 1)$ submatrices.

**(b)** How many submatrices does a $4 \times 5$ matrix possess?

**1.9–9**   Show that $\begin{bmatrix} A & | & B \\ \hline C & | & D \end{bmatrix}' = \begin{bmatrix} A' & | & C' \\ \hline B' & | & D' \end{bmatrix}$. Generalize.

**1.9–10**   If

$$P = \begin{bmatrix} a & b & c & d \\ -b & a & -d & c \\ -c & d & a & -b \\ -d & -c & b & a \end{bmatrix},$$

show simply (by appropriate partitioning) that

$$PP' = (a^2 + b^2 + c^2 + d^2)I_{(4)}.$$

## ADDENDA TO CHAPTER 1

These addenda to Chapter 1 are appended for the interested reader and do not constitute a part of the formal development of our subject. In them we attempt to indicate the astonishing pervasiveness of the Cayley product of matrices in applications. We also look briefly at a few other matters connected with Cayley multiplication.

### 1.1A  Linear Transformations

We have already seen, in the Prolegomenon, how a theory of matrices incorporating the Cayley product originated, with Arthur Cayley, in a study of linear transformations. By identifying the linear transformation

$$y_1 = a_{11}x_1 + a_{12}x_2 + \cdots + a_{1p}x_p$$
$$y_2 = a_{21}x_1 + a_{22}x_2 + \cdots + a_{2p}x_p \,,$$
$$\cdot \quad \cdot \quad \cdot \quad \cdot \quad \cdot$$
$$y_n = a_{n1}x_1 + a_{n2}x_2 + \cdots + a_{np}x_p$$

which maps points $(x_1, x_2, \cdots, x_p)$ of a $p$-dimensional space onto points $(y_1, y_2, \cdots, y_n)$ of an $n$-dimensional space, with the matrix $A = [a_{ij}]_{(n,p)}$, an algebra of matrices is induced from the corresponding algebra of linear transformations. If the preceding linear transformation (or mapping) is followed by the linear transformation

$$u_1 = b_{11}y_1 + b_{12}y_2 + \cdots + b_{1n}y_n$$
$$u_2 = b_{21}y_1 + b_{22}y_2 + \cdots + b_{2n}y_n \,,$$
$$\cdot \quad \cdot \quad \cdot \quad \cdot \quad \cdot \quad \cdot$$
$$u_m = b_{m1}y_1 + b_{m2}y_2 + \cdots + b_{mn}y_n$$

which maps points $(y_1, y_2, \cdots, y_n)$ of $n$ space onto points $(u_1, u_2, \cdots, u_m)$ of $m$ space, it can be shown, by eliminating the $y$'s, that points $(x_1, x_2, \cdots, x_p)$ of $p$ space are mapped onto points $(u_1, u_2, \cdots, u_m)$ of $m$ space directly by the linear transformation

$$u_1 = c_{11}x_1 + c_{12}x_2 + \cdots + c_{1p}x_p$$
$$u_2 = c_{21}x_1 + c_{22}x_2 + \cdots + c_{2p}x_p,$$
$$\cdot \quad \cdot \quad \cdot \quad \cdot \quad \cdot \quad \cdot \quad \cdot$$
$$u_m = c_{m1}x_1 + c_{m2}x_2 + \cdots + c_{mp}x_p$$

where, if we set $B = [b_{ij}]_{(m,n)}$ and $C = [c_{ij}]_{(m,p)}$, we have $C = BA$, in which $BA$ denotes the Cayley product of the matrices $B$ and $A$. In fact, if we introduce the column vectors

$$X = \{x_1, x_2, \cdots, x_p\}, \qquad Y = \{y_1, y_2, \cdots, y_n\}, \qquad U = \{u_1, u_2, \cdots, u_m\},$$

the reader can easily show that the three linear transformations above can be compactly written, in matrix notation, as

$$Y = AX, \qquad U = BY, \qquad U = CX,$$

respectively. Eliminating $Y$ from the first two of these matrix equations, we find

$$U = BY = B(AX) = (BA)X,$$

from which it follows that $C = BA$.

So strong and intimate is the tie-up between the theory of matrices and the theory of linear transformations that many textbooks on matrix theory develop the subject in conjunction with an algebraic study of linear transformations. This accounts for the many texts bearing such titles as *Matrix Theory and Linear Transformations* and *Linear Algebra and Matrix Theory*.

### 1.2A    Bilinear, Quadratic, and Hermitian Forms

We have noted, in the Prolegomenon, how a bilinear form can be characterized by a certain associated matrix. Thus the bilinear form

$$a_{11}x_1y_1 + a_{12}x_1y_2 + \cdots + a_{1n}x_1y_n$$
$$+ a_{21}x_2y_1 + a_{22}x_2y_2 + \cdots + a_{2n}x_2y_n$$
$$+ \quad \cdots \qquad \cdots \qquad \cdots \qquad \cdots$$
$$+ a_{m1}x_my_1 + a_{m2}x_my_2 + \cdots + a_{mn}x_my_n$$

is completely determined by the matrix $A = [a_{ij}]_{(m,n)}$. Indeed, if we denote the column vectors

$$\{x_1, x_2, \cdots, x_m\} \quad \text{and} \quad \{y_1, y_2, \cdots, y_n\}$$

by $X$ and $Y$, the reader can easily verify that the entire form can be written, in matrix notation, as the $1 \times 1$ matrix $X'AY$.

An important problem in the algebra of bilinear forms is the reduction of the form $X'AY$ to some simple standard type by subjecting $X$ and $Y$ to linear transformations $X = PU$ and $Y = QV$, where $P$ and $Q$ are numerical square matrices with nonvanishing determinants and

$$U = \{u_1, u_2, \cdots, u_m\}, \qquad V = \{v_1, v_2, \cdots, v_n\}.$$

Under such transformations, the original bilinear form $X'AY$ in $X$ and $Y$ becomes the bilinear form

$$(PU)'A(QV) = U'(P'AQ)V$$

in $U$ and $V$. Two bilinear forms related in this way are said to be *equivalent*. It is natural, then, to call the two associated matrices $A$ and $P'AQ$ *equivalent*, and the reduction problem mentioned above becomes that of determining two square matrices $P$ and $Q$, with nonvanishing determinants, such that $P'AQ$ is in some sort of simple standard form. We shall consider this problem later in the book, and there also take up other matrix matters that were originally motivated by the algebraic study of bilinear forms.

In the Prolegomenon we noted that a quadratic form can also be characterized by a certain associated matrix. For example, the quadratic form

$$\begin{aligned}
f \equiv \; & b_{11}x_1{}^2 &&+ b_{12}x_1x_2 &&+ \cdots + b_{1n}x_1x_n \\
& + b_{21}x_2x_1 &&+ b_{22}x_2{}^2 &&+ \cdots + b_{2n}x_2x_n \\
& + \cdots && \cdots && \cdots \qquad \cdots \\
& + b_{n1}x_nx_1 &&+ b_{n2}x_nx_2 &&+ \cdots + b_{nn}x_n{}^2
\end{aligned}$$

is completely determined by the square matrix $B = [b_{ij}]_{(n)}$. As in the case of the bilinear form given above, the reader can easily verify that

$$f = X'BX,$$

where $X = \{x_1, x_2, \cdots, x_n\}$. Now in Theorem 1.6.8 we showed that we may write $B$, uniquely, as $B = A + K$, where $A$ is symmetric and $K$ is skew-symmetric. It follows that

$$f = X'(A + K)X = X'AX + X'KX.$$

But, since $K = [k_{ij}]$ is skew-symmetric, $k_{ij} = -k_{ji}$ and

$$X'KX = \sum_{i,j=1}^{n} x_i k_{ij} x_j = 0.$$

It follows that $f = X'BX = X'AX$, where $A$ is symmetric. Since symmetric matrices possess many useful properties, it is customary to call the unique symmetric matrix $A$, rather than the matrix $B$, the *matrix* of the quadratic form. As in the case of bilinear forms, the algebraic study of quadratic forms has motivated large parts of matrix theory. For example, if $X$ is subjected to the linear transformation $X = PU$, where $P$ is square with nonvanishing determinant and $U = \{u_1, u_2, \cdots, u_n\}$, the quadratic form $X'BX$ becomes the quadratic form $U'(P'BP)U$. The two quadratic forms $X'BX$ and $U'(P'BP)U$, and also the two associated matrices $B$ and $P'BP$, are said to be *congruent*. Geometrically, the quadratic equation $X'BX = 0$ represents a hyperconicoid in projective $(n-1)$-dimensional space (a conic if $n = 3$, a conicoid if $n = 4$), and a linear transformation on $X$ with nonvanishing determinant is a projective transformation. It follows that the two hyperconicoids $X'BX = 0$ and $U'(P'BP)U = 0$ are *projectively congruent*, and it is apparent that the ability to simplify a quadratic form by a linear transformation has important geometrical implications. The study of congruent matrices has become a significant part of matrix theory and we shall consider the subject later in the book.

Another important kind of algebraic form is the so-called *Hermitian form*. This is a form that can be represented in matrix notation by the $1 \times 1$ matrix $X'A\overline{X}$, where $X = \{x_1, x_2, \cdots, x_n\}$ and $A$ is an $n \times n$ Hermitian matrix (that is, $A = A^*$). It is interesting—and perhaps at first a bit surprising—that the evaluation of a Hermitian form, obtained by replacing $X$ by a column vector of complex numbers, is always real. That this is so, however, was established in Theorem 1.8.9.

### 1.3A   A Matrix Approach to Complex Numbers

In the Prolegomenon we noted Sir William Rowan Hamilton's elegant representation of complex numbers as ordered pairs of real numbers subjected to certain definitions of addition and multiplication. Such a treatment removes all the mysticism that shrouds the classical approach, by which a complex number is considered as a strange hybrid of the form $a + bi$ in which $a$ and $b$ are real numbers and $i$ possesses the curious property that $i^2 = -1$. It is interesting that one can also give a (2 × 2)-matrix representation of complex numbers which accomplishes the same end. We merely outline the treatment and leave the completion of details to the interested reader. The treatment is based upon the fact that both the complex number $a + bi$ and the 2 × 2 matrix

$$\text{(1)} \qquad \begin{bmatrix} a & b \\ -b & a \end{bmatrix}$$

are completely determined by the two real numbers $a$ and $b$. Now, recalling that $a + bi = c + di$ if and only if $a = c$ and $b = d$,

$$(a + bi) + (c + di) = (a + c) + (b + d)i,$$

$$(a + bi)(c + di) = (ac - bd) + (ad + bc)i,$$

and noting that

$$\begin{bmatrix} a & b \\ -b & a \end{bmatrix} = \begin{bmatrix} c & d \\ -d & c \end{bmatrix}$$

if and only if $a = c$ and $b = d$,

$$\begin{bmatrix} a & b \\ -b & a \end{bmatrix} + \begin{bmatrix} c & d \\ -d & c \end{bmatrix} = \begin{bmatrix} a + c & b + d \\ -(b + d) & a + c \end{bmatrix},$$

$$\begin{bmatrix} a & b \\ -b & a \end{bmatrix} \begin{bmatrix} c & d \\ -d & c \end{bmatrix} = \begin{bmatrix} ac - bd & ad + bc \\ -(ad + bc) & ac - bd \end{bmatrix},$$

we see that we may define complex numbers to be matrices of the form (1), with equality, addition, and multiplication defined by matrix equality, addition, and Cayley multiplication. With these definitions it is easy to show that addition and multiplication of complex numbers are commutative and associative, and that multiplication is distributive over addition. The real numbers are embedded among the complex numbers by the association

$$\begin{bmatrix} a & 0 \\ 0 & a \end{bmatrix} \leftrightarrow a,$$

and, by Theorem 1.4.6, we may actually replace the matrix

$$\begin{bmatrix} a & 0 \\ 0 & a \end{bmatrix}$$

by the scalar $a$. Denoting the particular matrix

$$\begin{bmatrix} 0 & 1 \\ -1 & 0 \end{bmatrix}$$

by the symbol $i$, we then have

$$\begin{bmatrix} a & b \\ -b & a \end{bmatrix} = \begin{bmatrix} a & 0 \\ 0 & a \end{bmatrix} + \begin{bmatrix} 0 & b \\ -b & 0 \end{bmatrix}$$

$$= \begin{bmatrix} a & 0 \\ 0 & a \end{bmatrix} + b \begin{bmatrix} 0 & 1 \\ -1 & 0 \end{bmatrix} = a + bi,$$

where

$$i^2 = \begin{bmatrix} 0 & 1 \\ -1 & 0 \end{bmatrix}^2 = \begin{bmatrix} -1 & 0 \\ 0 & -1 \end{bmatrix} = -1,$$

and we have identified the matrix form of a complex number with the older classical form of a complex number.

### 1.4A    A Business Application†

In Example 9 of the Prolegomenon we considered a building contractor who constructs ranch style, Cape Cod style, and Colonial style houses. We assumed that 50, 70, and 60 units of steel are needed in the construction of the three styles of houses, respectively; and similarly, 200, 180, 250 units of wood; 160, 120, 80 units of glass; 70, 90, 50 units of paint; and 170, 210, 130 units of labor, respectively. This information was represented by the $3 \times 5$ matrix

$$A = \begin{array}{c} \\ \\ \end{array} \begin{array}{ccccc} \text{steel} & \text{wood} & \text{glass} & \text{paint} & \text{labor} \\ \begin{bmatrix} 50 & 200 & 160 & 70 & 170 \\ 70 & 180 & 120 & 90 & 210 \\ 60 & 250 & 80 & 50 & 130 \end{bmatrix} & & & & \end{array} \begin{array}{c} \text{ranch} \\ \text{Cape Cod} \\ \text{Colonial} \end{array}$$

Now let us suppose that a unit of steel costs \$30, a unit of wood costs \$16, a unit of glass costs \$10, a unit of paint costs \$2, and a unit of labor costs \$20. This information of the cost in dollars per unit of each of the needed raw materials may be represented by the matrix

$$B = \begin{bmatrix} 30 \\ 16 \\ 10 \\ 2 \\ 20 \end{bmatrix} \begin{array}{l} \text{steel} \\ \text{wood} \\ \text{glass} \\ \text{paint} \\ \text{labor} \end{array}$$

† Suggested by an example in J. G. Kemeny, J. L. Snell, G. L. Thompson, *Introduction to Finite Mathematics* (Prentice-Hall, Inc., 1957), p. 191.

Finally, suppose the contractor has orders for 5 ranch style houses, 7 Cape Cod houses, and 12 Colonial houses. We may represent this information by the matrix

$$C = [5, \quad 7, \quad 12].$$

(with column labels: ranch, Cape Cod, Colonial)

Then (as the reader can easily verify), the Cayley product $CA$ tells the contractor the number of units of each raw material he must order; the Cayley product $AB$ gives the cost of each of the three styles of houses; the Cayley product $CAB$ yields the total cost of materials and labor for the construction of all 24 houses. We note that

$$CA = [1460, \quad 5260, \quad 2600, \quad 1580, \quad 3880],$$

$$AB = \begin{bmatrix} 9{,}840 \\ 10{,}560 \\ 9{,}300 \end{bmatrix},$$

$$CAB = [234{,}720].$$

That is, the contractor must order 1460 units of steel, 5260 units of wood, 2600 units of glass, 1580 units of paint, and 3880 units of labor. The total cost of these materials and labor is $9840 for a ranch style house, $10,560 for a Cape Cod house, and $9300 for a Colonial house. The cost to the contractor for the construction of all 24 houses is $234,720.

### 1.5A   Enumeration of $k$-Stage Routes

Consider five cities, $C_1, C_2, C_3, C_4, C_5$, connected by roads as indicated in Figure 5. We shall call a route connecting two cities a *k-stage route* if one must pass through $k - 1$ cities while traversing the route. Thus the route ($C_1$ to $C_3$, $C_3$ to $C_5$, $C_5$ to $C_4$) is a 3-stage route from $C_1$ to $C_4$. With no trouble one can construct the matrix

$$M(1) = \begin{bmatrix} 0 & 1 & 1 & 1 & 0 \\ 1 & 1 & 1 & 0 & 1 \\ 1 & 1 & 0 & 1 & 1 \\ 1 & 0 & 1 & 0 & 1 \\ 0 & 1 & 1 & 1 & 0 \end{bmatrix}$$

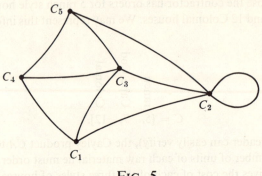

**FIG. 5**

in which $m(1)_{ij}$ is the number of 1-stage routes from $C_i$ to $C_j$. By simple counting, one similarly finds

$$M(2) = \begin{bmatrix} 3 & 2 & 2 & 1 & 3 \\ 2 & 4 & 3 & 3 & 2 \\ 2 & 3 & 4 & 2 & 2 \\ 1 & 3 & 2 & 3 & 1 \\ 3 & 2 & 2 & 1 & 3 \end{bmatrix}$$

in which $m(2)_{ij}$ is the number of 2-stage routes from $C_i$ to $C_j$. In general, we shall let $M(k)$ denote the matrix in which $m(k)_{ij}$ is the number of $k$-stage routes from $C_i$ to $C_j$. It is clear that the construction of $M(10)$, say, by the actual counting of the 10-stage routes, would be difficult, tedious, confusing, and probably unsystematic. It is therefore interesting to learn that $M(k) = [M(1)]^k$ and that Cayley multiplication of matrices thus furnishes an elegant and systematic procedure for solving the general problem. The proof of the general relation $M(k) = [M(1)]^k$ (by mathematical induction, say) is easy and is left to the interested reader.

The entire discussion above also holds for networks in which the roads are *directed*. Thus, if we have the network of directed roads illustrated in Figure 6, the associated matrices $M(1)$ and $M(2)$ become

$$M(1) = \begin{bmatrix} 0 & 0 & 1 & 0 & 0 \\ 1 & 1 & 0 & 0 & 0 \\ 0 & 1 & 0 & 1 & 1 \\ 1 & 0 & 0 & 0 & 1 \\ 0 & 1 & 0 & 0 & 0 \end{bmatrix}, \quad M(2) = \begin{bmatrix} 0 & 1 & 0 & 1 & 1 \\ 1 & 1 & 1 & 0 & 0 \\ 2 & 2 & 0 & 0 & 1 \\ 0 & 1 & 1 & 0 & 0 \\ 1 & 1 & 0 & 0 & 0 \end{bmatrix}.$$

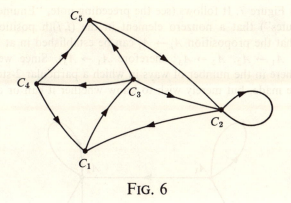

$$\text{F{\small IG}. 6}$$

Here, as in the preceding situation, $M(2) = [M(1)]^2$, and in general, $M(k) = [M(1)]^k$.

### 1.6A   Application to Mathematical Systems†

An intriguing application of the enumeration of the $k$-stage routes in a network can be made to a study of abstract mathematical systems. As an illustration, suppose we are given a set of primitive terms and the set $A_1, \cdots, A_6$ of six conditions involving them. Also, let us be given the set of postulates $A_1 \to A_2$ (read "$A_1$ implies $A_2$"), $A_2 \to A_3$, $A_2 \to A_4$, $A_4 \to A_5$, $A_6 \to A_3$, $A_1 \to A_6$, $A_5 \to A_6$. We may represent this postulate system by the matrix

$$M = \begin{bmatrix} 1 & 1 & 0 & 0 & 0 & 1 \\ 0 & 1 & 1 & 1 & 0 & 0 \\ 0 & 0 & 1 & 0 & 0 & 0 \\ 0 & 0 & 0 & 1 & 1 & 0 \\ 0 & 0 & 0 & 0 & 1 & 1 \\ 0 & 0 & 1 & 0 & 0 & 1 \end{bmatrix},$$

where $m_{ij} = 1$ if and only if we are given $A_i \to A_j$; since $A_i \to A_i$ for all $i$, we see that 1's also appear everywhere along the principal diagonal. This matrix is the matrix representing the 1-stage routes in the directed network

† See F. D. Parker, "Boolean matrices and logic," *Mathematics Magazine*, Vol. 37, No. 1 (January 1964), pp. 33–38.

pictured in Figure 7. It follows (see the preceding note, "Enumeration of $k$-stage routes") that a nonzero element in the $(i,j)$th position of $M^2$ indicates that the proposition $A_i \rightarrow A_j$ can be established in at most two steps (like $A_1 \rightarrow A_2$, $A_2 \rightarrow A_3$; therefore $A_1 \rightarrow A_3$). Since we are not interested here in the number of ways in which a particular 2-step deduction can be made, but merely wish to know whether it can or cannot be

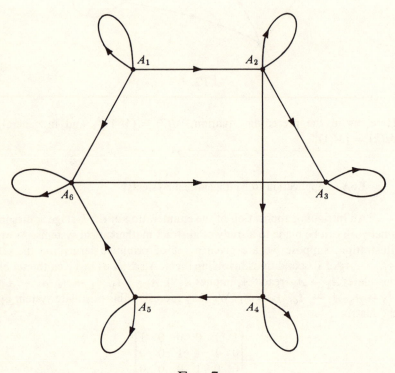

FIG. 7

made, we reduce all nonzero elements in $M^2$ to 1's. This can be automatically accomplished if in computing $M^2$ as the Cayley product of $M$ and $M$, we subject the scalars of $M$ to the *Boolean* laws: $0 \cdot 0 = 1 \cdot 0 = 0 \cdot 1 = 0$, $1 \cdot 1 = 1$, $1 + 1 = 1 + 0 = 0 + 1 = 1$, $0 + 0 = 0$. Adopting these Boolean laws, it follows that, in general, a unit element in the $(i,j)$th position of $M^k$ indicates that the proposition $A_i \rightarrow A_j$ can be established in at most $k$ steps. Since, clearly, no proposition of the form $A_i \rightarrow A_j$

would, in our example based on precisely six initial conditions, require more than five steps, it follows that all deducible implications of the form $A_i \to A_j$ are indicated by the positions of the 1's in $M^5$.

Suppose we have two postulate sets of the sort under consideration involving the same primitive terms and the same set of initial conditions $A_1, \cdots, A_6$. Suppose the two systems are represented by the matrices $M$ and $N$, respectively. Then, clearly, the two postulate sets are equivalent if and only if $M^5 = N^5$, and it is thus possible to use our matrix approach to determine the equivalence or nonequivalence of any two postulate sets employing the same primitive terms and initial conditions. Also, it is no surprise to the electrical engineer, who uses matrices to analyze switching circuits, that a simple electrical system can be built for a given mathematical system such as we are considering, and that the theorems of the form $A_i \to A_j$ can be discovered by watching a panel of lights.

Let $S$ be the set of all $n \times n$ matrices with 1's along the principal diagonal and with 0's and 1's elsewhere, where these scalars are subject to the Boolean laws of addition and multiplication given above. It then follows that $S$ is closed under matrix addition and multiplication, that addition is commutative and associative, that multiplication is associative and both right- and left-distributive over addition. It also follows that $I_{(n)}$, which is a member of $S$, serves as both an additive and a multiplicative identity; that is, if $M$ is any member of $S$, then $M + I_{(n)} = I_{(n)} + M = M$ and $MI_{(n)} = I_{(n)}M = M$. This suggests the study of abstract systems consisting of a nonempty set $S$ of elements in which two binary operations $(+)$ and $(\cdot)$ are defined satisfying the following postulates:

**P1**: For all $a, b$ in $S$, $a + b = b + a$.

**P2**: For all $a, b, c$ in $S$, $a + (b + c) = (a + b) + c$.

**P3**: For all $a, b, c$ in $S$, $a(bc) = (ab)c$.

**P4**: For all $a, b, c$ in $S$, $a(b + c) = (ab) + (ac), (b + c)a = (ba) + (ca)$.

**P5**: There exists an element $e$ of $S$ such that, for all $a$ in $S$, $a + e = a$ and $ae = ea = a$.

## 1.7A   Jordan and Lie Products of Matrices

The reader will recall that, in the Prolegomenon, it was mentioned that there are "products" of matrices other than the Cayley product. Here we note two of these products, one of which appears in connection with the mathematical study of quantum theory, and the other in a matrix representation of the Lie algebras associated with the study of continuous groups.

The first product is the *Jordan product*, named after the physicist Pascual Jordan (contemporary), one of the founders of modern quantum mechanics. This product, which we shall indicate by $A * B$, is defined for any two $n \times n$ $\mathscr{C}$ matrices $A$ and $B$ as the arithmetic mean of the two Cayley products $AB$ and $BA$; that is,

$$A * B = (AB + BA)/2.$$

It is easily seen that this product is commutative but, in general, nonassociative. The (special) *Jordan algebra* of $n \times n$ $\mathscr{C}$ matrices for which addition and scalar multiplication are defined as in Section 1.2, but for which matrix multiplication is defined as above, can easily be shown to satisfy the following laws (in which $A$, $B$, $C$ are arbitrary $n \times n$ $\mathscr{C}$ matrices and $k$ is an arbitrary complex number):

> **J1**: $A * B = B * A$,
>
> **J2**: $(kA) * B = A * (kB) = k(A * B)$,
>
> **J3**: $A * (B + C) = (A * B) + (A * C)$,
>
> **J4**: $(B + C) * A = (B * A) + (C * A)$,
>
> **J5**: $A * (B * A^2) = (A * B) * A^2$, where $A^2 = A * A = AA$.

The last relation is a special associative law. The initial study of *nonassociative algebras* satisfying the preceding five laws occurs in a paper by Jordan, published in 1933; the name *Jordan algebra* was introduced by A. A. Albert in 1946.

The second product is the *Lie product*, named after the Norwegian mathematician Marius Sophus Lie (1842–1899), who did important inaugural work in the study of continuous groups. This product, which we shall indicate by $A \times B$, is defined for any two $n \times n$ $\mathscr{C}$ matrices $A$ and $B$ as the difference of the two Cayley products $AB$ and $BA$; that is,

$$A \times B = AB - BA.$$

It is seen that this product is, in general, both noncommutative and nonassociative. The set of all $n \times n$ $\mathscr{C}$ matrices for which addition and scalar multiplication are defined as in Section 1.2, but for which matrix multiplication is defined as above, furnishes a matrix representation of a *Lie algebra*. The reader might care to verify the following laws, which hold in this nonassociative matrix algebra ($A, B, C$ represent arbitrary $n \times n$ $\mathscr{C}$ matrices and $k$ represents an arbitrary complex number):

**L1**: $A \times B = -(B \times A)$,

**L2**: $(kA) \times B = A \times (kB) = k(A \times B)$,

**L3**: $A \times (B + C) = (A \times B) + (A \times C)$,

**L4**: $(B + C) \times A = (B \times A) + (C \times A)$,

**L5**: $A \times (B \times C) + B \times (C \times A) + C \times (A \times B) = O$.

The last relation is known as the *Jacobi identity*. As a further interesting fact, it is easy to show that:

**L6**: If $A$ and $B$ are skew-symmetric, then $A \times B$ is skew-symmetric.

The Jordan and Lie matrix products are connected by identities. For example, it is not difficult to show that for any $n \times n$ $\mathscr{C}$ matrices $A, B, C$,

$$A \times (B * B) = 2[(A \times B) * B],$$

$$A \times (B \times C) = 4[(A * B) * C - (A * C) * B].$$

The Cayley, Jordan, and Lie products are connected by the identity

$$AB = (A * B) + (A \times B)/2.$$

Finally, let us consider the correspondence

$$\begin{bmatrix} 0 & u_1 & -u_2 \\ -u_1 & 0 & u_3 \\ u_2 & -u_3 & 0 \end{bmatrix} \leftrightarrow [u_1, u_2, u_3]$$

between a $3 \times 3$ skew-symmetric matrix and an associated vector of order 3. The matrix is called the *skew-symmetric matrix of the vector*, and the vector is called the *vector of the skew-symmetric matrix*. Recalling from elementary vector analysis the *scalar*, *dot*, or *inner product*,

$$[u_1, u_2, u_3] \cdot [v_1, v_2, v_3] = u_1 v_1 + u_2 v_2 + u_3 v_3,$$

of two third-order vectors, and the *vector*, *cross*, or *outer product*,

$$[u_1, u_2, u_3] \times [v_1, v_2, v_3] = [u_2 v_3 - v_2 u_3, u_3 v_1 - v_3 u_1, u_1 v_2 - v_1 u_2],$$

of the same two vectors, one can easily establish the following:

Let $A$ be the skew-symmetric matrix of the vector $U = [u_1, u_2, u_3]$, let $B$ be the skew-symmetric matrix of the vector $V = [v_1, v_2, v_3]$, and let

$k$ be any scalar. Then

    (a) $A + B$ is the skew-symmetric matrix of the vector $U + V$,

    (b) $kA$ is the skew-symmetric matrix of the vector $kU$,

    (c) $A \times B$ is the skew-symmetric matrix of the vector $U \times V$,

    (d) $U \cdot V = -(\text{tr } AB)/2$.

A very satisfying program is the establishment of the basic identities of vector algebra by the use of matrix algebra and the relations listed above.† Relation (c) furnishes an extension of the notion of the vector product of two third-order vectors to a vector product of any two vectors of order $n(n - 1)/2$.

### 1.8A  Square Roots of Matrices

If $z$ is a complex number and $n$ a positive integer, then it is well known that $z$ possesses exactly $n$ $n$th roots in the set of complex numbers. This raises the question as to whether some similar state of affairs exists for, say, square $\mathscr{C}$ matrices. That the situation is here considerably more complicated is attested to by the following three facts:

**I**  *Some square $\mathscr{C}$ matrices have an infinite number of square roots.* Since, for any complex number $k$,

$$\begin{bmatrix} k & 1 + k \\ 1 - k & -k \end{bmatrix}^2 = \begin{bmatrix} 1 & 0 \\ 0 & 1 \end{bmatrix},$$

it follows that $I_{(2)}$ has an infinite number of square roots.

**II**  *Some square $\mathscr{C}$ matrices have no square roots.* Suppose

$$\begin{bmatrix} 0 & 1 \\ 0 & 0 \end{bmatrix} = \begin{bmatrix} a & b \\ c & d \end{bmatrix}^2 = \begin{bmatrix} a^2 + bc & b(a + d) \\ c(a + d) & d^2 + bc \end{bmatrix}.$$

Then we must have

(1)                  $b(a + d) = 1,$

(2)                  $c(a + d) = 0,$

(3)                  $a^2 + bc = 0,$

(4)                  $d^2 + bc = 0.$

† See, for example, J. T. Schwartz, *Introduction to Matrices and Vectors*, McGraw-Hill Book Co., Inc., Section 6-4.

From (1) it follows that

(5) $$a + d \neq 0.$$

Therefore, from (2), we see that $c = 0$, and thence, from (3) and (4), that $a = d = 0$. But this contradicts (5). It follows that the matrix

$$\begin{bmatrix} 0 & 1 \\ 0 & 0 \end{bmatrix}$$

has no square root.

**III**   *Some square $\mathscr{C}$ matrices have only a (positive) finite number of square roots.*

By straightforward algebra the reader can show that the matrix

$$\begin{bmatrix} 3 & -4 \\ 1 & -1 \end{bmatrix}$$

has only the matrices

$$\pm \begin{bmatrix} 2 & -2 \\ 1/2 & 0 \end{bmatrix}$$

for square roots.

The preceding facts were to be suspected, since

$$\begin{bmatrix} a & b \\ c & d \end{bmatrix}^2 = \begin{bmatrix} a^2 + bc & b(a + d) \\ c(a + d) & d^2 + bc \end{bmatrix} = \begin{bmatrix} r & s \\ u & v \end{bmatrix}$$

implies that

$$a^2 + bc = r, \qquad b(a + d) = s, \qquad c(a + d) = u, \qquad d^2 + bc = v,$$

and this system of equations may, for given $r$, $s$, $u$, $v$, possess no, a (positive) finite number, or an infinite number of solutions. And, of course, a similar statement can be made concerning $n$th roots of any square $\mathscr{C}$ matrix.

The investigation of $n$th roots of matrices furnishes a somewhat involved study that is, of course, linked with the general study of matrix equations. The reader can find, in some of the textbook literature, methods of finding such roots, when they exist.

## 1.9A   Primitive Factorization of Matrices

The set $S$ of positive integers greater than 1 has the remarkable property that there exists a subset $P$ of $S$, namely, the set of prime numbers, such that the set $S - P$ is an infinite set and any member of $S$ can be expressed as a product of members of $P$. The prime numbers are thus building bricks from which all positive integers greater than 1 can be constructed multiplicatively. One naturally wonders if in the set $T$ of all $n \times n$ $\mathscr{C}$ matrices, say, there similarly exists a subset $Q$ such that $T - Q$ is infinite and any member of $T$ can be expressed as the Cayley product of members of $Q$. We state, without proof, two interesting theorems along this line.

**I  THEOREM** (J. Wellstein, 1909).    *Every* n × n $\mathscr{C}$ *matrix can be written as a product of $\mathscr{C}$ matrices of the following types*: $E_{ij}(k)$, *obtained by replacing the* (i, j)*th element of* $I_{(n)}$ *by the complex number* k; $\Gamma$, *obtained by cyclicly permuting the rows of* $I_{(n)}$,

**II  THEOREM** (C. Cellitti, 1914).    *Every* 2 × 2 *matrix with integral elements can be written as a product of powers of*

$$\begin{bmatrix} 1 & 1 \\ 0 & 1 \end{bmatrix}, \qquad \begin{bmatrix} 1 & 0 \\ 1 & 1 \end{bmatrix},$$

*and matrices of the form*

$$\begin{bmatrix} a & 0 \\ 0 & 1 \end{bmatrix},$$

*where* a *is an integer.*

Essentially the first theorem will be established in Chapter 2.

# 2. EQUIVALENCE

*2.1. Row equivalence of matrices. Problems.  2.2. Nonsingular matrices. Problems.  2.3. Column equivalence of matrices.  2.4. Equivalence of matrices. Problems.  2.5. Linear dependence and independence of a set of vectors.  2.6. Row rank and column rank of a matrix. Problems.  2.7. Rank of a matrix. Problems.  2.8. Application to the solution of systems of linear equations. Problems.  2.9. Linearly independent solutions of systems of linear equations. Problems.  ADDENDA.  2.1A. Left and right inverses.  2.2A. Some further methods of matrix inversion.  2.3A. Linear dependence and independence of a set of matrices.  2.4A. Lines in a plane and planes in space.  2.5A. An affine classification of conics and conicoids according to the ranks of their associated matrices.  2.6A. Kronecker product of matrices.  2.7A. Direct sum of matrices.*

## 2.1  Row equivalence of matrices

We commence with some motivation for subsequent discussion by considering, for the moment, the simultaneous solution of the pair of linear equations

$$3x - y = 9,$$
$$x + 2y = 3.$$

Employing the method of elimination, one might first multiply the first equation by 2 and add to this the second equation, to obtain $7x = 21$, and then divide by 7 (that is, multiply by $1/7$) to obtain $x = 3$. Next one might subtract the equation $x = 3$ from the second equation (or, in other words, add $-1$ times the equation $x = 3$ to the second equation), to obtain $2y = 0$, and then divide the resulting equation by 2 (that is, multiply by $1/2$), to obtain $y = 0$. We repeat these steps in the first column below.

$$
\begin{array}{c|c|c}
\begin{aligned} 3x - y &= 9 \\ x + 2y &= 3 \end{aligned} &
\begin{bmatrix} 3 & -1 \\ 1 & 2 \end{bmatrix}\begin{bmatrix} x \\ y \end{bmatrix} = \begin{bmatrix} 9 \\ 3 \end{bmatrix} &
\left[\begin{array}{cc|c} 3 & -1 & 9 \\ 1 & 2 & 3 \end{array}\right] \\[2mm]
\hline
\begin{aligned} 6x - 2y &= 18 \\ x + 2y &= 3 \end{aligned} &
\begin{bmatrix} 6 & -2 \\ 1 & 2 \end{bmatrix}\begin{bmatrix} x \\ y \end{bmatrix} = \begin{bmatrix} 18 \\ 3 \end{bmatrix} &
\left[\begin{array}{cc|c} 6 & -2 & 18 \\ 1 & 2 & 3 \end{array}\right] \\[2mm]
\hline
\begin{aligned} 7x &= 21 \\ x + 2y &= 3 \end{aligned} &
\begin{bmatrix} 7 & 0 \\ 1 & 2 \end{bmatrix}\begin{bmatrix} x \\ y \end{bmatrix} = \begin{bmatrix} 21 \\ 3 \end{bmatrix} &
\left[\begin{array}{cc|c} 7 & 0 & 21 \\ 1 & 2 & 3 \end{array}\right] \\[2mm]
\hline
\begin{aligned} x &= 3 \\ x + 2y &= 3 \end{aligned} &
\begin{bmatrix} 1 & 0 \\ 1 & 2 \end{bmatrix}\begin{bmatrix} x \\ y \end{bmatrix} = \begin{bmatrix} 3 \\ 3 \end{bmatrix} &
\left[\begin{array}{cc|c} 1 & 0 & 3 \\ 1 & 2 & 3 \end{array}\right] \\[2mm]
\hline
\begin{aligned} x &= 3 \\ 2y &= 0 \end{aligned} &
\begin{bmatrix} 1 & 0 \\ 0 & 2 \end{bmatrix}\begin{bmatrix} x \\ y \end{bmatrix} = \begin{bmatrix} 3 \\ 0 \end{bmatrix} &
\left[\begin{array}{cc|c} 1 & 0 & 3 \\ 0 & 2 & 0 \end{array}\right] \\[2mm]
\hline
\begin{aligned} x &= 3 \\ y &= 0 \end{aligned} &
\begin{bmatrix} 1 & 0 \\ 0 & 1 \end{bmatrix}\begin{bmatrix} x \\ y \end{bmatrix} = \begin{bmatrix} 3 \\ 0 \end{bmatrix} &
\left[\begin{array}{cc|c} 1 & 0 & 3 \\ 0 & 1 & 0 \end{array}\right] \\[2mm]
\hline
& \begin{bmatrix} x \\ y \end{bmatrix} = \begin{bmatrix} 3 \\ 0 \end{bmatrix} &
\end{array}
$$

The strategy of the solution was successively to convert the original pair of equations into *equivalent* pairs (that is, into pairs having precisely the same simultaneous solution as the original pair), finally obtaining the pair $x = 3$, $y = 0$, which constitutes the sought solution. The permissible operations on the two equations, which always led to an equivalent pair, are: (1) to multiply either equation by a nonzero constant, (2) to add to either equation a constant multiple of the other equation. By applying an adroit combination of these operations, we finally arrived at a pair of equations from which the sought solution could be read by inspection.

The second column in the steps given above merely reproduces the first column in a matrix form for each pair of equations. It is to be noted that now the basic operations are performed on the rows of the numerical matrices involved. That is, we multiply either row by a nonzero constant or we add to either row a constant multiple of the other row.

In the third column we do away with the essentially extraneous column vector $\{x, y\}$ and perform the basic operations on the rows of the $2 \times 3$ partitioned matrix of coefficients. When, by the basic operations we reduce the partitioned matrix to the form $[I_{(2)}|S]$, then the column vector $S$ constitutes the sought simultaneous solution of the original pair of linear equations.

We finally note that interchanging the two equations, and hence also interchanging the two rows of the associated partitioned matrix, is certainly also quite permissible, as long as we end up with a partitioned matrix of the form $[I_{(2)}|S]$. All this leads us to introduce the following definitions and accompanying notation.

**2.1.1** DEFINITIONS AND NOTATION. Let $A$ be an $m \times n$ matrix and denote the successive rows of $A$ by $A_i$, $i = 1, 2, \cdots, m$. The *elementary row operations* on matrix $A$ are:

(1) $o(j, k)$, the interchange of rows $A_j$ and $A_k$,

(2) $o(ck)$, the multiplication of row $A_k$ by the nonzero scalar $c$,

(3) $o(j + ak)$, the addition of $a$ times row $A_k$ to row $A_j$, $j \neq k$.

If $A$ and $B$ are two $m \times n$ matrices, then we say $A$ is *row-equivalent* to $B$ if and only if $B$ is obtainable from $A$ by the successive application of finitely many elementary row operations; we write $A \overset{\text{RE}}{=} B$.

**2.1.2** THEOREM. *The inverses of the elementary row operations* $o(j, k)$, $o(ck)$, $o(j + ak)$ *are the elementary row operations* $o(j, k)$, $o(c^{-1}k)$, $o(j - ak)$, *respectively.*

We leave the easy proof to the reader.

**2.1.3** THEOREM. *Row equivalence of matrices is reflexive, symmetric, and transitive.*

If $A$ is any matrix, then $A \overset{\text{RE}}{=} A$, for we may obtain $A$ from $A$ by the elementary row operation $o(1 \cdot 1)$. Again, if $A \overset{\text{RE}}{=} B$, then $B \overset{\text{RE}}{=} A$, for $B$ may be converted into $A$ by the sequence of elementary row operations composed of the inverses of the elementary row operations, taken in reverse order, that convert $A$ into $B$. Finally, if $A \overset{\text{RE}}{=} B$ and $B \overset{\text{RE}}{=} C$, then $A \overset{\text{RE}}{=} C$, for $C$ may be obtained from $A$ by the elementary row operations that convert $A$ into $B$, followed by the elementary row operations that convert $B$ into $C$.

*Note.* If $A$ is row-equivalent to $B$, we may, because of the symmetric property of the relationship, say "$A$ and $B$ are row-equivalent."

**2.1.4** DEFINITION. A matrix is said to be in *modified triangular form*, or *reduced echelon form*, if

(1) all the nonzero rows (if any) precede all the zero rows (if any),

(2) in any nonzero row, after the first row, the number of zeros preceding the first nonzero element is greater than the number of such zeros in the preceding row,

(3) the first nonzero element in each nonzero row is 1, and

(4) the first nonzero element in each nonzero row is the only nonzero element in its column.

*Example*

$$\begin{bmatrix} 0 & 1 & -2 & 3 & 0 & 0 & 1 & 2 \\ 0 & 0 & 0 & 0 & 1 & 0 & 2 & 4 \\ 0 & 0 & 0 & 0 & 0 & 1 & 3 & -1 \\ 0 & 0 & 0 & 0 & 0 & 0 & 0 & 0 \\ 0 & 0 & 0 & 0 & 0 & 0 & 0 & 0 \end{bmatrix}$$

**2.1.5** THEOREM. *An* m × n *matrix* A *is row-equivalent to an* m × n *matrix in modified triangular form.*

Either the first column of $A$ contains only zeros or, for some $k$, $a_{k1} \neq 0$. In this latter case, subject $A$ to the elementary row operations $o(1, k)$, $o(1/a_{k1} \cdot 1)$, and, if $a_{j1} \neq 0$ for $j \neq k$, $o(j - a_{j1} \cdot 1)$. It follows that $A$ is row-equivalent to a matrix of one of the forms

$$[O_{(m,1)} \quad C_{(m,n-1)}], \qquad \begin{bmatrix} 1 & D_{(1,n-1)} \\ O_{(m-1,1)} & E_{(m-1,n-1)} \end{bmatrix}.$$

In the first case, repeat the foregoing process with matrix $C$; in the second case, repeat the process with matrix $E$. If, in the latter case, $E$ is replaced by a matrix with 1 in its first row and column, a suitable multiple of this row in the whole matrix can be added to the first row to make the first element of $D$ zero without changing the first column of the whole matrix. Continuing this way, we finally achieve the desired modified triangular form.

*Example*

$$\begin{bmatrix} 1 & 2 & -1 & 4 \\ 3 & 2 & 0 & 2 \\ 0 & 1 & 3 & 2 \\ 3 & 3 & 3 & 4 \end{bmatrix} \xrightarrow[\substack{o(2-3\cdot1) \\ o(4-3\cdot1)}]{} \begin{bmatrix} 1 & 2 & -1 & 4 \\ 0 & -4 & 3 & -10 \\ 0 & 1 & 3 & 2 \\ 0 & -3 & 6 & -8 \end{bmatrix}$$

$$\xrightarrow[o(2,3)]{} \begin{bmatrix} 1 & 2 & -1 & 4 \\ 0 & 1 & 3 & 2 \\ 0 & -4 & 3 & -10 \\ 0 & -3 & 6 & -8 \end{bmatrix} \xrightarrow[\substack{o(1-2\cdot2) \\ o(3+4\cdot2) \\ o(4+3\cdot2)}]{} \begin{bmatrix} 1 & 0 & -7 & 0 \\ 0 & 1 & 3 & 2 \\ 0 & 0 & 15 & -2 \\ 0 & 0 & 15 & -2 \end{bmatrix}$$

$$\xrightarrow[o(1/15\cdot3)]{}\begin{bmatrix}1 & 0 & -7 & 0\\0 & 1 & 3 & 2\\0 & 0 & 1 & -2/15\\0 & 0 & 15 & -2\end{bmatrix}\xrightarrow[\substack{o(1+7\cdot3)\\o(2-3\cdot3)\\o(4-15\cdot3)}]{}\begin{bmatrix}1 & 0 & 0 & -14/15\\0 & 1 & 0 & 36/15\\0 & 0 & 1 & -2/15\\0 & 0 & 0 & 0\end{bmatrix}$$

**2.1.6** DEFINITION AND NOTATION. A matrix obtained by applying an elementary row operation to an identity matrix $I$ is called an *elementary row transformation matrix*, or, in abbreviated form, an *ERT matrix*. We shall denote them by the self-explanatory symbolism $R(j, k)$, $R(ck)$, $R(j + ak)$.

*Examples.* If $I$ is of order three, then

$$R(2, 3) = \begin{bmatrix}1 & 0 & 0\\0 & 0 & 1\\0 & 1 & 0\end{bmatrix}, \qquad R(2\cdot3) = \begin{bmatrix}1 & 0 & 0\\0 & 1 & 0\\0 & 0 & 2\end{bmatrix},$$

$$R(1 - 2\cdot3) = \begin{bmatrix}1 & 0 & -2\\0 & 1 & 0\\0 & 0 & 1\end{bmatrix}.$$

**2.1.7** LEMMA. *If* A *and* B *are suitably conformable matrices, then* $(AB)_i = A_iB$.

The easy proof is left to the reader.

**2.1.8** THEOREM. *The application of an elementary row operation to matrix* A *can be effected by premultiplication of* A *by a corresponding ERT matrix.*

1. Applying Lemma 2.1.7, we note that

$$[R(j, k)A]_j = [R(j, k)]_j A = I_k A = (IA)_k = A_k;$$

$$[R(j, k)A]_k = [R(j, k)]_k A = I_j A = (IA)_j = A_j;$$

$$[R(j, k)A]_i = [R(j, k)]_i A = I_i A = (IA)_i = A_i, \qquad i \neq j, k.$$

Therefore $R(j, k)A$ is equivalent to applying elementary row operation $o(j, k)$ to $A$.

2. Again,

$$[R(ck)A]_k = [R(ck)]_k A = cI_k A = c(IA)_k = cA_k;$$

$$[R(ck)A]_i = [R(ck)]_i A = I_i A = (IA)_i = A_i, \qquad i \neq k.$$

Therefore $R(ck)A$ is equivalent to applying elementary row operation $o(ck)$ to $A$.

3. Finally,

$$[R(j+ak)A]_j = [R(j+ak)]_j A = (I_j + aI_k)A = I_j A + aI_k A = A_j + aA_k;$$

$$[R(j+ak)A]_i = [R(j+ak)]_i A = I_i A = A_i, \qquad i \neq j.$$

Therefore $R(j+ak)A$ is equivalent to applying elementary row operation $o(j+ak)$ to $A$.

*Note.* This theorem suggests the *inverse golden rule*, "Do unto $I$ as you would do unto $A$."

**2.1.9** COROLLARY. *If* A *and* B *are two* m × n *row-equivalent matrices, then* B = SA, *where* S *is a product of* mth-order ERT *matrices.*

## PROBLEMS

**2.1–1**   Prove Theorem 2.1.2.

**2.1–2**   Reduce $\begin{bmatrix} 3 & -2 & 1 & 6 \\ 2 & 5 & -3 & -2 \\ 4 & -9 & 5 & 14 \end{bmatrix}$ to modified triangular form.

**2.1–3**   Reduce $\begin{bmatrix} 4 & 7 & -14 & 10 \\ 2 & 3 & -4 & -4 \\ 1 & 1 & 1 & 6 \end{bmatrix}$ to modified triangular form.

**2.1–4**   Reduce the transpose of the matrix in Problem 2.1–2 to modified triangular form.

**2.1–5**   Prove that two matrices in modified triangular form are row-equivalent if and only if they are equal, and thus sharpen Theorem 2.1.5 to: "An $m \times n$ matrix $A$ is row equivalent to a *unique* $m \times n$ matrix in modified triangular form."

**2.1–6**   Show that $R(j - ak) = R(-1k)R(j + ak)R(-1k)$.

**2.1–7**   Let $E_{(n)}(r, s)$ denote the $n \times n$ matrix having a 1 at the intersection of its $r$th row and $s$th column, and having 0's everywhere else. Show that, where all matrices are assumed to be $n \times n$,

(a)  $R(j, k) = I - E(j, j) + E(k, j) - E(k, k) + E(j, k)$,

(b)  $R(ck) = I + (c - 1)E(k, k)$,

(c)  $R(j + ak) = I + aE(j, k)$,

**(d)** $E(r, s)$ is idempotent if $r = s$ and is nilpotent of index 2 if $r \neq s$.

**2.1–8** Show that

**(a)** $R(j + ak) = R(a^{-1}k)R(j + k)R(ak)$,

**(b)** $R(j, k) = R(-1k)R(k + j)R(-1j)R(j + k)R(-1k)R(k + j)$,

thus showing that ERT matrices of the types $R(j + ak)$ and $R(j, k)$ can be defined in terms of ERT matrices of the types $R(ck)$ and $R(j + k)$. It follows that any elementary row operation can be performed by a finite sequence of elementary row operations of the types $o(ck)$ and $o(j + k)$.

**2.1–9** Prove Lemma 2.1.7.

**2.1–10** Prove Corollary 2.1.9.

**2.1–11** Factor $\begin{bmatrix} 1 & -1 & 0 \\ -1 & 2 & 0 \\ 0 & 0 & 1 \end{bmatrix}$ into a product of ERT matrices.

**2.1–12** If, in the product $AB$, $A$ and $B$ are in modified triangular form, does it necessarily follow that $AB$ is in modified triangular form?

## 2.2 Nonsingular matrices

In this section we introduce the important concept of *nonsingular matrix*.

**2.2.1** DEFINITIONS. A matrix $A$ is said to be *nonsingular* if a matrix $B$ exists such that $BA = AB = I$. If $B$ exists, it is called an *inverse* of $A$. If $A$ has no inverse, $A$ is said to be *singular*.

**2.2.2** THEOREM. *A nonsingular matrix* A *is square and has a unique inverse.*

That $A$ is square follows from Theorem 1.3.5. Now suppose $B$ and $C$ are inverses of $A$. Then $C = CI = C(AB) = (CA)B = IB = B$.

**2.2.3** NOTATION. If $A$ is a nonsingular matrix, we denote its unique inverse by $A^{-1}$.

Some authors denote the inverse of a nonsingular matrix $A$ by $A^I$.

**2.2.4** THEOREM. *If* A *and* B *are nonsingular matrices of the same order, then* AB *is a nonsingular matrix and* $(AB)^{-1} = B^{-1}A^{-1}$.

Since $A$ and $B$ are nonsingular, $A^{-1}$ and $B^{-1}$ exist. Then

$$(B^{-1}A^{-1})(AB) = B^{-1}[A^{-1}(AB)] = B^{-1}[(A^{-1}A)B]$$
$$= B^{-1}(IB) = B^{-1}B = I$$

and

$$(AB)(B^{-1}A^{-1}) = [(AB)B^{-1}]A^{-1} = [A(BB^{-1})]A^{-1}$$
$$= (AI)A^{-1} = AA^{-1} = I.$$

It follows that $AB$ is nonsingular and that $(AB)^{-1} = B^{-1}A^{-1}$.

The reader should find no trouble in extending Theorem 2.2.4 to the more general situation of Corollary 2.2.5, nor should he find trouble in establishing the easy but useful Theorems 2.2.6 and 2.2.7.

**2.2.5** COROLLARY. *If* A, B, $\cdots$, M, N *are nonsingular matrices of the same order, then* AB $\cdots$ MN *is a nonsingular matrix and* $(AB \cdots MN)^{-1}$ $= N^{-1}M^{-1} \cdots B^{-1}A^{-1}$.

**2.2.6** THEOREM. *If* A *is nonsingular, then* A$^{-1}$ *is nonsingular and* $(A^{-1})^{-1} = A$.

**2.2.7** THEOREM. *If* A *is nonsingular, then* A$'$ *is nonsingular and* $(A')^{-1} = (A^{-1})'$.

**2.2.8** THEOREM. *ERT matrices are nonsingular.*
It is easily verified that the inverse of $R(j, k)$ is $R(j, k)$, the inverse of $R(ck)$ is $R(c^{-1}k)$, and the inverse of $R(j + ak)$ is $R(j - ak)$.

**2.2.9** THEOREM. *If* A $\overset{RE}{=}$ B, *then there exists a nonsingular matrix* S *such that* B $=$ SA *and* A $=$ S$^{-1}$B.
By Corollary 2.1.9, $B = SA$, where $S$ is a product of ERT matrices. But $S$ is nonsingular by Theorem 2.2.8 and Corollary 2.2.5.

**2.2.10** THEOREM. *An nth-order matrix* A *is row-equivalent to* I$_{(n)}$ *if and only if it is nonsingular.*
Suppose $A$ is nonsingular. Let $A \overset{RE}{=} B$, where $B$ is in modified

triangular form. Then $B = SA$, where $S$ is nonsingular (by Theorem 2.2.9). Therefore $B$ is nonsingular (by Theorem 2.2.4) and $B^{-1}$ exists.

We now show that $b_{kk} \neq 0$ for all $k$. For suppose the contrary, and recall that the first nonzero element in the $i$th row of $B$ is preceded by at least $i - 1$ zeros. If $b_{kk} = 0$ and $k = n$, the $n$th row of $B$ must be a row of zeros. If $b_{kk} = 0$ and $k < n$, the $(k + 1)$st row of $B$ must have at least $k + 1$ zeros preceding its first nonzero element, and at least one row after the $k$th must be a row of zeros. In any case, then, $B$ will contain a row of zeros. Then $BB^{-1} = I$ will have the corresponding row a row of zeros. But this is contrary to the definition of $I$, and it follows that $b_{kk} \neq 0$ for all $k$.

Recalling the structure of a matrix in modified triangular form, it now follows that $b_{kk} = 1$ for all $k$ and $b_{ij} = 0$ when $i \neq j$. That is, $B = I$.

Conversely, suppose $A \overset{\text{RE}}{=} I$. Then $I = SA$, where $S$ is nonsingular (by Theorem 2.2.9), and $A = S^{-1}I = S^{-1}$ is nonsingular (by Theorem 2.2.6).

**2.2.11** THEOREM. *A matrix* A *is nonsingular if and only if it can be written as a product of ERT matrices.*

Suppose $A$ is nonsingular. Then $A \overset{\text{RE}}{=} I$. Therefore $A = SI = S$, where $S$ is a product of ERT matrices.

Conversely, suppose $A = S$, where $S$ is a product of ERT matrices. Then $A$ is nonsingular (by Theorem 2.2.8 and Corollary 2.2.5).

**2.2.12** THEOREM. *If* B = SA, *where* S *is nonsingular, then* A $\overset{\text{RE}}{=}$ B. (This is the converse of Theorem 2.2.9.)

$S$ is a product of ERT matrices (by Theorem 2.2.11). It follows that $B$ is obtainable from $A$ by a finite sequence of elementary row operations. Therefore $A \overset{\text{RE}}{=} B$.

**2.2.13** THEOREM. *If a square matrix* A *is reducible to* $I_{(n)}$ *by a sequence of elementary row operations, then the same sequence of elementary row operations performed on* $I_{(n)}$ *produces* $A^{-1}$.

We have $(R_s \cdots R_2 R_1)A = I_{(n)}$, where $R_1, \cdots, R_s$ are ERT matrices. Now $A^{-1}$ exists (by Theorem 2.2.10) and

$$(R_s \cdots R_2 R_1)(AA^{-1}) = I_{(n)}A^{-1},$$

whence $(R_s \cdots R_2 R_1)I_{(n)} = A^{-1}$, and the theorem follows.

*Example.* Find the inverse of $A = \begin{bmatrix} 2 & -2 & 4 \\ 2 & 3 & 2 \\ -1 & 1 & -1 \end{bmatrix}$.

Theorem 2.2.13 says that a sequence of elementary row operations that converts a nonsingular matrix $A$ into $I$ will convert $I$ into $A^{-1}$. This suggests that we consider the partitioned matrix $[A|I]$ and choose a sequence of elementary row operations on this partitioned matrix that will convert the submatrix $A$ into $I$; then the submatrix $I$ will automatically be converted into $A^{-1}$. For the matrix $A$ given above, we have

$$[A|I] = \begin{bmatrix} 2 & -2 & 4 & 1 & 0 & 0 \\ 2 & 3 & 2 & 0 & 1 & 0 \\ -1 & 1 & -1 & 0 & 0 & 1 \end{bmatrix} \xleftrightarrow[o(1/2 \cdot 1)]{}$$

$$\begin{bmatrix} 1 & -1 & 2 & 1/2 & 0 & 0 \\ 2 & 3 & 2 & 0 & 1 & 0 \\ -1 & 1 & -1 & 0 & 0 & 1 \end{bmatrix} \xleftrightarrow[\substack{o(2-2\cdot1)\\o(3+1\cdot1)}]{}$$

$$\begin{bmatrix} 1 & -1 & 2 & 1/2 & 0 & 0 \\ 0 & 5 & -2 & -1 & 1 & 0 \\ 0 & 0 & 1 & 1/2 & 0 & 1 \end{bmatrix} \xleftrightarrow[o(1/5 \cdot 2)]{}$$

$$\begin{bmatrix} 1 & -1 & 2 & 1/2 & 0 & 0 \\ 0 & 1 & -2/5 & -1/5 & 1/5 & 0 \\ 0 & 0 & 1 & 1/2 & 0 & 1 \end{bmatrix} \xleftrightarrow[o(1+1\cdot2)]{}$$

$$\begin{bmatrix} 1 & 0 & 8/5 & 3/10 & 1/5 & 0 \\ 0 & 1 & -2/5 & -1/5 & 1/5 & 0 \\ 0 & 0 & 1 & 1/2 & 0 & 1 \end{bmatrix} \xleftrightarrow[\substack{o(1-8/5\cdot3)\\o(2+2/5\cdot3)}]{}$$

$$\begin{bmatrix} 1 & 0 & 0 & -1/2 & 1/5 & -8/5 \\ 0 & 1 & 0 & 0 & 1/5 & 2/5 \\ 0 & 0 & 1 & 1/2 & 0 & 1 \end{bmatrix} = [I|A^{-1}].$$

## PROBLEMS

**2.2–1**  Prove Corollary 2.2.5.

**2.2–2**  Establish Theorem 2.2.6.

**2.2–3**  Establish Theorem 2.2.7.

**2.2–4**  Find the inverse of $\begin{bmatrix} 1 & -1 & 0 \\ -1 & 2 & 0 \\ 0 & 0 & 1 \end{bmatrix}$.

**2.2–5**  Find the inverse of $\begin{bmatrix} 1 & 1 & 2 & 1 \\ 2 & 3 & 4 & 1 \\ 3 & 3 & 3 & 1 \\ 1 & 2 & 3 & 1 \end{bmatrix}$.

**2.2–6**  Let $A = \begin{bmatrix} a + ib & c + id \\ -c + id & a - ib \end{bmatrix}$, where $a^2 + b^2 + c^2 + d^2 = 1$. Find $A^{-1}$.

**2.2–7**  If $A$ is nonsingular and $n$ is a positive integer, we define $A^{-n} = (A^{-1})^n$. We further define $A^0 = I$, where $A$ is any square matrix, singular or not. Show that the two laws of exponents, $A^m A^n = A^{m+n}$ and $(A^m)^n = A^{mn}$, hold for all integral values of $m$ and $n$ when $A$ is nonsingular.

**2.2–8**  If $A = \begin{bmatrix} \cosh x & \sinh x \\ \sinh x & \cosh x \end{bmatrix}$, show that $A^n = \begin{bmatrix} \cosh nx & \sinh nx \\ \sinh nx & \cosh nx \end{bmatrix}$ for all integral values of $n$.

**2.2–9**  If $A$ is nonsingular and symmetric, show that $A^{-1}$ is symmetric.

**2.2–10**  Prove that $A_{(n)}$ is nonsingular if and only if a matrix $B$ exists such that $B(AX) = X$ for all $n$-dimensional column vectors $X$.

**2.2–11**  If $S = I + A + A^2 + \cdots + A^m$ and $I - A$ is nonsingular, show that $S = (I - A^{m+1})(I - A)^{-1}$.

**2.2–12**  Given that $A^{-1}$ exists, determine $X$ so that $\begin{bmatrix} A^{-1} & O \\ X & A^{-1} \end{bmatrix}$ is the inverse of $\begin{bmatrix} A & O \\ B & A \end{bmatrix}$.

**2.2–13**  Find, by means of Problem 2.2–12, the inverses of

$$\begin{bmatrix} 1 & 0 & 0 & 0 \\ 1 & 1 & 0 & 0 \\ 0 & 0 & 1 & 0 \\ 0 & 0 & 1 & 1 \end{bmatrix} \quad \text{and} \quad \begin{bmatrix} 1 & 0 & 0 & 0 \\ 1 & 1 & 0 & 0 \\ 1 & 1 & 1 & 0 \\ 1 & 1 & 1 & 1 \end{bmatrix}.$$

**2.2–14**  If $A$ is nonsingular, show that $A^{-1} = (A'A)^{-1}A'$, and thus show how the inversion of any nonsingular matrix $A$ may be reduced to the inversion of a symmetric nonsingular matrix.

**2.2–15**  (a) Show that the inverse of a nonsingular diagonal matrix

$$\text{diag}(k_1, k_2, \cdots, k_n)$$

is the diagonal matrix

$$\text{diag}(1/k_1, 1/k_2, \cdots, 1/k_n).$$

(b) If $A_1, A_2, \cdots, A_s$ are nonsingular matrices, show that the inverse of the direct sum of $A_1, \cdots, A_s$ is the direct sum of the inverses of $A_1, \cdots, A_s$.

**2.2–16**  Suppose nonsingular matrices $A$ and $B$ are symmetric and commute. Show that

(a) $A^{-1}$ and $B$, $A$ and $B^{-1}$, $A^{-1}$ and $B^{-1}$ also commute,

(b) $A^{-1}B$, $AB^{-1}$, $A^{-1}B^{-1}$ are symmetric.

**2.2–17**  Suppose $A$ is nonsingular and that $AB = BA$. Show that (see Problem 2.2–7):

(a) If $S^2 = B$, then $(A^{-1}SA)^2 = B$.

(b) If $r$ is any integer, $A^r B = BA^r$.

(c) If $B$ is nonsingular and $r$ and $s$ are any integers, $A^r B^s = B^s A^r$.

**2.2–18**  Let $A$ be a square matrix such that

$$A^k + c_{k-1}A^{k-1} + \cdots + c_0 I = O, \qquad c_0 \neq 0, \quad k > 1.$$

Show that $A$ is nonsingular and that

$$A^{-1} = -(1/c_0)(A^{k-1} + c_{k-1}A^{k-2} + \cdots + c_1 I).$$

**2.2–19**  If $A$ and $B$ are $n \times n$ matrices and $B$ is nonsingular, show that

$$BA^k B^{-1} = (BAB^{-1})^k$$

for every positive integer $k$.

**2.2–20**  If $A$ is a nonsingular $\mathscr{C}$ matrix, show that: (a) $\bar{A}^{-1} = \overline{A^{-1}}$, (b) $(A^*)^{-1} = (A^{-1})^*$.

## 2.3  Column equivalence of matrices

Analogous to the elementary row operations on a matrix are the elementary column operations on a matrix.

**2.3.1**  Definitions and notation.  Let $A$ be an $m \times n$ matrix and denote the successive columns of $A$ by $A'_i$, $i = 1, 2, \cdots, n$. The *elementary column operations* on matrix $A$ are

(1) $o'(j, k)$, the interchange of columns $A'_j$ and $A'_k$,

(2) $o'(ck)$, the multiplication of column $A'_k$ by the nonzero scalar $c$, and

(3) $o'(j + ak)$, the addition of $a$ times column $A'_k$ to column $A'_j$, $j \neq k$.

If $A$ and $B$ are two $m \times n$ matrices, then we say $A$ is *column-equivalent* to $B$ if and only if $B$ is obtainable from $A$ by the successive application of finitely many elementary column operations; we write $A \overset{\text{CE}}{=} B$.

Though we can establish theorems involving elementary column operations by paralleling the proofs of the corresponding theorems about elementary row operations, the following simple relation frequently yields shorter and more elegant demonstrations.

**2.3.2**   THEOREM.   *If* $A \overset{CE}{=} B$, *then* $A' \overset{RE}{=} B'$, *and conversely.*

For an elementary column operation on $A$ is an elementary row operation on $A'$, and conversely.

**2.3.3**   THEOREM.   *Column equivalence of matrices is reflexive, symmetric, and transitive.*

This follows from Theorems 2.3.2 and 2.1.3.

*Note.* It follows that if $A$ is column-equivalent to $B$, we may say "*A and B are column-equivalent.*"

**2.3.4**   DEFINITION AND NOTATION.   A matrix obtained by applying an elementary column operation to an identity matrix $I$ is called an *elementary column transformation matrix* or, in abbreviated form, an *ECT matrix.* We shall denote these by the self-explanatory symbolism $C(j, k)$, $C(ck)$, $C(j + ak)$.

**2.3.5**   THEOREM.   *The application of an elementary column operation to matrix* A *can be effected by postmultiplication of* A *by a corresponding ECT matrix.*

Let $B$ be obtained from $A$ by an elementary column operation $o'$. Then $B'$ is obtained from $A'$ by the corresponding row operation $o$. That is, $B' = RA'$, where $R$ is the appropriate ERT matrix. Therefore $B = AR' = AC$, where $C$ is the ECT matrix corresponding to the ERT matrix $R$.

**2.3.6**   COROLLARY.   *If* A *and* B *are two* m × n *column-equivalent matrices, then* B = AT, *where* T *is a product of* nth-*order ECT matrices.*

**2.3.7**   THEOREM.   *An ECT matrix is also an ERT matrix.*

The reader can show that $C(j, k) = R(j, k)$, $C(ck) = R(ck)$, $C(j + ak) = R(k + aj)$.

**2.3.8**   THEOREM.   *ECT matrices are nonsingular.*
Proved by Theorems 2.3.7 and 2.2.8.

**2.3.9**   THEOREM.   $A \stackrel{CE}{=} B$ *if and only if there exists a nonsingular matrix* $T$ *such that* $B = AT$.

By Theorem 2.3.2, $A \stackrel{CE}{=} B$ if and only if $A' \stackrel{RE}{=} B'$. By Theorems 2.2.9 and 2.2.12, $A' \stackrel{RE}{=} B'$ if and only if there exists a nonsingular matrix $S$ such that $B' = SA'$. But there exists a nonsingular matrix $S$ such that $B' = SA'$ if and only if there exists a nonsingular matrix $T$ (in fact, $T = S'$) such that $A = BT$. The theorem now follows.

**2.3.10**   THEOREM.   *An* nth-*order matrix* $A$ *is column-equivalent to* $I_{(n)}$ *if and only if it is nonsingular.*

By Theorem 2.3.2, $A \stackrel{CE}{=} I_{(n)}$ if and only if $A' \stackrel{RE}{=} I'_{(n)} = I_{(n)}$. By Theorem 2.3.10, $A' \stackrel{RE}{=} I_{(n)}$ if and only if $A'$ is nonsingular. But $A'$ is nonsingular if and only if $A$ is nonsingular. The theorem now follows.

By paralleling the proofs of Theorems 2.2.11 and 2.2.13, the reader may easily establish the following two concluding theorems.

**2.3.11**   THEOREM.   *A matrix* $A$ *is nonsingular if and only if it can be written as a product of ECT matrices.*

**2.3.12**   THEOREM.   *If a square matrix* $A$ *is reducible to* $I_{(n)}$ *by a sequence of elementary column operations, then the same sequence of elementary column operations performed on* $I_{(n)}$ *produces* $A^{-1}$.

## 2.4   Equivalence of matrices

An extension (actually a combination) of the concepts of row equivalence and column equivalence of matrices is the important concept of *equivalence* of matrices.

**2.4.1**   DEFINITION AND NOTATION.   If matrix $B$ can be obtained from matrix $A$ by the successive application of finitely many elementary row and column operations, we say that $A$ is *equivalent* to $B$, and we write $A \stackrel{E}{=} B$.

**2.4.2** THEOREM. *If* A *and* B *are* m × n *matrices, then* A $\overset{E}{=}$ B *if and only if* B = SAT, *where* S *and* T *are* m × m *and* n × n *nonsingular matrices, respectively.*

Suppose $A \overset{E}{=} B$. Then we may obtain $B$ from $A$ by a finite sequence of elementary row and column operations. It follows (by Theorems 2.1.8 and 2.3.5) that $B = SAT$, where $S$ is a product of appropriate ERT matrices and $T$ is a product of appropriate ECT matrices. But (by Theorems 2.2.8, 2.3.8, 2.2.5) $S$ and $T$ are nonsingular.

Conversely, suppose $B = SAT$, where $S$ and $T$ are nonsingular matrices. Then (by Theorems 2.2.11 and 2.3.11) $S$ can be expressed as a product of ERT matrices and $T$ as a product of ECT matrices. It follows (by Theorems 2.1.8 and 2.3.5) that we may obtain $B$ from $A$ by a finite sequence of elementary row and column operations. That is, $A \overset{E}{=} B$.

**2.4.3** THEOREM. *Equivalence of matrices is reflexive, symmetric, and transitive.*

We first note that if $A$ is any $m \times n$ matrix, then $A \overset{E}{=} A$, since $A = I_{(m)}AI_{(n)}$. Next, suppose $A \overset{E}{=} B$. Then (by Theorem 2.4.2) $B = SAT$, where $S$, $T$, $S^{-1}$, $T^{-1}$ are nonsingular. Therefore $S^{-1}BT^{-1} = S^{-1}(SAT)T^{-1} = (S^{-1}S)A(TT^{-1}) = A$, and (by Theorem 2.4.2) $B \overset{E}{=} A$.

Finally, suppose $A \overset{E}{=} B$ and $B \overset{E}{=} C$. Then (by Theorem 2.4.2) $B = SAT$, $C = UBV$, where $S$, $T$, $U$, $V$ are nonsingular. Therefore $C = U(SAT)V = (US)A(TV)$ and (by Theorem 2.4.2) $C \overset{E}{=} A$, since $US$ and $TV$ are nonsingular (by Theorem 2.2.4).

**2.4.4** THEOREM. *Every nonzero* m × n *matrix* A *is equivalent to an* m × n *partitioned matrix of the form*

$$\begin{bmatrix} I_{(r)} & O_{(r,n-r)} \\ O_{(m-r,r)} & O_{(m-r,n-r)} \end{bmatrix}.$$

Since $A \neq O$, there exists an element $a_{ij} \neq 0$. By elementary row and column operations we carry $a_{ij}$ into the top left corner. Now multiplying the first row by $1/a_{ij}$, we get 1 in the top left corner. By additions of suitable multiples of the first row to the other rows, we introduce zeros everywhere in the first column except for the 1 at the top. By additions of suitable multiples of the first column to the other columns, we then introduce zeros everywhere in the first row except for the 1 at the left. $A$

has been carried into a matrix of the form

$$\begin{bmatrix} 1 & O_{(1,n-1)} \\ O_{(m-1,1)} & B_{(m-1,n-1)} \end{bmatrix}.$$

Treating $B$, if it is nonzero, in the same way, and continuing the process, we obtain the desired form.

**2.4.5** DEFINITION. Matrices of the form $\begin{bmatrix} I_{(r)} & O \\ O & O \end{bmatrix}$ are called *canonical matrices.*

**2.4.6** COROLLARY. *If* A *is a nonzero* m × n *matrix, then there exist* m × m *and* n × n *nonsingular matrices* S *and* T, *respectively, such that* SAT *is a canonical matrix.*

## PROBLEMS

**2.3–1**  Prove Corollary 2.3.6.

**2.3–2**  Complete the proof of Theorem 2.3.7.

**2.3–3**  Supply proofs for Theorems 2.3.11 and 2.3.12.

**2.3–4**  Find the inverses of

(a) $\begin{bmatrix} 1 & -1 & 0 \\ -1 & 2 & 0 \\ 0 & 0 & 1 \end{bmatrix}$, (b) $\begin{bmatrix} 1 & 1 & 2 & 1 \\ 2 & 3 & 4 & 1 \\ 3 & 3 & 3 & 1 \\ 1 & 2 & 3 & 1 \end{bmatrix}$,

by the method of Theorem 2.3.12.

**2.4–1**  Prove Corollary 2.4.6.

**2.4–2**  Reduce $\begin{bmatrix} 1 & 2 & -1 & 3 \\ 2 & 4 & -4 & 7 \\ -1 & -2 & -1 & -2 \end{bmatrix}$ to an equivalent canonical matrix.

**2.4–3**  Find matrices $A$ and $B$ such that

$$A \begin{bmatrix} 2 & 2 & -6 \\ 1 & -2 & -2 \end{bmatrix} B$$

is a canonical matrix.

**2.4–4** (a) Prove that an $n \times n$ matrix is nonsingular if and only if it is equivalent to the $n \times n$ identity matrix.

(b) Prove that an $n \times n$ matrix is nonsingular if and only if it is row-equivalent to the $n \times n$ identity matrix.

**2.4–5** Let $A$ be a symmetric $n \times n$ $\mathscr{C}$ matrix. Show that there exists a nonsingular matrix $P$ such that $PAP'$ is a canonical matrix.

**2.4–6** If $A \overset{\text{E}}{=} B$, does it necessarily follow that:

(a) $A' \overset{\text{E}}{=} B'$?

(b) $A^2 \overset{\text{E}}{=} B^2$?

(c) $AB \overset{\text{E}}{=} BA$?

**2.4–7** Enumerate the number of $m \times n$ canonical matrices.

**2.4–8** (a) Show that the product of two conformable canonical matrices is a canonical matrix.

(b) Show by a counterexample that if $A$ and $B$ are conformable for multiplication and if $A \overset{\text{E}}{=} R$ and $B \overset{\text{E}}{=} S$, where $R$ and $S$ are canonical matrices, we do not necessarily have $AB \overset{\text{E}}{=} RS$.

**2.4–9** If $A_1 \overset{\text{E}}{=} A_2$ and $B_1 \overset{\text{E}}{=} B_2$, show that

$$\begin{bmatrix} A_1 & O \\ O & B_1 \end{bmatrix} \overset{\text{E}}{=} \begin{bmatrix} A_2 & O \\ O & B_2 \end{bmatrix}.$$

**2.4–10** (a) Let $A$ be an $n \times n$ nonsingular matrix. Show that one can find nonsingular matrices $P$ and $Q$ such that

$$\left[ \begin{array}{c|c} A & I_{(n)} \\ \hline I_{(n)} & O_{(n)} \end{array} \right] \overset{\text{E}}{=} \left[ \begin{array}{c|c} I_{(n)} & P \\ \hline Q & O_{(n)} \end{array} \right]$$

and show that $A^{-1} = QP$.

(b) If $A$ is symmetric and nonsingular, prove that there exists a nonsingular matrix $P$ such that $A^{-1} = P'P$.

**2.4–11** Show that not every matrix can be reduced to a canonical matrix by elementary row operations alone.

## 2.5 Linear dependence and independence of a set of vectors

At this point two important questions naturally arise: (1) Is the canonical matrix to which a given matrix is equivalent unique? (2) If so, what determines the value of $r$ in the canonical matrix? To answer these questions, we introduce the valuable concept of linear dependence and independence of a set of vectors.

**2.5.1**   DEFINITIONS.   Let $V_1, \cdots, V_p$ be $p$ $n$th-order row (column) vectors. If there exist scalars $c_1, \cdots, c_p$, not all zero, such that

$$c_1 V_1 + \cdots + c_p V_p = O,$$

then the vectors $V_1, \cdots, V_p$ are said to be *linearly dependent*; otherwise these vectors are said to be *linearly independent*.

**2.5.2**   DEFINITION.   A vector $V$ is said to be a *linear combination* of the vectors $V_1, \cdots, V_p$ if and only if there exist scalars $d_1, \cdots, d_p$ such that

$$V = d_1 V_1 + \cdots + d_p V_p.$$

**2.5.3**   THEOREM.   *If* p *vectors are linearly independent, then any* q < p *of these vectors are linearly independent.*

Denote the $p$ vectors by $V_1, \cdots, V_p$, and suppose (renumbering the vectors if necessary) that $V_1, \cdots, V_q$ are linearly dependent. Then there exist scalars $c_1, \cdots, c_q$, not all zero, such that

$$c_1 V_1 + \cdots + c_q V_q = O.$$

It then follows that

$$c_1 V_1 + \cdots + c_q V_q + c_{q+1} V_{q+1} + \cdots + c_p V_p = O,$$

where $c_{q+1} = \cdots = c_p = 0$, and $V_1, \cdots, V_p$ are linearly dependent. This contradiction proves the theorem.

**2.5.4**   THEOREM.   *A single vector is linearly dependent if and only if it is a zero vector.*

For if $V$ is a vector, $c$ a scalar, and $cV = O$, then either $c = 0$ or $V = O$.

**2.5.5**   THEOREM.   *Let* r *denote the number of vectors in a maximum linearly independent subset of a set of* p *nonzero linearly dependent vectors* $V_1, \cdots, V_p$. *Then* 0 < r < p *and each of the remaining* p − r *vectors in the set is a linear combination of the* r *linearly independent vectors.*

Clearly, $r < p$ and, by Theorem 2.5.4, $r > 0$. Let, then, renumbering the vectors if necessary, $V_1, \cdots, V_r$ be a maximum subset of linearly independent vectors and let $V$ denote any one of the other $p - r$ vectors. Then $V, V_1, \cdots, V_r$ are linearly dependent and we have

$$cV + c_1 V_1 + \cdots + c_r V_r = O,$$

where the $c$'s are not all zero. Now if $c = 0$, then at least one of $c_1, \cdots, c_r$ must be nonzero, and $V_1, \cdots, V_r$ would be linearly dependent, which is contrary to hypothesis. It follows that $c \neq 0$ and that

$$V = -(c_1/c)V_1 - \cdots - (c_r/c)V_r,$$

a linear combination of $V_1, \cdots, V_r$.

### 2.6   Row rank and column rank of a matrix

We now apply the theory of linear dependence and independence of a set of vectors to the set of rows and to the set of columns of a matrix.

**2.6.1**   DEFINITIONS.   The maximum number of linearly independent rows of a matrix is called the *row rank* of the matrix, and the maximum number of linearly independent columns of a matrix is called the *column rank* of the matrix.

**2.6.2**   THEOREM.   *Row-equivalent matrices have the same row rank.*

Let $A$ be an $m \times n$ matrix of row rank $r$. We shall show that elementary row operations performed on $A$ do not increase the row rank of $A$. This is obvious if $r = m$. We therefore suppose $r < m$.

Interchanging two rows of $A$ clearly does not alter the row rank of $A$.

Multiplying row $A_k$ of $A$ by $c \neq 0$ does not increase the row rank of $A$. For consider any $r + 1$ rows of the new matrix. If these $r + 1$ rows do not contain row $cA_k$, they are linearly dependent, since they are the same rows of $A$. Suppose these $r + 1$ rows do contain row $cA_k$, and designate the corresponding $r + 1$ rows of $A$ by $A_{i_1}, A_{i_2}, \cdots, A_{i_r}, A_k$. These rows are linearly dependent; that is, there exist scalars $c_1, \cdots, c_r$, $d$ not all zero, such that

$$c_1 A_{i_1} + c_2 A_{i_2} + \cdots + c_r A_{i_r} + d A_k = O.$$

Then

$$(cc_1)A_{i_1} + (cc_2)A_{i_2} + \cdots + (cc_r)A_{i_r} + d(cA_k) = O,$$

where $cc_1, cc_2, \cdots, cc_r$, $d$ are not all zero, and $A_{i_1}, A_{i_2}, \cdots, A_{i_r}, cA_k$ are linearly dependent.

Adding a nonzero multiple $a$ of row $A_k$ to row $A_j$, $j \neq k$, does not increase the row rank of $A$. For consider any $r + 1$ rows of the new matrix. If these $r + 1$ rows do not contain row $A_j + aA_k$, they are linearly dependent, since they are the same $r + 1$ rows of $A$. Suppose these $r + 1$ rows do contain row $A_j + aA_k$ and denote these rows by $A_{i_1}, A_{i_2}, \cdots, A_{i_r}, A_j + aA_k$. Now rows $A_{i_1}, A_{i_2}, \cdots, A_{i_r}, A_j$ are linearly dependent, and rows $A_{i_1}, A_{i_2}, \cdots, A_{i_r}, A_k$ are linearly dependent. That is, there are scalars $c_1, \cdots, c_r, e$, not all zero, and scalars $d_1, \cdots, d_r, f$, not all zero, such that

$$c_1 A_{i_1} + \cdots + c_r A_{i_r} + e A_j = O,$$

$$d_1 A_{i_1} + \cdots + d_r A_{i_r} + f(aA_k) = O.$$

If $e = 0$ or $f = 0$, we have either

$$c_1 A_{i_1} + \cdots + c_r A_{i_r} + e(A_j + aA_k) = O$$

or

$$d_1 A_{i_1} + \cdots + d_r A_{i_r} + f(A_j + aA_k) = O,$$

and $A_{i_1}, \cdots, A_{i_r}, A_j + aA_k$ are linearly dependent, for if $e = 0$, not all the $c_i$'s are zero, and if $f = 0$, not all the $d_i$'s are zero. If $e \neq 0$ and $f \neq 0$, then

$$A_j = c'_1 A_{i_1} + \cdots + c'_r A_{i_r},$$

where $c'_i = -c_i/e$, and

$$aA_k = d'_1 A_{i_1} + \cdots + d'_r A_{i_r},$$

where $d'_i = -d_i/f$. It follows that

$$(c'_1 + d'_1)A_{i_1} + \cdots + (c'_r + d'_r)A_{i_r} - (A_j + aA_k) = O,$$

and $A_{i_1}, \cdots, A_{i_r}, A_j + aA_k$ are again linearly dependent.

Now let $B$ be obtained from $A$ by a finite sequence of elementary row operations, and let row rank $B = s$. By the preceding, $s \leq r$. But $A$ can also be obtained from $B$ by a finite sequence of elementary row operations, whence $r \leq s$. It follows that $r = s$, and the theorem is established.

**2.6.3** THEOREM. *If* S *is nonsingular, then row rank* SA = *row rank* A.

For, by Theorem 2.2.12, $SA \overset{\text{RE}}{=} A$.

**2.6.4** THEOREM. *Row rank of* $\begin{bmatrix} I_{(r)} & O \\ O & O \end{bmatrix}$ *is* r.

The reader can supply the easy proof.

**2.6.5** THEOREM. *If* S *is nonsingular and of order* n, *then row rank* S = n.

For $S \overset{\text{RE}}{=} I_{(n)}$, and row rank $I_{(n)} = n$.

**2.6.6** THEOREM. *Column-equivalent matrices have the same column rank*.

Suppose $A \overset{\text{CE}}{=} B$. Then $A' \overset{\text{RE}}{=} B'$ and row rank $A' =$ row rank $B'$. But row rank $A' =$ column rank $A$, and row rank $B' =$ column rank $B$. Therefore column rank $A =$ column rank $B$.

**2.6.7** THEOREM. *If* T *is nonsingular, then column rank* AT = *column rank* A.

For, by Theorem 2.3.9, $AT \overset{\text{CE}}{=} A$.

## PROBLEMS

**2.5–1** Prove that if among $p$ $n$th-order like vectors (that is, all the vectors are row vectors or all are column vectors) there is a subset of $q < p$ vectors that are linearly dependent, then the vectors of the entire set are linearly dependent.

**2.5–2** (a) Prove that two $n$th-order like vectors are linearly dependent if and only if one is a scalar multiple of the other.

(b) Prove that the vectors of a set of $n$th-order like vectors that contains the zero vector are linearly dependent.

(c) Show that a set of $n$th-order like vectors that contains two equal vectors is a linearly dependent set of vectors.

**2.5–3** If the $n$th-order like vectors $V_1, \cdots, V_k$ are linearly independent, show that the $n$th-order like vectors $V_1, \cdots, V_k, V$ are linearly dependent if and only if $V$ can be expressed as a linear combination of $V_1, \cdots, V_k$.

**2.5–4** (a) Show that the *n unit vectors* constituting the rows of $I_{(n)}$ are linearly independent.

(b) Show that any $n$th-order row vector can be expressed as a linear combination of the preceding $n$ unit vectors.

**2.5–5** Prove that any set of more than $n$ $n$th-order like vectors must be linearly dependent.

**2.5–6**  Show that if $V_1, \cdots, V_k$ are linearly independent vectors and if $V = a_1 V_1 + \cdots + a_k V_k$, then the vectors $V - V_1, \cdots, V - V_k$ are linearly independent if and only if $a_1 + \cdots + a_k \neq 1$.

**2.5–7**  If $V_1, \cdots, V_k$ are linearly independent $n$th-order like vectors, show that

$$a_1 V_1 + \cdots + a_k V_k = b_1 V_1 + \cdots + b_k V_k$$

if and only if $a_1 = b_1, \cdots, a_k = b_k$.

**2.5–8**  Determine whether the following sets of vectors are linearly dependent or independent:

(**a**) $[0, 1, 2]$, $[1, -2, 1]$, $[1, 1, 1]$.

(**b**) $[1, 0, 2]$, $[2, 0, 1]$, $[1, 1, 1]$.

(**c**) $[0, 1, 4]$, $[2, 1, 5]$, $[2, 0, 1]$.

**2.5–9**  Find maximal subsets of linearly independent vectors of the following sets of vectors:

(**a**) $[1, 0, 1]$, $[2, 0, 1]$, $[3, 0, 1]$, $[3, 0, 0]$.

(**b**) $[6, 3, -9]$, $[3, -2, 1]$, $[2, 1, -3]$, $[5, 2, -11]$, $[3, 1, -8]$.

(**c**) $[1, 1 + i, i]$, $[i, -i, 1 - i]$, $[1 + 2i, 1 - i, 2 - i]$.

(**d**) $[1, 1 + i, i]$, $[i, -i, 1 - i]$, $[0, 1 - 2i, 2 - i]$.

**2.5–10**  (**a**) Let $(x_i, y_i)$, $i = 1, 2, 3$, represent the rectangular Cartesian coordinates of three points in a plane. Show that the three points are collinear if and only if the three vectors $[x_i, y_i, 1]$ are linearly dependent.

(**b**) Let $(x_i, y_i, z_i)$, $i = 1, 2, 3, 4$, represent the rectangular Cartesian coordinates of four points in space. Show that the four points are coplanar if and only if the four vectors $[x_i, y_i, z_i, 1]$ are linearly dependent.

(**c**) Let $a_i x + b_i y + c_i = 0$, $i = 1, 2, 3$, represent the Cartesian equations of three lines in a plane. Show that the three lines are concurrent or codirectional if and only if the three vectors $[a_i, b_i, c_i]$ are linearly dependent.

**2.6–1**  Prove Theorem 2.6.4.

**2.6–2**  Prove that in an upper triangular matrix the columns having nonzero diagonal elements are linearly independent.

**2.6–3**  If the $n$th-order like vectors $V_1, \cdots, V_n$ are linearly independent, show that the vectors $W_1, \cdots, W_n$, where

$$W_i = a_{i1} V_1 + \cdots + a_{in} V_n,$$

are linearly independent if and only if $[a_{ij}]$ is nonsingular.

**2.6–4**  Prove that the $k$ $n$th-order column vectors $V_1, \cdots, V_k$ are linearly independent if and only if the equation

$$[V_1, \cdots, V_k]X_{(k,1)} = O$$

has only the trivial solution $X = O$.

**2.6–5** Show that the $n$th-order like vectors $V_1, \cdots, V_k$ are linearly dependent if and only if $a_1 V_1, \cdots, a_k V_k$, where no $a_i$ is zero, are linearly dependent.

## 2.7 Rank of a matrix

It is very interesting (and perhaps somewhat unanticipated) that, no matter what the relative dimensions of a matrix, the maximum number of linearly independent rows and the maximum number of linearly independent columns are always equal. We now prove this fact, and thereby introduce the important concept of *rank* of a matrix.

**2.7.1** THEOREM. *The row rank and the column rank of a matrix are equal.*

This is obviously true if the matrix is a zero matrix. Let $A$ be nonzero. Then there exist (by Theorem 2.4.6) nonsingular matrices $S$ and $T$ such that

$$SAT = C = \begin{bmatrix} I_{(r)} & O \\ O & O \end{bmatrix},$$

of row and column rank $r$. Now (by Theorem 2.6.3) row rank $A$ = row rank $SA$ = row rank $CT^{-1} = r_1$, say. But, partitioning $T^{-1}$ into the submatrix $T_1$ of its first $r$ rows and the submatrix $T_2$ of its remaining rows,

$$CT^{-1} = \begin{bmatrix} I_{(r)} & O \\ O & O \end{bmatrix}\begin{bmatrix} T_1 \\ T_2 \end{bmatrix} = \begin{bmatrix} T_1 \\ O \end{bmatrix},$$

where the $r$ rows of $T_1$ are linearly independent (being $r$ rows of the nonsingular matrix $T^{-1}$). It follows that $r_1 = r$. Now consider $T'A'S' = C'$. As above, row rank $A'$ = row rank $T'A'$ = row rank $C'(S')^{-1} = r$. That is, column rank $A = r$ = row rank $A$.

**2.7.2** DEFINITION AND NOTATION. The common value of the row rank and the column rank of a matrix $A$ is called the *rank* of $A$ and will be denoted by $\rho(A)$ or $\rho[A]$.

**2.7.3** THEOREM. *A given matrix* A *is equivalent to a unique canonical matrix* C.

By the proof of Theorem 2.7.1, we must have $\rho(C) = \rho(A)$. But there is a unique canonical matrix of given rank and given dimensions.

**2.7.4** THEOREM. *Two* m × n *matrices are equivalent if and only if they have the same rank.*

Suppose $A \overset{E}{=} B$. Then $A \overset{E}{=} C$, a canonical matrix whose rank is the rank $r$ of $A$. Also, $B \overset{E}{=} C$ and $\rho(B) = r$.

Conversely, suppose $\rho(A) = \rho(B) = r$. Then $A$ and $B$ are equivalent to the same canonical matrix $C$, and are therefore equivalent to one another.

**2.7.5** THEOREM. *The rank of a matrix in modified triangular form equals the number of its nonzero rows.*

Let $A_1, \cdots, A_r$ denote the nonzero rows and consider a linear relation of the form

$$c_1 A_1 + c_2 A_2 + \cdots + c_r A_r = O.$$

Then $c_k = 0$, $1 \leqq k \leqq r$; for if $A_k$ has its first nonzero element in the $i_k$th position, the vector

$$c_1 A_1 + c_2 A_2 + \cdots + c_r A_r$$

has $c_k$ in the $i_k$th position. It follows that $A_1, A_2, \cdots, A_r$ are linearly independent.

**2.7.6** THEOREM. *The rank of a product of matrices cannot exceed the rank of any factor.*

It suffices to prove the theorem for the product $AB$ of two matrices $A$ and $B$. Let $r = \rho(A)$. Then there exist nonsingular matrices $S$ and $T$ such that

$$SAT = C = \begin{bmatrix} I_{(r)} & O \\ O & O \end{bmatrix}.$$

Therefore $AB = S^{-1}CT^{-1}B = S^{-1}(CT^{-1}B)$. Since $S^{-1}$ is nonsingular, $\rho(AB) = \rho(CT^{-1}B)$, by Theorem 2.6.3. But $CT^{-1}B = C(T^{-1}B)$ has for its first $r$ rows the first $r$ rows of $T^{-1}B$, the remaining rows, if any, being zero rows. Hence $\rho(CT^{-1}B)$ is at most $r$. It follows that $\rho(AB) \leq r = \rho(A)$.

Also, $\rho(AB) = \rho(AB)' = \rho(B'A') \leq \rho(B') = \rho(B)$.

**2.7.7**  THEOREM.  *The rank of a sum of matrices cannot exceed the sum of the ranks of the matrices.*

It suffices to prove the theorem for the sum $A + B$ of two $m \times n$ matrices $A$ and $B$. To this end we note that

$$\begin{bmatrix} I_{(m)} & I_{(m)} \\ O_{(m)} & O_{(m)} \end{bmatrix} \begin{bmatrix} A \\ B \end{bmatrix} = \begin{bmatrix} A + B \\ O \end{bmatrix} = C,$$

say. It follows, by Theorem 2.7.6 and by considering row ranks, that

$$\rho(A + B) = \rho(C) \leq \rho \begin{bmatrix} A \\ B \end{bmatrix} \leq \rho(A) + \rho(B).$$

**2.7.8**  THEOREM.  *If* A *and* B *are two* n $\times$ n *matrices, then* $\rho(\mathrm{AB}) \geq \rho(\mathrm{A}) + \rho(\mathrm{B}) - \mathrm{n}$.

Let $\rho(A) = r$. Then there exist nonsingular matrices $P$ and $Q$ such that

$$PAQ = \begin{bmatrix} I_{(r)} & O \\ O & O \end{bmatrix}$$

and

$$A = P^{-1} \begin{bmatrix} I_{(r)} & O \\ O & O \end{bmatrix} Q^{-1}.$$

Define matrix $C$ by

$$C = P^{-1} \begin{bmatrix} O & O \\ O & I_{(n-r)} \end{bmatrix} Q^{-1}.$$

Then

$$A + C = P^{-1} \left( \begin{bmatrix} I_{(r)} & O \\ O & O \end{bmatrix} + \begin{bmatrix} O & O \\ O & I_{(n-r)} \end{bmatrix} \right) Q^{-1}$$

$$= P^{-1} \begin{bmatrix} I_{(r)} & O \\ O & I_{(n-r)} \end{bmatrix} Q^{-1} = P^{-1} Q^{-1},$$

and $A + C$ is nonsingular. Also, since

$$C \overset{\mathrm{E}}{=} \begin{bmatrix} O & O \\ O & I_{n-(r)} \end{bmatrix},$$

$\rho(C) = n - r$. We now have

$$\rho(B) = \rho[(A + C)B] \qquad \text{(since } A + C \text{ is nonsingular)}$$

$$= \rho(AB + CB)$$

$$\leqq \rho(AB) + \rho(CB) \qquad \text{(by Theorem 2.7.7)}$$

$$\leqq \rho(AB) + \rho(C) \qquad \text{(by Theorem 2.7.6)}$$

$$= \rho(AB) + n - r.$$

It follows that $\rho(AB) \geqq \rho(A) + \rho(B) - n$.

**2.7.9**   DEFINITION AND NOTATION.   If $A$ is a square matrix of order $n$ and rank $r$, then $n - r$ is called the *nullity* of $A$ and will be denoted by $v(A)$.

**2.7.10**   SYLVESTER'S LAW OF NULLITY.†   *The nullity of the product of two square matrices is at least as great as the nullity of either factor, and at most as great as the sum of the nullities of the two factors.*

The reader can easily supply a proof, using Definition 2.7.9 and Theorems 2.7.6 and 2.7.8.

**2.7.11**   COROLLARY.   *If* A *and* B *are* nth-*order square matrices such that* AB = O, *then* $\rho(A) + \rho(B) \leqq$ n.

# PROBLEMS

**2.7–1**   If $A$ and $B$ are $n$th-order row and column vectors, respectively, what limitations are there on the ranks of $AB$ and $BA$?

**2.7–2**   Show why it suffices to prove Theorem 2.7.6 for the product $AB$ of two matrices $A$ and $B$ and to prove Theorem 2.7.7 for the sum $A + B$ of two matrices $A$ and $B$.

**2.7–3**   Suppose $A_{(m,n)}$, $m \leqq n$, is real. (a) If $\rho(A) = m$, show that $AA'$ is nonsingular and that the elements on the principal diagonal of $AA'$ are nonnegative. (b) If $\rho(A) < m$, show that $AA'$ is singular.

† Found by J. J. Sylvester in 1884.

**2.7–4**  Show that $A$, $A^*$, $A^*A$, $AA^*$ all have the same rank.

**2.7–5**  From a square matrix of order $n$ and rank $r$, $s$ rows (or columns) are selected. Prove that the rank of the submatrix thus obtained cannot be less than $r + s - n$.

**2.7–6**  Prove that if columns $j_1, j_2, \cdots, j_r$ of matrix $B$ are linearly dependent (independent), then columns $j_1, j_2, \cdots, j_r$ of matrix $AB$, where $A$ is nonsingular, are also linearly dependent (independent).

**2.7–7**  Let

$$M = \begin{bmatrix} A_1 & O & O \\ O & A_2 & O \\ O & O & A_3 \end{bmatrix},$$

where $A_1$, $A_2$, $A_3$ are square matrices. Show that

$$\rho(M) = \rho(A_1) + \rho(A_2) + \rho(A_3).$$

**2.7–8**  Show that the locus of points $(x, y, z)$ in Cartesian 3-space such that $\rho \begin{bmatrix} x & y & z \\ 1 & x & y \end{bmatrix} = 1$ is the twisted cubic space curve with parametric equations $x = t$, $y = t^2$, $z = t^3$.

**2.7–9**  Show that a matrix of rank $r$ can be expressed as the sum of $r$ matrices each of rank 1.

**2.7–10**  Show that any $k + 1$ linear combinations of $k$ $n$th-order like vectors are linearly dependent.

**2.7–11**  Prove Theorem 2.7.10 and Corollary 2.7.11.

**2.7–12**  Show that two $n \times n$ matrices are equivalent if and only if they have the same nullity.

**2.7–13**  If $A$ is an $n \times n$ idempotent matrix, show that $\rho(A) + \rho(I_{(n)} - A) = n$.

**2.7–14**  (a) Show that $\rho \begin{bmatrix} A & O \\ O & O \end{bmatrix} = \rho(A)$.

(b) Show that the rank of a matrix is never less than the rank of any of its submatrices.

**2.7–15**  Using Problem 2.7–14, show that the conclusion of Theorem 2.7.8 still holds when $A$ is $m \times n$ and $B$ is $n \times p$.

**2.7–16**  Find the ranks of $A$, $A^2$, $A^3$, $A^4$, where

$$A = \begin{bmatrix} 0 & 1 & 0 & 0 \\ 0 & 0 & 1 & 0 \\ 0 & 0 & 0 & 1 \\ 0 & 0 & 0 & 0 \end{bmatrix}.$$

## 2.8 Application to the solution of systems of linear equations

At the start of the present chapter we motivated the study of equivalence of matrices with the important problem of finding a simultaneous solution of a system of linear equations. We now return to the motivating problem and apply to it some of the theory developed so far in the chapter. We commence by introducing some convenient notation and definitions.

**2.8.1** NOTATION AND DEFINITIONS. We shall write the system of linear equations

$$a_{11}x_1 + a_{12}x_2 + \cdots + a_{1n}x_n = c_1,$$
$$a_{21}x_1 + a_{22}x_2 + \cdots + a_{2n}x_n = c_2,$$
$$\cdots \qquad \cdots \qquad \cdots \qquad \cdots \qquad \cdots$$
$$a_{m1}x_1 + a_{m2}x_2 + \cdots + a_{mn}x_n = c_m,$$

in the matrix form

$$AX = C,$$

where $A = [a_{ij}]_{(m,n)}$, $X = \{x_1, \cdots, x_n\}$, $C = \{c_1, \cdots, c_m\}$. The partitioned matrix $[A|C]$ will be denoted by $A^+$. $A$ is called the *matrix*, and $A^+$ the *augmented matrix*, of the system of equations. For convenience we shall frequently refer to the matrix equation $AX = C$ as "the system $AX = C$," meaning, of course, the allied system of scalar equations.

*Example.* For the system of equations

$$x_1 - x_2 + 2x_3 + x_4 = 2,$$
$$3x_1 + 2x_2 \qquad + x_4 = 1,$$
$$4x_1 + x_2 + 2x_3 + 2x_4 = 3,$$

we have

$$X = \{x_1, x_2, x_3, x_4\}, \qquad C = \{2, 1, 3\},$$

$$A = \begin{bmatrix} 1 & -1 & 2 & 1 \\ 3 & 2 & 0 & 1 \\ 4 & 1 & 2 & 2 \end{bmatrix}, \qquad A^+ = \begin{bmatrix} 1 & -1 & 2 & 1 & 2 \\ 3 & 2 & 0 & 1 & 1 \\ 4 & 1 & 2 & 2 & 3 \end{bmatrix}.$$

**2.8.2** THEOREM. *If A is nonsingular, the system* $AX = C$ *has the unique solution* $X = A^{-1}C$.

Suppose $X$ is a solution of the system. Then $AX = C$. Since $A$ is nonsingular, we may premultiply both sides of this equation by $A^{-1}$, obtaining $X = A^{-1}C$. That is, if the system has a solution, that solution must be $A^{-1}C$. But $A^{-1}C$ is a solution, since $A(A^{-1}C) = (AA^{-1})C = IC = C$.

*Example.* Solve the system of equations

$$2x_1 - 2x_2 + 4x_3 = 1,$$

$$2x_1 + 3x_2 + 2x_3 = 2,$$

$$-x_1 + x_2 - x_3 = 3.$$

Since (see the example following Theorem 2.2.13)

$$A^{-1} = (1/10)\begin{bmatrix} -5 & 2 & -16 \\ 0 & 2 & 4 \\ 5 & 0 & 10 \end{bmatrix},$$

we have

$$X = (1/10)\begin{bmatrix} -5 & 2 & -16 \\ 0 & 2 & 4 \\ 5 & 0 & 10 \end{bmatrix}\begin{bmatrix} 1 \\ 2 \\ 3 \end{bmatrix} = \begin{bmatrix} -49/10 \\ 8/5 \\ 7/2 \end{bmatrix}.$$

**2.8.3** DEFINITION. Two systems of equations are said to be *equivalent* if every solution of each is a solution of the other.

**2.8.4** THEOREM. *If* S *is an* m × m *nonsingular matrix, then the systems* AX = C *and* SAX = SC *are equivalent.*

Suppose $AY = C$, then $SAY = SC$. Conversely, suppose $SAY = SC$, then $S^{-1}SAY = S^{-1}SC$, or $AY = C$.

**2.8.5** DEFINITIONS. A system of equations is said to be *consistent* if it has a solution; otherwise the system is said to be *inconsistent*.

**2.8.6** THEOREM. *The system* AX = C *is consistent if and only if* rank $A^+$ = r = rank A, *in which case* r *of the unknowns may be expressed as linear functions of the* c's *and of the remaining* n − r *unknowns, to which arbitrary values may be assigned.*

Let $S$ be a nonsingular matrix such that $SA^+$ is in modified triangular form. Now $SA^+ = [SA|SC]$, and $SA$ is also in modified triangular form. It follows that $SA$ has precisely $m - r$ zero rows at the bottom.

Suppose the system $AX = C$ has a solution $Y$. Then $AY = C$, $SAY = SC$. Since $SA$ has $m - r$ zero rows at the bottom, so does $SAY$, and hence also $SC$.

Conversely, suppose $SC$ has $m - r$ zero rows at the bottom, so that

$$SC = \{b_1, \cdots, b_r, 0, \cdots, 0\}.$$

Let the first nonzero elements (which are 1's) in the rows of $SA$ lie in columns $k_1, k_2, \cdots, k_r$ and denote the $(ij)$th element of $SA$ by $p_{ij}$. Then the matrix equation $SAX = SC$ yields the following set of $r$ scalar equations:

$$x_{k_1} + p_{1,k_1+1}x_{k_1+1} + \cdots + p_{1n}x_n = b_1,$$
where $p_{1k_j} = 0$ for $j = 2, \cdots, r$;

$$x_{k_2} + p_{2,k_2+1}x_{k_2+1} + \cdots + p_{2n}x_n = b_2,$$
where $p_{2k_j} = 0$ for $j = 3, \cdots, r$;

. . . . . . . . . . . . . .

$$x_{k_r} + p_{r,k_r+1}x_{k_r+1} + \cdots + p_{rn}x_n = b_r.$$

It follows that the system $SAX = SC$, and hence (by Theorem 2.8.4) the system $AX = C$, has a solution in which $x_{k_1}, x_{k_2}, \cdots, x_{k_r}$ are linear functions of the remaining $x_j$'s and the constants $b_i$. Since the $b_i$'s are elements of $SC$, each $b_i$ is a linear combination of the $c$'s. Hence $x_{k_1}, \cdots, x_{k_r}$ are linear functions of the $c$'s and the remaining unknowns, the latter being able to assume arbitrary values.

At this stage of the argument we have shown that the system $AX = C$ is consistent if and only if the bottom $m - r$ rows of $SC$ are zero rows, and that in such a situation we have a solution of the kind described in the statement of the theorem.

It remains to prove that $SC$ has the bottom $m - r$ rows as zero rows if and only if rank $A^+ = r =$ rank $A$, or since $SA^+$ and $SA$ have the same ranks as $A^+$ and $A$, if and only if rank $(SA^+) = r =$ rank $(SA)$. But this follows immediately, since $SA^+ = [SA|SC]$ and $SA^+$ and $SA$ are both in modified triangular form.

*Example.*   Solve the system of equations

$$x_1 - x_2 + 2x_3 + x_4 = 2,$$
$$3x_1 + 2x_2 \qquad + x_4 = 1,$$
$$4x_1 + x_2 + 2x_3 + 2x_4 = 3.$$

The system may be expressed by the matrix equation

$$\begin{bmatrix} 1 & -1 & 2 & 1 \\ 3 & 2 & 0 & 1 \\ 4 & 1 & 2 & 2 \end{bmatrix} \begin{bmatrix} x_1 \\ x_2 \\ x_3 \\ x_4 \end{bmatrix} = \begin{bmatrix} 2 \\ 1 \\ 3 \end{bmatrix}.$$

By elementary row operations we transform $A^+$, and hence $A$ along with it, into modified triangular form:

$$\begin{bmatrix} 1 & -1 & 2 & 1 & 2 \\ 3 & 2 & 0 & 1 & 1 \\ 4 & 1 & 2 & 2 & 3 \end{bmatrix} \underset{\substack{o(2-3\cdot1) \\ o(3-4\cdot1)}}{\longleftrightarrow} \begin{bmatrix} 1 & -1 & 2 & 1 & 2 \\ 0 & 5 & -6 & -2 & -5 \\ 0 & 5 & -6 & -2 & -5 \end{bmatrix} \underset{\substack{o(1/5\cdot2) \\ o(1/5\cdot3)}}{\longleftrightarrow}$$

$$\begin{bmatrix} 1 & -1 & 2 & 1 & 2 \\ 0 & 1 & -6/5 & -2/5 & -1 \\ 0 & 1 & -6/5 & -2/5 & -1 \end{bmatrix} \underset{\substack{o(1+1\cdot2) \\ o(3-1\cdot2)}}{\longleftrightarrow} \begin{bmatrix} 1 & 0 & 4/5 & 3/5 & 1 \\ 0 & 1 & -6/5 & -2/5 & -1 \\ 0 & 0 & 0 & 0 & 0 \end{bmatrix}.$$

Since rank $A^+$ = rank $A = 2$, the system of equations is consistent. Moreover it is equivalent to the system

$$\begin{bmatrix} 1 & 0 & 4/5 & 3/5 \\ 0 & 1 & -6/5 & -2/5 \\ 0 & 0 & 0 & 0 \end{bmatrix} \begin{bmatrix} x_1 \\ x_2 \\ x_3 \\ x_4 \end{bmatrix} = \begin{bmatrix} 1 \\ -1 \\ 0 \end{bmatrix},$$

which yields

$$x_1 = -(4/5)x_3 - (3/5)x_4 + 1,$$

$$x_2 = (6/5)x_3 + (2/5)x_4 - 1.$$

We note that two of the variables, $x_1$ and $x_2$, are expressed as linear functions of the remaining two variables, $x_3$ and $x_4$, and that arbitrary values may be assigned to these last two variables.

**2.8.7** DEFINITIONS. The system $AX = C$ is said to be *homogeneous* or *nonhomogeneous* according as $C = O$ or $C \neq O$.

**2.8.8** THEOREM. *A homogeneous system* AX = O *has a solution* X = O (known as the *trivial solution*).

**2.8.9** THEOREM. *A homogeneous system* AX = O, *where* A *is* m × n, *has a nontrivial solution if and only if rank* A < n.

If rank $A = r < n$, then $n - r > 0$ and, by Theorem 2.8.6, $r$ of the unknowns may be expressed as linear functions of the $n - r > 0$ other unknowns, and these $n - r$ other unknowns may be assigned arbitrary values.

If $r = n$, then, in the proof of Theorem 2.8.6,

$$SA = I_{(n)} \quad \text{or} \quad \begin{bmatrix} I_{(n)} \\ O \end{bmatrix},$$

and the equation $SAX = O$ shows that we must have $X = O$. It follows that if $AX = O$ has a nontrivial solution, we must have $r \neq n$; that is, we must have $r < n$.

**2.8.10** Corollary. *The homogeneous system* $AX = O$, *where* A *is square, has a nontrivial solution if and only if* A *is singular.*

## PROBLEMS

**2.8–1**   Prove Corollary 2.8.10.

**2.8–2**   Show that the system of equations

$$x_1 + 2x_2 - x_3 = 3,$$
$$3x_1 - x_2 + 2x_3 = 1,$$
$$2x_1 - 2x_2 + 3x_3 = 2,$$
$$x_1 - x_2 + x_3 = -1,$$

is consistent, and solve the system.

**2.8–3**   Show that the system of equations

$$5x_1 + 3x_2 + 7x_3 - 4 = 0,$$
$$3x_1 + 26x_2 + 2x_3 - 9 = 0,$$
$$7x_1 + 2x_2 + 10x_3 - 5 = 0,$$

is consistent, and solve the system.

**2.8–4**   Show that the system of equations

$$2x_1 + 6x_2 \qquad\qquad + 11 = 0,$$
$$6x_1 + 20x_2 - 6x_3 + 3 = 0,$$
$$6x_2 - 18x_3 + 1 = 0,$$

is inconsistent.

**2.8–5**    Solve the system of equations

$$x_1 + 3x_2 - 2x_3 = 0,$$
$$2x_1 - x_2 + 4x_3 = 0,$$
$$x_1 - 11x_2 + 14x_3 = 0.$$

**2.8–6**    Show that the only real value of $a$ for which the following system of equations has a nontrivial solution is 6:

$$x_1 + 2x_2 + 3x_3 = ax_1,$$
$$3x_1 + x_2 + 2x_3 = ax_2,$$
$$2x_1 + 3x_2 + x_3 = ax_3.$$

**2.8–7**    Find for what values of $a$ and $b$ the system of equations

$$x_1 + x_2 + x_3 = 6,$$
$$x_1 + 2x_2 + 3x_3 = 10,$$
$$x_1 + 2x_2 + ax_3 = b,$$

has (**a**) no solution, (**b**) a unique solution, (**c**) an infinite number of solutions.

**2.8–8**    Find the values of $b$ for which the system of equations

$$x_1 + x_2 + x_3 = 1,$$
$$x_1 + 2x_2 + 4x_3 = b,$$
$$x_1 + 4x_2 + 10x_3 = b^2,$$

has a solution, and solve the system completely in each case.

**2.8–9**    Solve the system of equations

$$ax_1 + 2x_2 - 2x_3 = 1,$$
$$4x_1 + 2ax_2 - x_3 = 2,$$
$$6x_1 + 6x_2 + ax_3 = 3,$$

considering, in particular, the case where $a = 2$.

**2.8–10**    Suppose $A$ is an $m \times n$ matrix of rank $r$, which can be partitioned into the form

$$A = \begin{bmatrix} A_1 & B_1 \\ A_2 & B_2 \end{bmatrix},$$

where $A_1$ is nonsingular and of order $r$. Show that

$$X = \{x_1, \cdots, x_n\} = \begin{bmatrix} -A_1^{-1}B_1X_2 \\ X_2 \end{bmatrix},$$

where $X_2 = \{x_{r+1}, \cdots, x_n\}$ is arbitrary, is a solution of $AX = O$.

**2.8–11**   (a) Show that if the column vectors $X_1, \cdots, X_k$ are solutions of $AX = O$, then so also is any linear combination $Y$ of $X_1, \cdots, X_k$.

(b) Show that if the column vectors $X_1, \cdots, X_k$ are solutions of $AX = O$, and if the column vector $Z$ is a solution of $AX = C$, then $Y + Z$, where $Y$ is any linear combination of $X_1, \cdots, X_k$, is also a solution of $AX = C$.

**2.8–12**   (a) Let $A$ be an $n \times n$ matrix and let $B'_1, B'_2, \cdots, B'_n$ denote the columns of an $n \times n$ matrix $B$. Show that if $\rho[A|B'_1] = \rho[A|B'_2] = \cdots = \rho[A|B'_n] = \rho(A)$, then $\rho[A|B] = \rho(A)$.

(b) Show that if $A, B, X$ are $n \times n$ matrices, the matrix equation $AX = B$ has a solution if and only if $\rho[A|B] = \rho(A)$. Show, in particular, that if $A$ is nonsingular, the matrix equation has the unique solution $X = A^{-1}B$.

**2.8–13**   If $A$ is an $m \times n$ matrix and $X$ is an $n$th-order column vector, show that the system $AX = O$ always has a nontrivial solution if $m < n$.

**2.8–14**   Let $P$ be an $m \times n$ matrix, $Y$ an $m$th-order row vector, and $Q$ an $n$th-order row vector. Show how the solution of the matrix equation $YP = Q$ can be reduced to the solution of a matrix equation of the type $AX = C$ considered in Section 2.8.

## 2.9   Linearly independent solutions of systems of linear equations

In this section we determine the maximum number of linearly independent solutions possessed by a consistent system $AX = C$, and we show how such a maximum set may be found.

**2.9.1**   THEOREM.   *A consistent system of linear equations, whose matrix is* m $\times$ n *and of rank* r, *has exactly* n $-$ r *linearly independent solutions if the system is homogeneous, and exactly* n $-$ r $+$ 1 *linearly independent solutions if the system is nonhomogeneous.*

In the proof of Theorem 2.8.6, we have seen (renumbering the unknowns if necessary) that all solutions of the consistent system $AX = C$

may be expressed in the form

$$x_1 = \sum_{j=r+1}^{n} d_{1j}x_j + b_1,$$

(1) $$x_2 = \sum_{j=r+1}^{n} d_{2j}x_j + b_2,$$

$$\cdot \quad \cdot \quad \cdot \quad \cdot \quad \cdot \quad \cdot \quad \cdot \quad \cdot$$

$$x_r = \sum_{j=r+1}^{n} d_{rj}x_j + b_r,$$

where the $d$'s and the $b$'s are constants and $x_{r+1}, \cdots, x_n$ may assume arbitrary values. It follows that if $r = n$, there is only the one solution $\{b_1, b_2, \cdots, b_n\}$. If the system is nonhomogeneous, at least one of the $b$'s is different from zero and we have precisely one linearly independent solution. If the system is homogeneous, then, by Theorem 2.8.9, all the $b$'s must be zero, and we have no linearly independent solution. Thus the theorem is certainly true if $r = n$.

If $r < n$, consider the $n \times (n - r + 2)$ matrix

$$R = \begin{bmatrix} x'_1 & b_1 & d_{1,r+1} + b_1 & d_{1,r+2} + b_1 & \cdots & d_{1,n} + b_1 \\ x'_2 & b_2 & d_{2,r+1} + b_2 & d_{2,r+2} + b_2 & \cdots & d_{2,n} + b_2 \\ \cdots & \cdots & \cdots & \cdots & \cdots & \cdots \\ x'_r & b_r & d_{r,r+1} + b_r & d_{r,r+2} + b_r & \cdots & d_{r,n} + b_r \\ x'_{r+1} & 0 & 1 & 0 & \cdots & 0 \\ x'_{r+2} & 0 & 0 & 1 & \cdots & 0 \\ \cdots & \cdots & \cdots & \cdots & \cdots & \cdots \\ x'_n & 0 & 0 & 0 & \cdots & 1 \end{bmatrix}$$

and denote the $i$th column by $R'_i$. Then $R'_2$ is a solution of the given system of equations (obtained by setting $x_{r+1} = x_{r+2} = \cdots = x_n = 0$); $R'_3$ is a solution (obtained by setting $x_{r+1} = 1$, $x_{r+2} = x_{r+3} = \cdots = x_n = 0$); $\cdots$; $R'_{n-r+2}$ is a solution (obtained by setting $x_{r+1} = x_{r+2} = \cdots = x_{n-1} = 0, x_n = 1$).

Note that the last $n - r$ columns of $R$ are linearly independent, since the matrix of these columns has its last $n - r$ rows linearly independent. If the $b$'s are not all zero, then actually the last $n - r + 1$ columns of $R$ are linearly independent because the matrix of these columns contains $n - r + 1$ linearly independent rows, namely, the last $n - r$ along with any one containing a $b_k \neq 0$.

Suppose $R'_1$ is any solution of the system of equations. Now consider

$$R'_1 - \sum_{j=r+1}^{n} x'_j(R'_{j-r+2} - R'_2) - R'_2$$

$$= R'_1 - x'_{r+1}(R'_3 - R'_2) - x'_{r+2}(R'_4 - R'_2) - \cdots$$
$$- x'_n(R'_{n-r+2} - R'_2) - R'_2$$

$$= R'_1 - x'_{r+1}\{d_{1,r+1}, d_{2,r+1}, \cdots, d_{r,r+1}, 1, 0, \cdots, 0\}$$
$$- x'_{r+2}\{d_{1,r+2}, d_{2,r+2}, \cdots, d_{r,r+2}, 0, 1, \cdots, 0\}$$
$$\cdot \quad \cdot \quad \cdot \quad \cdot \quad \cdot \quad \cdot \quad \cdot \quad \cdot \quad \cdot \quad \cdot \quad \cdot \quad \cdot \quad \cdot \quad \cdot$$
$$- x'_n\{d_{1,n}, d_{2,n}, \cdots, d_{r,n}, 0, 0, \cdots, 1\} - R'_2.$$

Let us find the $i$th element of this column vector. If $i \leqq r$ we have

$$x'_i - x'_{r+1}d_{i,r+1} - x'_{r+2}d_{i,r+2} - \cdots - x'_n d_{i,n} - b_i$$

$$= x'_i - \sum_{j=r+1}^{n} d_{ij}x'_j - b_i = 0,$$

by (1). If $i = r + k$, $k > 0$, we have

$$x'_{r+k} - x'_{r+k} = 0.$$

It follows that

$$R'_1 - \sum_{j=r+1}^{n} x'_j(R'_{j-r+2} - R'_2) - R'_2 = O,$$

and $R'_1$ is a linear combination of $R'_2, R'_3, \cdots, R'_{n-r+2}$. Therefore, if the $b$'s are not all zero (that is, if the given system of equations is nonhomogeneous), the system has exactly $n - r + 1$ linearly independent solutions. If the $b$'s are all zero (that is, if the given system of equations is homogeneous), the system has exactly $n - r$ linearly independent solutions.

*Example.* Find a maximum set of linearly independent solutions of the system

$$x_1 - x_2 + 2x_3 + x_4 = 2,$$
$$3x_1 + 2x_2 + x_4 = 1,$$
$$4x_1 + x_2 + 2x_3 + 2x_4 = 3.$$

This is the system considered in the example following Theorem 2.8.6. There we found

$$x_1 = -(4/5)x_3 - (3/5)x_4 + 1,$$
$$x_2 = (6/5)x_3 + (2/5)x_4 - 1.$$

Thus we have $n = 4$ and $r = 2$, whence the maximum number of linearly independent solutions is (since the system is nonhomogeneous) $n - r + 1 = 3$. Now here we have $b_1 = 1$, $b_2 = -1$, $d_{13} = -4/5$, $d_{23} = 6/5$, $d_{14} = -3/5$, $d_{24} = 2/5$. It follows that

$$d_{13} + b_1 = 1/5, \quad d_{23} + b_2 = 1/5,$$

$$d_{14} + b_1 = 2/5, \quad d_{24} + b_2 = -3/5,$$

and as a maximum set of linearly independent solutions we may take the last three columns of the associated matrix

$$R = \begin{bmatrix} x'_1 & 1 & 1/5 & 2/5 \\ x'_2 & -1 & 1/5 & -3/5 \\ x'_3 & 0 & 1 & 0 \\ x'_4 & 0 & 0 & 1 \end{bmatrix}.$$

Theorem 2.9.1 enables us to prove some theorems concerning ranks of matrices in a very convenient way, which we now illustrate by reproving and extending some of the theorems of Section 2.7.

**2.9.2** THEOREM. *The rank of the product of two matrices cannot exceed the rank of either factor.*

Let $A$ and $B$ be $m \times n$ and $n \times p$ matrices and consider the systems $(AB)X = O$ and $BX = O$. By Theorem 2.9.1, the first system has exactly $p - \rho(AB)$ linearly independent solutions and the second system has exactly $p - \rho(B)$ linearly independent solutions. Since any solution of the second system is obviously a solution of the first system, but not necessarily conversely, it follows that $p - \rho(B) \leq p - \rho(AB)$, or $\rho(AB) \leq \rho(B)$. It then further follows that $\rho(AB) = \rho(AB)' = \rho(B'A') \leq \rho(A') = \rho(A)$.

**2.9.3** THEOREM. *If A and B are m × n and n × p matrices such that* AB = O, *then* $\rho(A) + \rho(B) \leq n$.

Since each column of $B$ is a solution of the system $AX = O$, the number $\rho(B)$ of linearly independent columns of $B$ cannot exceed the number $n - \rho(A)$ of linearly independent solutions of the system $AX = O$. That is, $\rho(B) \leq n - \rho(A)$, or $\rho(A) + \rho(B) \leq n$.

**2.9.4** DEFINITION. Let $P$ and $Q$ be $m \times n$ and $n \times p$ matrices and let $\rho(Q) = r$, $\rho(PQ) = s$. Then (by Theorem 2.9.2) $s \leq r$, whence $p - s \geq$

$p - r$. By Theorem 2.9.1, the system $QX = O$ has exactly $p - r$ linearly independent solutions, say $X_1, \cdots, X_{p-r}$. These solutions are clearly also solutions of the system $(PQ)X = O$. By Theorem 2.9.1, the system $(PQ)X = O$ has exactly $p - s$ linearly independent solutions. For such a set we may take the set $X_1, \cdots, X_{p-r}$, augmented by $(p - s) - (p - r) = r - s = t$, say, further appropriate vectors $Y_1, \cdots, Y_t$. Then $Y_1, \cdots, Y_t$ are linearly independent solutions of the system $(PQ)X = O$ but are not solutions of the system $QX = O$. We shall refer to such a set $Y_1, \cdots, Y_t$ as an *extension of solutions of* QX = O *to solutions of* (PQ)X = O.

**2.9.5**   LEMMA.   *Let* P *and* Q *be* m × n *and* n × p *matrices and let* $Y_1, \cdots, Y_t$ *be an extension of solutions of* QX = O *to solutions of* (PQ)X = O. *Then* $QY_1, \cdots, QY_t$ *are linearly independent.*

For suppose

$$c_1 Q Y_1 + \cdots + c_t Q Y_t = O.$$

Then

$$Q(c_1 Y_1 + \cdots + c_t Y_t) = O,$$

and $c_1 Y_1 + \cdots + c_t Y_t$ is a solution of the system $QX = O$. Suppose (see Theorem 2.9.1) $X_1, \cdots, X_{p-r}$, where $r = \rho(Q)$, constitute a maximal linearly independent set of solutions of the system $QX = O$. Then we must have

$$c_1 Y_1 + \cdots + c_t Y_t = a_1 X_1 + \cdots + a_{p-r} X_{p-r}.$$

But, since $X_1, \cdots, X_{p-r}, Y_1, \cdots, Y_t$ constitute a linearly independent set of solutions of the system $(PQ)X = O$, it follows that all the $c$'s and all the $a$'s must be zero, and (since all the $c$'s must be zero) the lemma is established.

**2.9.6**   THE FROBENIUS INEQUALITY.[†]   *If* A, B, C *are* m × n, n × p, p × q *matrices, then* $\rho(AB) + \rho(BC) \leqq \rho(B) + \rho(ABC)$.

Let $Y_1, \cdots, Y_t$ be an extension of solutions of $(BC)X = O$ to solutions of $(ABC)X = O$. Then, by Lemma 2.9.5, the vectors $BCY_1, \cdots, BCY_t$ are linearly independent. Now $Z_1 = CY_1, \cdots, Z_t = CY_t$ are also linearly independent, for if

$$a_1 Z_1 + \cdots + a_t Z_t = O,$$

then

† Given by Georg Frobenius in 1911.

$$a_1(BCY_1) + \cdots + a_t(BCY_t) = O,$$

whence $a_1 = \cdots = a_t = 0$. It follows that $Z_1, \cdots, Z_t$ is a set of $t$ linearly independent vectors, each of which is a solution of the system $(AB)X = O$ but not of the system $BX = O$. Since, by Theorem 2.9.1, the systems $BX = O$, $(AB)X = O$, $(BC)X = O$, and $(ABC)X = O$ have, respectively, exactly $p - \rho(B), p - \rho(AB), q - \rho(BC), q - \rho(ABC)$ linearly independent solutions, we have

$$\rho(BC) - \rho(ABC) = [q - \rho(ABC)] - [q - \rho(BC)]$$
$$= t \leqq [p - \rho(AB)] - [p - \rho(B)] = \rho(B) - \rho(AB).$$

That is, $\rho(AB) + \rho(BC) \leqq \rho(B) + \rho(ABC)$.

The Frobenius inequality is remarkably comprehensive. Setting $C = O$ and $A = O$ in turn, we obtain $\rho(AB) \leqq \rho(B)$ and $\rho(BC) \leqq \rho(B)$, which gives Theorem 2.9.2. If we set $B = I$, we obtain $\rho(A) + \rho(C) \leqq n + \rho(AC)$, which is the extension of Theorem 2.7.8 given in Problem 2.7–15. Combining these results, we have a slight extension of Sylvester's law of nullity. Setting $B = I$ and $AC = O$, we obtain $\rho(A) + \rho(C) \leqq n$, which is Theorem 2.9.3.

## PROBLEMS

**2.9–1**   Exhibit a maximum set of linearly independent solutions for each of the following systems of equations:

(a)  $3x_1 - 2x_2 + x_3 + 6 = 0,$
     $2x_1 + 5x_2 - 3x_3 - 2 = 0,$
     $4x_1 - 9x_2 + 5x_3 + 14 = 0.$

(b)  $4x_1 + 7x_2 - 14x_3 = 10,$
     $2x_1 + 3x_2 - 4x_3 = -4,$
     $x_1 + x_2 + x_3 = 6.$

(c)  $4x_1 - x_2 + x_3 = 5,$
     $2x_1 - 3x_2 + 5x_3 = 1,$
     $x_1 + x_2 - 2x_3 = 2,$
     $5x_1 - x_3 = 2.$

(d)  $2x_1 - x_2 - 2x_3 = 0,$
     $x_1 - 2x_2 + x_3 = 0,$
     $2x_1 - 3x_2 - x_3 = 0.$

**(e)** $6x_1 + 4x_2 + 3x_3 - 84x_4 = 0,$
$\quad x_1 + 2x_2 + 3x_3 - 48x_4 = 0,$
$\quad 4x_1 - 4x_2 - \phantom{3}x_3 - 24x_4 = 0,$
$\quad x_1 - 2x_2 + \phantom{3}x_3 - 12x_4 = 0.$

**2.9–2**  If $A$ is any square matrix and $p$, $q$, $r$ are arbitrary positive integers, show that

$$\rho(A^{p+q}) + \rho(A^{q+r}) \leqq \rho(A^q) + \rho(A^{p+q+r})$$

and then deduce, as particular cases, that

$$\rho(A^{p+1}) + \rho(A^{p+q}) \leqq \rho(A^p) + \rho(A^{p+q+1}),$$

$$\rho(A^{p+1}) + \rho(A^{2p-1}) \leqq \rho(A^p) + \rho(A^{2p}),$$

$$2\rho(A^3) \leqq \rho(A^2) + \rho(A^4).$$

**2.9–3**  If $A$ is an $n \times n$ matrix, show that $2\rho(A) \leqq n + \rho(AA')$.


# ADDENDA TO CHAPTER 2


The following items, given only in skeleton form and largely connected with the concepts of rank and of the inverse of a matrix, are inserted here for those interested students who might care to develop some of them into short expository papers.


## 2.1A   Left and Right Inverses

This item concerns itself with a generalization of the concept of the inverse of a nonsingular matrix.


**I**   DEFINITIONS AND NOTATION.   An $m \times n$ matrix $A$ is said to have matrix $A^L$ as a *left inverse* if $A^L A = I$, and $A$ is then said to be *nonsingular on the left*; in this case $I$ must be $n \times n$ and $A^L$ must be $n \times m$. Similarly, $A$ is said to have $A^R$ as a *right inverse* if $AA^R = I$, and $A$ is then said to be *nonsingular on the right*; in this case $I$ must be $m \times m$ and $A^R$ must be $n \times m$. Left and right inverses of a matrix $A$ that are not also right and left inverses, respectively, of $A$ are called *one-sided inverses* of $A$.


**II**   THEOREM.   *One-sided inverses are not necessarily unique.*

**III** THEOREM. *If* A *has both a left inverse* $A^L$ *and a right inverse* $A^R$, *then* A *is nonsingular and* $A^L = A^R = A^{-1}$.

**IV** THEOREM. *If* A *and* B *are conformable for multiplication and if* A *has a left inverse* $A^L$ *and* B *a left inverse* $B^L$, *then* AB *has* $B^L A^L$ *for a left inverse. Also, if* A *has right inverse* $A^R$ *and* B *a right inverse* $B^R$, *then* AB *has* $B^R A^R$ *for a right inverse.*

**V** THEOREM. *If* A *has a left (right) inverse, then* A' *has a right (left) inverse.*

**VI** THEOREM. *If* A *is square and has a left inverse* $A^L$ *(or a right inverse* $A^R$), *then* A *is nonsingular and* $A^L$ *(or* $A^R$) $= A^{-1}$.

**VII** THEOREM. *The row (column) rank of a matrix is not altered by premultiplying (postmultiplying) the matrix by a matrix that is nonsingular on the left (right).*

### 2.2A   Some Further Methods of Matrix Inversion

Following Theorem 2.2.13, we illustrated a method of finding the inverse of a nonsingular matrix $A$ by use of elementary row operations alone. The procedure lay in converting the partitioned matrix $[A|I]$, by elementary row operations, into a matrix of the form $[I|P]$. Then $P = A^{-1}$. We shall refer to this method as Method 1.

The problem of inverting a nonsingular matrix is such an important one that a large number of alternative methods have been devised and examined. In the present item we indicate a few of these alternative methods; still other methods will be considered later in the text after sufficient theory has been developed.

**I** METHOD 2. Problem 2.4–10 (a) suggests a method of inverting a nonsingular matrix $A$ by the use of both elementary row and elementary column operations. The procedure lies in converting the partitioned matrix

$$\begin{bmatrix} A & | & I \\ I & | & O \end{bmatrix},$$

by elementary row and elementary column operations, into the partitioned form

$$\left[\begin{array}{c|c} I & P \\ \hline Q & O \end{array}\right].$$

Then $A^{-1} = QP$. It is to be noted that Method 1 is a special case of Method 2. Nevertheless, Method 1 is worthy of special consideration, for though Method 2 may be faster, Method 1 is more readily coded for high-speed computing machines. The reader is invited to find by Method 2 the inverse of

$$\begin{bmatrix} 2 & -2 & 4 \\ 2 & 3 & 2 \\ -1 & 1 & -1 \end{bmatrix},$$

the matrix whose inverse was found following Theorem 2.2.13 by Method 1.

**II**   Method 3.   Let $A$ be an $n \times n$ matrix and consider an $n \times n$ matrix $X$ such that $AX = I$. If we denote the columns of $X$ by $X'_1, \cdots, X'_n$ and those of $I$ by $I'_1, \cdots, I'_n$, the solution of the $n$ linear systems

$$AX'_1 = I'_1, \cdots, AX'_n = I'_n$$

will yield matrix $X$. But, if $A$ is nonsingular, $X'_1, \cdots, X'_n$ are uniquely determined and $X = A^{-1}$.

Taking

$$A = \begin{bmatrix} 2 & -2 & 4 \\ 2 & 3 & 2 \\ -1 & 1 & -1 \end{bmatrix}, \qquad X = \begin{bmatrix} x_1 & y_1 & z_1 \\ x_2 & y_2 & z_2 \\ x_3 & y_3 & z_3 \end{bmatrix},$$

the reader is invited to find $X = A^{-1}$ by Method 3. The problem of inverting matrix $A$ is here reduced to the problem of solving linear systems of equations. Since much research has been expended on efficient solutions of this latter problem, a number of techniques are available for finding $X = A^{-1}$.

**III**   Method 4 (*method of partitioning*).   Let us partition a non-singular matrix $A$ into four submatrices

$$A = \begin{bmatrix} \alpha_{11} & \alpha_{12} \\ \alpha_{21} & \alpha_{22} \end{bmatrix},$$

where $\alpha_{11}$ is $(n-1) \times (n-1)$. Denote the inverse of $A$ by $B$ and similarly partition $B$ as

$$B = \begin{bmatrix} \beta_{11} & \beta_{12} \\ \beta_{21} & \beta_{22} \end{bmatrix},$$

where $\beta_{11}$ is $(n-1) \times (n-1)$. Then we have

$$\begin{bmatrix} \alpha_{11} & \alpha_{12} \\ \alpha_{21} & \alpha_{22} \end{bmatrix} \begin{bmatrix} \beta_{11} & \beta_{12} \\ \beta_{21} & \beta_{22} \end{bmatrix} = I,$$

or

(1) $$\alpha_{11}\beta_{11} + \alpha_{12}\beta_{21} = I_{(n-1)},$$

(2) $$\alpha_{11}\beta_{12} + \alpha_{12}\beta_{22} = O_{(1,n-1)},$$

(3) $$\alpha_{21}\beta_{11} + \alpha_{22}\beta_{21} = O_{(n-1,1)},$$

(4) $$\alpha_{21}\beta_{12} + \alpha_{22}\beta_{22} = I_{(1)}.$$

Assuming that $\alpha_{11}$ is nonsingular we find, from (2),

(5) $$\beta_{12} = -\alpha_{11}^{-1}\alpha_{12}\beta_{22}.$$

Substituting this expression for $\beta_{12}$ in (4) and solving for $\beta_{22}$, we find

(6) $$\beta_{22} = (\alpha_{22} - \alpha_{21}\alpha_{11}^{-1}\alpha_{12})^{-1} = D,$$

say, provided $\alpha_{22} - \alpha_{21}\alpha_{11}^{-1}\alpha_{12} \neq O_{(1)}$. We now find, from (5),

(7) $$\beta_{12} = -\alpha_{11}^{-1}\alpha_{12}D.$$

From (1) we find

(8) $$\beta_{11} = \alpha_{11}^{-1} - \alpha_{11}^{-1}\alpha_{12}\beta_{21}.$$

Substituting this expression for $\beta_{11}$ in (3) we find

(9) $$\beta_{21} = (\alpha_{22} - \alpha_{21}\alpha_{11}^{-1}\alpha_{12})^{-1}\alpha_{21}\alpha_{11}^{-1} = -D(\alpha_{21}\alpha_{11}^{-1}).$$

We now finally find, from (8),

(10) $$\beta_{11} = \alpha_{11}^{-1} + (\alpha_{11}^{-1}\alpha_{12}D)(\alpha_{21}\alpha_{11}^{-1}).$$

Now, using (6), (7), (9), and (10), we can form

$$A^{-1} = B = \begin{bmatrix} \alpha_{11}^{-1} + (\alpha_{11}^{-1}\alpha_{12}D)(\alpha_{21}\alpha_{11}^{-1}) & -\alpha_{11}^{-1}\alpha_{12}D \\ -D(\alpha_{21}\alpha_{11}^{-1}) & D \end{bmatrix},$$

where $D = (\alpha_{22} - \alpha_{21}\alpha_{11}^{-1}\alpha_{12})^{-1}$, and it follows that $A^{-1}$ can be found if $\alpha_{11}^{-1}$ exists and can be found and if $\alpha_{22} - \alpha_{21}\alpha_{11}^{-1}\alpha_{12} \neq O$. To find

$\alpha_{11}{}^{-1}$, we use the same process applied to the $(n-1) \times (n-1)$ matrix $\alpha_{11}$, and so on.

The process can be used in a recursive fashion to find the inverse of any nonsingular $n \times n$ matrix $A = [a_{ij}]$. Define the sequence of matrices $S_1, S_2, \cdots, S_n$ by

$$S_1 = [a_{11}], \qquad S_2 = \left[\begin{array}{c|c} S_1 & a_{12} \\ \hline a_{21} & a_{22} \end{array}\right], \qquad S_3 = \left[\begin{array}{cc|c} & & a_{13} \\ & S_2 & a_{23} \\ \hline a_{31} & a_{32} & a_{33} \end{array}\right], \cdots,$$

$$S_n = A = \left[\begin{array}{cccc|c} & & & & a_{1n} \\ & & & & a_{2n} \\ & & S_{n-1} & & \cdots \\ & & & & a_{n-1,n} \\ \hline a_{n1} & a_{n2} & \cdots & a_{n,n-1} & a_{nn} \end{array}\right].$$

Find $S_1{}^{-1}$; then find (by the process) $S_2{}^{-1}$; then find (by the process) $S_3{}^{-1}, \cdots$; then finally find (by the process) $S_n{}^{-1} = A^{-1}$. The reader is invited to carry out the procedure for the nonsingular matrix

$$A = \begin{bmatrix} 2 & -2 & 4 \\ 2 & 3 & 2 \\ -1 & 1 & -1 \end{bmatrix}.$$

This method requires that inverses of certain smaller matrices exist. There are ways of coping with the situation where one or more of these smaller matrices are singular, but we shall not consider this matter here. The method (as also each of the other methods) appreciably simplifies if $A$ is symmetric, and in this connection it is interesting to note that the inversion of any nonsingular matrix can be reduced to the inversion of a symmetric nonsingular matrix, since $A^{-1} = (A'A)^{-1}A'$ and $A'A$ is symmetric.†

### 2.3A   Linear Dependence and Independence of a Set of Matrices

An interesting exercise is to extend Definitions 2.5.1 and 2.5.2 to a set $M_1, \cdots, M_p$ of $p$ $m \times n$ $\mathscr{C}$ matrices and then establish analogs of

† In connection with the method of partitioning, the interested student may like to consult: (1) R. A. Frazer, W. J. Duncan, A. R. Collar, *Elementary Matrices* (Cambridge University Press, 1950), pp. 112–118; (2) M. Lotkin and R. Remage, "Matrix inversion by partitioning," BRL Report 823, Aberdeen Proving Ground, Maryland, 1952; (3) M. Lotkin and R. Remage, "Scaling and error analysis for matrix inversion by partitioning," *Annals of Mathematical Statistics*, Vol. 24 (1953), pp. 428–439.

Theorems 2.5.3, 2.5.4, and 2.5.5. Imagine matrices $M_1, \cdots, M_p$ piled successively in layers on top of one another to form a three-dimensional rectangular array of numbers. We might call such a piling an $m \times n \times p$ *box matrix*. Formulate definitions of *width rank*, *length rank*, and *depth rank* of a box matrix. One can regard an $m \times n \times p$ box matrix as made up of $m\ p \times n$ *width layers*, or $n\ p \times m$ *length layers*, or $p\ m \times n$ *depth layers*. Define *elementary operations* on a set of layers and try to extend to box matrices as many of the concepts and theorems on rank and equivalence of rectangular matrices as you can.

### 2.4A    Lines in a Plane and Planes in Space

In Cartesian analysis a linear equation

$$ax + by = c,$$

where $a$, $b$, $c$ are real numbers, represents a line in the plane, and a linear equation

$$ax + by + cz = d,$$

where $a$, $b$, $c$, $d$ are real numbers, represents a plane in space. Associated with the pair of lines

$$a_{11}x + a_{12}y = d_1,$$

$$a_{21}x + a_{22}y = d_2,$$

are the two matrices

$$A = \begin{bmatrix} a_{11} & a_{12} \\ a_{21} & a_{22} \end{bmatrix}, \qquad D = \begin{bmatrix} a_{11} & a_{12} & d_1 \\ a_{21} & a_{22} & d_2 \end{bmatrix},$$

and associated with the triple of planes

$$a_{11}x + a_{12}y + a_{13}z = d_1,$$

$$a_{21}x + a_{22}y + a_{23}z = d_2,$$

$$a_{31}x + a_{32}y + a_{33}z = d_3,$$

are the two matrices

$$A = \begin{bmatrix} a_{11} & a_{12} & a_{13} \\ a_{21} & a_{22} & a_{23} \\ a_{31} & a_{32} & a_{33} \end{bmatrix}, \qquad D = \begin{bmatrix} a_{11} & a_{12} & a_{13} & d_1 \\ a_{21} & a_{22} & a_{23} & d_2 \\ a_{31} & a_{32} & a_{33} & d_3 \end{bmatrix}.$$

The various relations of the pair of lines to one another and of the triple of planes to one another can be neatly expressed in terms of the ranks of the associated matrices $A$ and $D$. The reader can show that we have the following:

*Pair of lines:*

| | | |
|---|---|---|
| $\rho(A) = 2,$ | $\rho(D) = 2.$ | The lines intersect in a point. |
| $\rho(A) = 1,$ | $\rho(D) = 2.$ | The lines are parallel. |
| $\rho(A) = 1,$ | $\rho(D) = 1.$ | The lines coincide. |

*Triple of planes:*

| | | |
|---|---|---|
| $\rho(A) = 3,$ | $\rho(D) = 3.$ | The planes intersect in a point. |
| $\rho(A) = 2,$ | $\rho(D) = 3.$ | The planes form a prism. |
| $\rho(A) = 1,$ | $\rho(D) = 3.$ | This case cannot arise. |
| $\rho(A) = 2,$ | $\rho(D) = 2.$ | The planes have a line in common. |
| $\rho(A) = 1,$ | $\rho(D) = 2.$ | The planes are parallel. |
| $\rho(A) = 1,$ | $\rho(D) = 1.$ | The planes coincide. |

### 2.5A   An Affine Classification of Conics and Conicoids According to the Ranks of Their Associated Matrices

A conic (real, imaginary, or degenerate) can be represented in nonhomogeneous coordinates by an equation of the form

$$ax^2 + 2hxy + by^2 + 2gx + 2fy + c = 0,$$

and in homogeneous coordinates by an equation of the form

$$ax^2 + by^2 + cz^2 + 2fyz + 2gzx + 2hxy = 0,$$

where the coefficients are real numbers. If we set

$$A = \begin{bmatrix} a & h \\ h & b \end{bmatrix}, \qquad D = \begin{bmatrix} a & h & g \\ h & b & f \\ g & f & c \end{bmatrix},$$

$$X = \{x, y\}, \qquad B = [g, f], \qquad Y = \{x, y, z\},$$

the two preceding equations can be written in the forms

(1) $$X'AX + 2BX + c = 0$$

and

(2) $$Y'DY = 0,$$

respectively. We shall refer to $A$ and $D$ as the *associated matrices* of the conic.

Now set

$$C = \{p, q\}, \qquad U = \{u, v\}, \qquad V = \{u, v, w\}.$$

Then a transformation of the form

$$X = C + SU,$$

where $S$ is a $2 \times 2$ nonsingular matrix, is called an *affine transformation*, and a transformation of the form

$$Y = RV,$$

where $R$ is a $3 \times 3$ nonsingular matrix, is called a *nonsingular collineation*. It is easy to show that an affine transformation is a special type of nonsingular collineation. We now prove an important theorem.

**I.** **Theorem.** *The ranks of the associated matrices of a conic are unchanged when the conic is subjected to an affine transformation.*

Setting $X = C + SU$ (where $S$ is nonsingular) in (1), we obtain

$$(C + SU)'A(C + SU) + 2B(C + SU) + c$$

$$= (C' + U'S')A(C + SU) + 2B(C + SU) + c$$

$$= U'(S'AS)U + 2(C'AS + BS)U + (C'AC + 2BC + c)$$

$$= 0.$$

Since $S$ is nonsingular, $\rho(S'AS) = \rho(A)$, and the rank of the first associated matrix has not changed.

Now the affine transformation $X = C + SU$ is a special case of a nonsingular collineation $Y = RV$. Setting $Y = RV$ in (2), we obtain

$$V'(R'DR)V = 0.$$

Since $R$ is nonsingular, $\rho(R'DR) = \rho(D)$, and the rank of the second associated matrix has not changed.

It is shown in any analytic geometry text that, by means of a rigid motion (which is a special type of affine transformation), we may transform the equation of any given conic into a corresponding standard form. By actually forming the associated matrices of these standard forms, we can then find the ranks of these matrices (which ranks we have shown to be invariant under affine transformations). In doing this, we are able to make the following affine classification of conics according to the ranks of the associated matrices:

$\rho(A) = 2,\quad \rho(D) = 3.$    The conic is a real or imaginary ellipse, or a hyperbola.

$\rho(A) = 1,\quad \rho(D) = 3.$    The conic is a parabola.

$\rho(A) = 2,\quad \rho(D) = 2.$    The conic is a pair of distinct real or imaginary intersecting lines.

$\rho(A) = 1,\quad \rho(D) = 2.$    The conic is a pair of parallel lines.

$\rho(A) = 1,\quad \rho(D) = 1.$    The conic is a pair of coincident lines.

The reader is invited to fill in the details.

Stepping up one dimension, we have that a conicoid (real, imaginary, or degenerate) can be represented in nonhomogeneous coordinates by an equation of the form

$$ax^2 + by^2 + cz^2 + 2fyz + 2gzx + 2hxy + 2lx + 2my + 2nz + d = 0,$$

and in homogeneous coordinates by an equation of the form

$$ax^2 + by^2 + cz^2 + dw^2 + 2fyz + 2gzx + 2hxy$$
$$+ 2lxw + 2myw + 2nzw = 0,$$

where the coefficients are real numbers. The associated matrices here are

$$A = \begin{bmatrix} a & h & g \\ h & b & f \\ g & f & c \end{bmatrix}, \qquad D = \begin{bmatrix} a & h & g & l \\ h & b & f & m \\ g & f & c & n \\ l & m & n & d \end{bmatrix}.$$

By paralleling the foregoing treatment for conics, the reader may now obtain the following affine classification of conicoids according to the ranks of the associated matrices.

$\rho(A) = 3,\quad \rho(D) = 4.$    The conicoid is a real or imaginary ellipsoid, or a hyperboloid of one sheet, or a hyperboloid of two sheets.

$\rho(A) = 2,\quad \rho(D) = 4.$    The conicoid is an elliptic or a hyperbolic paraboloid.

$\rho(A) = 1,\quad \rho(D) = 4.$    This case cannot arise.

$\rho(A) = 3,\quad \rho(D) = 3.$    The conicoid is a real or imaginary cone.

$\rho(A) = 2,\quad \rho(D) = 3.$    The conicoid is an elliptic, a hyperbolic, or an imaginary cylinder.

$\rho(A) = 1, \quad \rho(D) = 3.$     The coinicoid is a parabolic cylinder.

$\rho(A) = 2, \quad \rho(D) = 2.$     The conicoid is a pair of distinct real or imaginary intersecting planes.

$\rho(A) = 1, \quad \rho(D) = 2.$     The conicoid is a pair of parallel planes.

$\rho(A) = 1, \quad \rho(D) = 1.$     The conicoid is a pair of coincident planes.

### 2.6A    Kronecker Product of Matrices

A useful way of combining *any* two rectangular matrices to obtain a third rectangular matrix is by forming their *Kronecker product*. Let $A_{(m,n)} = [a_{ij}]_{(m,n)}$ and $B_{(p,q)}$ be any two rectangular matrices; then the Kronecker product of $A$ and $B$, here denoted by $A \otimes B$, is the $mp \times nq$ partitioned matrix

$$\begin{bmatrix} a_{11}B & a_{12}B & \cdots & a_{1n}B \\ a_{21}B & a_{22}B & \cdots & a_{2n}B \\ \cdots & \cdots & \cdots & \cdots \\ a_{m1}B & a_{m2}B & \cdots & a_{mn}B \end{bmatrix}.$$

This product is also known as the *direct product* and the *tensor product* of $A$ and $B$.

Some elementary facts about the Kronecker product are as follows:

(1) If $B$ and $C$ have the same dimensions, then $A \otimes (B + C) = (A \otimes B) + (A \otimes C)$ and $(B + C) \otimes A = (B \otimes A) + (C \otimes A)$.

(2) $(A \otimes B)' = A' \otimes B'$.

(3) $\overline{A \otimes B} = \overline{A} \otimes \overline{B}$ and $(A \otimes B)^* = A^* \otimes B^*$.

(4) If $AC$ and $BD$ exist, then $(A \otimes B)(C \otimes D)$ exists and is equal to $(AC) \otimes (BD)$.

(5) If $A$ and $B$ are nonsingular, then so is $A \otimes B$, and $(A \otimes B)^{-1} = A^{-1} \otimes B^{-1}$.

(6) $\rho(A \otimes B) = \rho(A)\rho(B)$.

### 2.7A    Direct Sum of Matrices

The "direct sum" of a finite sequence of square matrices is defined in Definition 1.9.7, and several properties of the direct sum appear in subsequent problems. Here we redefine the direct sum of two square matrices, introduce an alternative notation, and gather together a number of elementary properties of the direct sum.

DEFINITION AND NOTATION.    If $A$ and $B$ are two square matrices of orders $m$ and $n$, respectively, the matrix

$$A \dotplus B = \begin{bmatrix} A & O \\ O & B \end{bmatrix},$$

of order $m + n$, is called their *direct sum*.

The following elementary properties of the direct sum can be established by the reader:

(1) $(A \dotplus B) \dotplus C = A \dotplus (B \dotplus C)$.

(2) If $k$ is a scalar, then $k(A \dotplus B) = (kA) \dotplus (kB)$.

(3) $(A \dotplus B) \dotplus (C \dotplus D) = (A \dotplus C) \dotplus (B \dotplus D)$.

(4) $(A \dotplus B)(C \dotplus D) = (AC) \dotplus (BD)$.

(5) $(A \dotplus B)' = A' \dotplus B'$.

(6) If $A$ and $B$ are nonsingular, then so is $A \dotplus B$, and $(A \dotplus B)^{-1} = A^{-1} \dotplus B^{-1}$.

(7) $\rho(A \dotplus B) = \rho(A) + \rho(B)$.

(8) $A \dotplus B \overset{\mathrm{E}}{=} B \dotplus A$.

(9) If $A \overset{\mathrm{E}}{=} C$ and $B \overset{\mathrm{E}}{=} D$, then $A \dotplus B \overset{\mathrm{E}}{=} C \dotplus D$.

(10) If $f(x)$ is a polynomial, then $f(A \dotplus B) = f(A) \dotplus f(B)$.

# 3. DETERMINANTS

*3.1. Permutations. 3.2. The notion of determinant. Problems. 3.3. Some elementary properties of determinants. Problems. 3.4. Cofactors. Problems. 3.5. Cyclic determinants and Vandermonde determinants. Problems. 3.6. Chio's expansion. Appendix. Problems. 3.7. Laplace's expansion. Problems. 3.8. The product theorem. Problems. 3.9. Determinant rank of a matrix. Problems. 3.10. Adjoint of a square matrix. Problems. ADDENDA 3.1A. A geometric study of permutations. 3.2A. Permutation matrices. 3.3A. Permanents. 3.4A. Postulational definitions of determinant. 3.5A. The sweep-out process for evaluating determinants. 3.6A. Pfaffians. 3.7A. Solution of systems of linear equations. 3.8A. Continuants. 3.9A. An application of determinants to triangles and tetrahedra. 3.10A. Quantitative aspect of linear independence of vectors. 3.11A. Sylvester's dialytic method of elimination.*

## 3.1 Permutations

In this section we develop the subject of permutations to an extent that will be sufficient for our purposes.

**3.1.1** DEFINITION. Any ordered arrangement $(i_1, i_2, \cdots, i_n)$ of the $n$ integers $1, 2, \cdots, n$ will be called a *permutation* of $1, 2, \cdots, n$.

*Example.* $(1, 2, 3)$, $(1, 3, 2)$, $(2, 3, 1)$, $(2, 1, 3)$, $(3, 1, 2)$, $(3, 2, 1)$ are the possible permutations of $1, 2, 3$.

**3.1.2** THEOREM.    *There are* n! *permutations of* 1, 2, $\cdots$, n.

In building up a permutation of 1, 2, $\cdots$, $n$, the first place may be filled in with any one of the $n$ integers; the second place may then be filled in with any one of the remaining $n - 1$ integers; the third place may then be filled in with any one of the remaining $n - 2$ integers; $\cdots$; the next to the last place may be filled in with either of the two remaining integers; the last place must be filled in with the sole remaining integer. There are thus

$$n(n - 1)(n - 2) \cdots (2)(1) = n!$$

possible permutations of 1, 2, $\cdots$, $n$.

**3.1.3** DEFINITIONS.    A situation in a permutation $(i_1, i_2, \cdots, i_n)$ of 1, 2, $\cdots$, $n$, in which $i_r$ precedes $i_s$ and $i_r > i_s$, is called an *inversion*. The permutation is said to be *odd* or *even*, according as it possesses an odd or an even number of inversions.

*Example.*    The permutation (4, 3, 5, 2, 1) of 1, 2, 3, 4, 5 has three inversions relative to the first element, two inversions relative to the second element, two inversions relative to the third element, one inversion relative to the fourth element, and no inversions relative to the last element. Thus the permutation possesses $3 + 2 + 2 + 1 = 8$ inversions, and is therefore even.

**3.1.4** DEFINITIONS.    The operation of interchanging any two distinct elements of a permutation $(i_1, i_2, \cdots, i_n)$ of 1, 2, $\cdots$, $n$ is called a *transposition*. If the two elements are adjacent, the transposition is called an *adjacent transposition*.

**3.1.5** THEOREM.    *A transposition converts an even* (*odd*) *permutation of* 1, 2, $\cdots$, n *into an odd* (*even*) *permutation of* 1, 2, $\cdots$, n.

Let $p$ denote the permutation $(i_1, \cdots, i_r, \cdots, i_s, \cdots, i_n)$ of 1, 2, $\cdots$, $n$, and let $q$ denote the permutation obtained from $p$ by the transposition that interchanges $i_r$ and $i_s$. If $i_r$ and $i_s$ are adjacent, the conclusion of the theorem certainly holds, for if $i_r > i_s$, their interchange will diminish the total number of inversions in the permutation by 1, and if $i_r < i_s$, their interchange will increase the total number of inversions in the permutation by 1. Now suppose $i_r$ and $i_s$ are separated in $p$ by $m$ elements. By $m + 1$ successive adjacent transpositions we may carry $i_r$ forward into the position originally occupied by $i_s$, and then by $m$ further successive adjacent transpositions, we may carry $i_s$ backward into the position originally occupied by $i_r$. That is, interchanging $i_r$ and $i_s$ can be effected by $2m + 1$ adjacent transpositions. Since each adjacent transposition either increases or decreases the total number of inversions by 1, the $2m + 1$ adjacent transpositions will increase or decrease the total number of inversions by an odd number, and the theorem follows.

**3.1.6** THEOREM. *Of the* n! *permutations of* 1, 2, $\cdots$, n, *where* n > 1, *exactly half are odd and half are even.*

Let $S$ denote the set of $n!$ permutations of 1, 2, $\cdots$, $n$ and let $S'$ denote the set of permutations obtained by subjecting each permutation of $S$ to a given fixed transposition. Then $S'$ is composed of the $n!$ permutations of 1, 2, $\cdots$, $n$, for if any two of the permutations of $S'$ are identical, the two corresponding permutations of $S$ would have to be identical. But, by Theorem 3.1.5, the odd (even) permutations of $S$ have become the even (odd) permutations of $S'$. It follows that there must be exactly as many odd permutations in $S$ as there are even ones, and the theorem is established.

## 3.2  The notion of determinant

There are several ways in which one may formulate a definition of the so-called *determinant* of a square matrix. In this section we choose the *explicit definition*. An equivalent *inductive definition* will appear in Section 3.4, and several *postulational definitions* will be suggested in an addendum to the chapter.

**3.2.1** NOTATION. Let $p$ be a permutation of 1, 2, $\cdots$, $n$. By $e_p$ we shall mean $+1$ or $-1$, according as $p$ is even or odd.

**3.2.2** DEFINITIONS AND NOTATION. Let $A = [a_{ij}]$ be an $n \times n$ matrix. By the *determinant* of $A$, written $|A|$, or $|a_{ij}|$, or $d(A)$, we mean

$$|A| = \sum_p e_p a_{i_1 1} a_{i_2 2} \cdots a_{i_n n},$$

where the subscript $p$ under the summation sign indicates that the summation is extended over all permutations $p = (i_1, i_2, \cdots, i_n)$ of 1, 2, $\cdots$, $n$.

The determinant of a square matrix is thus a certain scalar associated with the matrix. The integer $n$ is called the *order* of the determinant, and the numbers $e_p a_{i_1 1} a_{i_2 2} \cdots a_{i_n n}$ are called the *terms* of the determinant.

*Example*

$$\begin{vmatrix} a_{11} & a_{12} & a_{13} \\ a_{21} & a_{22} & a_{23} \\ a_{31} & a_{32} & a_{33} \end{vmatrix} = a_{11}a_{22}a_{33} + a_{21}a_{32}a_{13} + a_{31}a_{12}a_{23}$$

$$- a_{31}a_{22}a_{13} - a_{11}a_{32}a_{23} - a_{21}a_{12}a_{33},$$

since

$$e_{(1,2,3)} = e_{(2,3,1)} = e_{(3,1,2)} = 1$$

and

$$e_{(3,2,1)} = e_{(1,3,2)} = e_{(2,1,3)} = -1.$$

The following theorem is an immediate consequence of the definition of a determinant.

**3.2.3**  THEOREM.  *Each term of* $|A|$ *contains one and only one element from each row and each column of* A, *and every such product is plus or minus some term of* $|A|$.

# PROBLEMS

**3.1–1**  Show that a permutation of $1, 2, \cdots, n$, which has $k$ inversions, can be reduced to the natural order $(1, 2, \cdots, n)$ by $k$, but never fewer than $k$, adjacent transpositions.

**3.1–2**  Show that an even (odd) permutation of $1, 2, \cdots, n$ may be reduced to the natural order $(1, 2, \cdots, n)$ only by an even (odd) number of adjacent transpositions.

**3.1–3**  Show that a permutation of $1, 2, \cdots, n$ is even (odd) if and only if an even (odd) number of adjacent transpositions is required to reduce it to the natural order $(1, 2, \cdots, n)$.

**3.1–4**  Show that any permutation of $1, 2, \cdots, n$ can be reduced to the natural order $(1, 2, \cdots, n)$ by not more than $n(n-1)/2$ adjacent transpositions.

**3.1–5**  Prove that the permutation $(i_1, i_2, \cdots, i_n)$ of $1, 2, \cdots, n$ is even or odd according as the product

$$(i_n - i_{n-1})(i_n - i_{n-2}) \cdots (i_n - i_2)(i_n - i_1)$$
$$(i_{n-1} - i_{n-2}) \cdots (i_{n-1} - i_2)(i_{n-1} - i_1)$$
$$\cdots\cdots\cdots\cdots\cdots\cdots$$
$$(i_3 - i_2)(i_3 - i_1)$$
$$(i_2 - i_1)$$

is positive or negative.

**3.1–6** Prove that the number of inversions in a permutation of $1, 2, \cdots, n$ may be found either by counting the number of smaller integers following each integer of the permutation or by counting the number of larger integers preceding each integer of the permutation.

**3.1–7** Find the number of inversions in each of the following permutations of $1, 2, \cdots, n$:

(**a**) $(n, n - 1, \cdots, 2, 1)$;

(**b**) $(n - 1, n, n - 3, n - 2, \cdots, 1, 2)$, $n$ even;

(**c**) $(2, 4, \cdots, n - 1, 1, 3, \cdots, n)$, $n$ odd;

(**d**) $(1, 3, \cdots, n - 1, 2, 4, \cdots, n)$, $n$ even;

(**e**) $(3, 6, 9, \cdots, n, 1, 4, 7, \cdots, n - 2, 2, 5, 8, \cdots, n - 1)$, $n$ a multiple of 3;

(**f**) $(1, 4, 7, \cdots, n - 2, 2, 5, 8, \cdots, n - 1, 3, 6, 9, \cdots, n)$ $n$ a multiple of 3.

**3.2–1** Let $p, q, r$ denote the permutations

$$(i_1, i_2, \cdots, i_n), \qquad (j_{i_1}, j_{i_2}, \cdots, j_{i_n}), \qquad (j_1, j_2, \cdots, j_n)$$

of $1, 2, \cdots, n$. Prove that $e_p e_q = e_r$.

**3.2–2** (**a**) Show that the third-order determinant

$$\begin{vmatrix} a_{11} & a_{12} & a_{13} \\ a_{21} & a_{22} & a_{23} \\ a_{31} & a_{32} & a_{33} \end{vmatrix}$$

can be expanded by repeating the first two columns after the third column (as here indicated):

and then adding the three products of triples of elements on the diagonal lines marked ($+$) and subtracting the three products of triples of elements on the diagonal lines marked ($-$). [This mnemonic, or rule, for expanding a determinant of the third order is known as the *Sarrus rule*. It is stated in the second (1846) edition of P. J. E. Finck's *Éléments d'algèbre* with credit to P. F. Sarrus (1798–1861).]

(**b**) Show that the general rule suggested by part (a) is not valid for determinants of order different from 3.

**3.2–3** Prove Theorem 3.2.3.

**3.2–4** Show that $|a_{ij}|_{(n)}$ possesses $n!$ terms, and if $n > 1$, that half of these terms have plus signs and half have minus signs.

**3.2–5** Prove that the determinant of a square matrix possessing a row (column) of zeros is equal to zero.

**3.2–6** Show that

$$\begin{vmatrix} a_{11} & a_{12} & a_{13} \\ a_{21} & a_{22} & a_{23} \\ a_{31} & a_{32} & a_{33} \end{vmatrix} = a_{11} \begin{vmatrix} a_{22} & a_{23} \\ a_{32} & a_{33} \end{vmatrix} - a_{12} \begin{vmatrix} a_{21} & a_{23} \\ a_{31} & a_{33} \end{vmatrix} + a_{13} \begin{vmatrix} a_{21} & a_{22} \\ a_{31} & a_{32} \end{vmatrix}.$$

**3.2–7** Prove that $|a_{ij}|_{(n)}$, regarded as a polynomial in the letters $a_{ij}$, cannot be factored into a product of other polynomials in the $a_{ij}$'s.

**3.2–8** Find the signs needed to complete the following into terms of $|a_{ij}|_{(5)}$:

$$a_{31}a_{42}a_{13}a_{54}a_{25}, \qquad a_{31}a_{22}a_{53}a_{14}a_{45}, \qquad a_{43}a_{12}a_{54}a_{21}a_{35},$$

$$a_{12}a_{24}a_{33}a_{41}a_{55}, \qquad a_{31}a_{14}a_{22}a_{45}a_{53}, \qquad a_{24}a_{15}a_{33}a_{51}a_{42}.$$

**3.2–9** All elements of matrix $A = [a_{ij}]_{(n)}$ are zero except possibly for $a_{1n}, a_{2,n-1}, a_{3,n-2}, \cdots, a_{n-1,2}, a_{n1}$. Show that

$$|A| = (-1)^{n(n-1)/2} a_{1n}a_{2,n-1} \cdots a_{n-1,2}a_{n1}.$$

**3.2–10** Prove that it is impossible for all six terms of the determinant of a $3 \times 3$ $\mathscr{R}$ matrix to be positive.

## 3.3 Some elementary properties of determinants

Many of the elementary properties of determinants follow easily from the definition of a determinant given in the preceding section. We here state and prove as theorems a number of these properties. We commence with an important lemma.

**3.3.1** LEMMA. *If the product $a_{i_1 1}a_{i_2 2} \cdots a_{i_n n}$ has its factors reordered so that the first suffixes $i_1, i_2, \cdots, i_n$ come into natural order, the product becoming $a_{1 j_1}a_{2 j_2} \cdots a_{n j_n}$, then the two permutations $(i_1, i_2, \cdots, i_n)$ and $(j_1, j_2, \cdots, j_n)$ of $1, 2, \cdots, n$ are both even or both odd.*

The reordering of the factors can be accomplished by a succession of transpositions of these factors. But (see Theorem 3.1.5) each transposition of factors changes the number of inversions in the first suffixes by an odd number, and the number of inversions in the second suffixes by an odd number. It follows that the sum of the number of inversions in the first suffixes and the number of inversions in the second suffixes always remains even or always remains odd. But at the start, this sum is the number of inversions in $(i_1, i_2, \cdots, i_n)$, and at the end, this sum is the number of inversions in $(j_1, j_2, \cdots, j_n)$. Therefore the two permutations $(i_1, i_2, \cdots, i_n)$ and $(j_1, j_2, \cdots, j_n)$ are both even or both odd.

**3.3.2** THEOREM.  *If* A *is a square matrix, then* $|A'| = |A|$.
Let $A = [a_{ij}]$ and $A' = [b_{ij}]$. Then $b_{ij} = a_{ji}$ and

$$|A'| = \sum_p e_p b_{i_1 1} b_{i_2 2} \cdots b_{i_n n} = \sum_p e_p a_{1 i_1} a_{2 i_2} \cdots a_{n i_n}.$$

Now (by Theorem 3.2.3) $a_{1 i_1} a_{2 i_2} \cdots a_{n i_n}$ is, except possibly for sign, a term of $|A|$. If we write the product $a_{1 i_1} a_{2 i_2} \cdots a_{n i_n}$ in the form $a_{j_1 1} a_{j_2 2} \cdots a_{j_n n}$, by rearranging the factors so that the second suffixes come into natural order, and denote by $q$ the permutation $(j_1, j_2, \cdots, j_n)$, then (by Lemma 3.3.1) $e_q = e_p$, and hence $e_p a_{1 i_1} a_{2 i_2} \cdots a_{n i_n}$ is exactly a term of $|A|$. It follows that the terms of $|A'|$ are the terms of $|A|$, and $|A'| = |A|$.

**3.3.3** COROLLARY (The principle of duality of determinant theory). *Any theorem concerning the rows, columns, and value of an arbitrary determinant remains valid if the words "row" and "column" are everywhere interchanged in the statement of the theorem.*

**3.3.4** COROLLARY.  *If* A *is a square matrix, then*

$$|A| = \sum_q e_q a_{1 j_1} a_{2 j_2} \cdots a_{n j_n},$$

*where the summation is taken over all permutations* $q = (j_1, j_2, \cdots, j_n)$ *of* 1, 2, $\cdots$, n.

**3.3.5** THEOREM.  *If matrix* B *is obtained from square matrix* A *by the interchange of two rows or two columns, then* $|B| = -|A|$.
Let the *r*th and *s*th rows of $A$ be interchanged to give $B$. Then, in passing from the terms of $|A|$ to the corresponding terms of $|B|$, the natural order of the column suffixes is not altered, but the row suffixes

receive one transposition, namely, $r$ with $s$. Hence each term has undergone a change of sign. It follows that $|B| = -|A|$. The other part of the theorem follows by the principle of duality of determinant theory.

**3.3.6**    THEOREM.    *If two rows or two columns of a square $\mathscr{C}$ matrix* A *are identical, then* $|A| = 0$.

Suppose two rows of square $\mathscr{C}$ matrix $A$ are identical. Interchanging these two identical rows we find, by Theorem 3.3.5, $|A| = -|A|$. Since $A$ is a $\mathscr{C}$ matrix, it follows that $|A| = 0$. The other part of the theorem follows by the principle of duality of determinant theory.

*Note.* We have stressed that $A$ is a $\mathscr{C}$ matrix because later (in Chapter 6) we shall see that there exist $S$ matrices for which the preceding proof is invalid, though the theorem continues to hold. Since $|A| = -|A|$, we have $|A|(1 + 1) = 0$, whence either $|A| = 0$ or $1 + 1 = 0$. Now, in the complex number field, $1 + 1 \neq 0$, and so $|A| = 0$. But there exists a field $S$ wherein $1 + 1 = 0$. It then follows that for $S$ matrices, the foregoing proof is invalid.

**3.3.7**    THEOREM.    *If matrix* B *is obtained from square matrix* A *by multiplying the* i*th row, or the* j*th column, of* A *by a scalar* c, *then* $|B| = c|A|$.

Suppose $B$ is obtained from $A$ by multiplying the $i$th row of $A$ by $c$. Then, by Corollary 3.3.4,

$$|B| = \sum_q e_q a_{1j_1} a_{2j_2} \cdots c a_{ij_i} \cdots a_{nj_n}$$
$$= c \sum_q e_q a_{1j_1} a_{2j_2} \cdots a_{ij_i} \cdots a_{nj_n} = c|A|.$$

The other part of the theorem follows by the principle of duality of determinant theory.

**3.3.8**    COROLLARY.    *If square matrix* A *is of order* n *and* c *is a scalar, then* $|cA| = c^n|A|$.

**3.3.9**    COROLLARY.    *If square matrix* A *has a row (column) of zeros, then* $|A| = 0$.

**3.3.10**    THEOREM.    *If matrix* B *is obtained from square matrix* A *by replacing the* k*th row* $A_k$ *of* A *by* $A_k + cA_i$, *or the* k*th column* $A'_k$ *of* A *by* $A'_k + cA'_i$, *where* i ≠ k *and* c *is a scalar, then* $|B| = |A|$.

Suppose $B$ is obtained from $A$ by replacing the $k$th row $A_k$ of $A$ by $A_k + cA_i$. Then, by Corollary 3.3.4,

$$|B| = \sum_q e_q a_{1j_1} a_{2j_2} \cdots (a_{kj_k} + ca_{ij_k}) \cdots a_{nj_n}$$

$$= \sum_q e_q a_{1j_1} a_{2j_2} \cdots a_{kj_k} \cdots a_{nj_n} + c \sum_q e_q a_{1j_1} a_{2j_2} \cdots a_{ij_k} \cdots a_{nj_n}.$$

The first of the summations directly above is equal to $|A|$, and the second is equal to zero, since it is the value of a determinant whose $i$th and $k$th rows, $i \neq k$, are identical (see Theorem 3.3.6). It follows that $|B| = |A|$. The other part of the theorem follows by the principle of duality of determinant theory.

The elementary properties of determinants given in this section are often useful in evaluating a determinant. As an illustration, consider the evaluation of the determinant

$$D = \begin{vmatrix} a+d & 3a & b+2a & b+d \\ 2b & d+b & c-b & c-d \\ a+c & c-2d & d & a+3d \\ b-d & c-d & a+c & a+b \end{vmatrix}.$$

By Theorem 3.3.10, the value of $D$ is not changed by the following sequence of operations: column 1 $-$ column 2, column 1 $+$ column 3, column 1 $-$ column 4. But it will now be found that the first column is a column of zeros. Therefore, by Corollary 3.3.9, $D = 0$.

## PROBLEMS

**3.3–1**    Prove Corollaries 3.3.3 and 3.3.4.

**3.3–2**    Prove Corollaries 3.3.8 and 3.3.9.

**3.3–3**    Prove that the value of a determinant is unaltered when any row (column) is replaced by a linear combination of the other rows (columns).

**3.3–4**    Evaluate the following determinants by employing elementary properties of determinants:

(a) $\begin{vmatrix} 15 & 16 & 17 \\ 18 & 19 & 20 \\ 21 & 22 & 23 \end{vmatrix}$,    (b) $\begin{vmatrix} 1 & a & b+c \\ 1 & b & c+a \\ 1 & c & a+b \end{vmatrix}$,

**(c)** $\begin{vmatrix} x-y & y-z & z-x \\ y-z & z-x & x-y \\ z-x & x-y & y-z \end{vmatrix}$,     **(d)** $\begin{vmatrix} -ab & ac & ae \\ bd & -cd & de \\ bf & cf & -ef \end{vmatrix}$.

**3.3–5**   Evaluate each of the following determinants:

**(a)** $\begin{vmatrix} c & a & d & b \\ a & c & d & b \\ a & c & b & d \\ c & a & b & d \end{vmatrix}$,     **(b)** $\begin{vmatrix} 1 & p & q & r+s \\ 1 & q & r & s+p \\ 1 & r & s & p+q \\ 1 & s & p & q+r \end{vmatrix}$.

**3.3–6**   Prove that if $A$ is a Hermitian matrix, then the value of $|A|$ is a real number.

**3.3–7**   Evaluate $\begin{vmatrix} 1 & w & w^2 \\ w & w^2 & 1 \\ w^2 & 1 & w \end{vmatrix}$,     where $w^3 = 1$.

**3.3–8**   Evaluate $\begin{vmatrix} 0^2 & 1^2 & 2^2 & 3^2 \\ 1^2 & 2^2 & 3^2 & 4^2 \\ 2^2 & 3^2 & 4^2 & 5^2 \\ 3^2 & 4^2 & 5^2 & 6^2 \end{vmatrix}$.

**3.3–9**   Evaluate $\begin{vmatrix} a & 1 & 1 & 1 & 1 \\ 1 & a & 1 & 1 & 1 \\ 1 & 1 & a & 1 & 1 \\ 1 & 1 & 1 & a & 1 \\ 1 & 1 & 1 & 1 & a \end{vmatrix}$.

**3.3–10**   Show, by means of elementary properties of determinants, that

$$\begin{vmatrix} a+b & b+c & c+a \\ p+q & q+r & r+p \\ u+v & v+w & w+u \end{vmatrix} = 2 \begin{vmatrix} a & b & c \\ p & q & r \\ u & v & w \end{vmatrix}.$$

**3.3–11**   If $abc \neq 0$, show that

$$\begin{vmatrix} a^2 & bc & a^2 \\ b^2 & b^2 & ac \\ ab & c^2 & c^2 \end{vmatrix} = \begin{vmatrix} ac & bc & ab \\ bc & ab & ac \\ ab & ac & bc \end{vmatrix},$$

and thus show that $ab + bc + ca$ is a factor of the left-hand determinant.

**3.3–12**   Show that

$$\begin{vmatrix} a+b & c & c \\ a & b+c & a \\ b & b & c+a \end{vmatrix} = 2 \begin{vmatrix} 0 & b & a \\ b & 0 & c \\ a & c & 0 \end{vmatrix}.$$

**3.3–13**   Show that

$$\begin{vmatrix} 1 & a & a^2 \\ 1 & b & b^2 \\ 1 & c & c^2 \end{vmatrix} = (a - b)(b - c)(c - a).$$

**3.3–14**   Show that

$$\begin{vmatrix} 1 & a & a^3 \\ 1 & b & b^3 \\ 1 & c & c^3 \end{vmatrix} = (a - b)(b - c)(c - a)(a + b + c).$$

**3.3–15**   Show that

$$\begin{vmatrix} (b + c)^2 & ab & ac \\ ab & (c + a)^2 & bc \\ ac & bc & (a + b)^2 \end{vmatrix} = 2abc(a + b + c)^3.$$

## 3.4   Cofactors

The concept of cofactor is useful in the study of determinants; it leads to the inductive definition of a determinant, to the easy demonstration of various theorems, and to the valuable idea of the adjoint of a square matrix.

**3.4.1**   DEFINITION AND NOTATION.   By Theorem 3.2.3, an element $a_{ij}$ of a square matrix $A$ can occur at most once in any term of $|A|$. Collect all the terms of $|A|$ that contain the element $a_{ij}$. Then the sum of all these terms may be written as $a_{ij}A_{ij}$. $A_{ij}$ is called the *cofactor* of element $a_{ij}$.

*Example.*   An examination of the expansion of $|A_{(3)}|$ as given in the example following Definition 3.2.2 reveals that the cofactor $A_{12}$ of $a_{12}$ in $|A_{(3)}|$ is $a_{31}a_{23} - a_{21}a_{33}$.

**3.4.2**   THEOREM.   *If* A *is a square matrix of order* n, *then*

(1)
$$|A| = \sum_{j=1}^{n} a_{ij}A_{ij},$$

(2)
$$|A| = \sum_{i=1}^{n} a_{ij}A_{ij}.$$

Since each term of $|A|$ contains one and only one element from each row and each column of $A$,

$$|A| = a_{i1}A_{i1} + a_{i2}A_{i2} + \cdots + a_{in}A_{in} = \sum_{j=1}^{n} a_{ij}A_{ij}$$

and

$$|A| = a_{1j}A_{1j} + a_{2j}A_{2j} + \cdots + a_{nj}A_{nj} = \sum_{i=1}^{n} a_{ij}A_{ij}.$$

**3.4.3** THEOREM (alternative proof of Theorem 3.3.7). *If matrix* B *is obtained from square matrix* A *by multiplying the* i*th row, or the* j*th column, of* A *by a scalar* c, *then* $|B| = c|A|$.

Suppose $B$ is obtained from $A$ by multiplying the $i$th row of $A$ by $c$. Then [see Theorem 3.4.2 (1)]

$$|B| = \sum_{j=1}^{n} ca_{ij}A_{ij} = c \sum_{j=1}^{n} a_{ij}A_{ij} = c|A|.$$

A similar proof, using Theorem 3.4.2 (2), can be given if $B$ is obtained from $A$ by multiplying the $j$th column of $A$ by $c$.

**3.4.4** DEFINITION AND NOTATION. Let $A_{(i)(j)}$ denote the $(n-1) \times (n-1)$ submatrix of the $n \times n$ matrix $A$ obtained by deleting the $i$th row and the $j$th column of $A$. Then $|A_{(i)(j)}|$ is called the *minor* of element $a_{ij}$ of $A$.

**3.4.5** THEOREM. *If* A *is a square matrix of order* n, *then*

$$A_{ij} = (-1)^{i+j}|A_{(i)(j)}|.$$

We first prove that $A_{11} = |A_{(1)(1)}|$. We have the identity

$$a_{11}A_{11} \equiv \sum_{p'} e_{p'}a_{11}a_{i_2 2}a_{i_3 3} \cdots a_{i_n n},$$

where $p'$ runs over the $(n-1)!$ permutations $(1, i_2, i_3, \cdots, i_n)$ of $1, 2, \cdots, n$, in which the first element is fixed as 1. If we let $q$ denote the permutation $(i_2, i_3, \cdots, i_n)$ of $2, 3, \cdots, n$, then clearly $e_q = e_{p'}$, whence

$$a_{11}A_{11} \equiv a_{11} \sum_{q} e_q a_{i_2 2}a_{i_3 3} \cdots a_{i_n n} \equiv a_{11}|A_{(1)(1)}|.$$

It follows that $A_{11} = |A_{(1)(1)}|$.

We now prove that $A_{ij} = (-1)^{i+j}|A_{(i)(j)}|$. By $i-1$ successive interchanges of adjacent rows, and $j-1$ successive interchanges of adjacent columns, we may carry element $a_{ij}$ into the top left position without disturbing the relative positions of the elements in the other rows and columns. Denote the resulting matrix by $B$, and let $B_{11}$ denote the cofactor in $B$ of $b_{11} = a_{ij}$. Then, by the first part of our proof, $B_{11} = |A_{(i)(j)}|$, and the sum of the terms of $|B|$ containing $a_{ij}$ is $a_{ij}|A_{(i)(j)}|$. But

$$|B| = (-1)^{i-1+j-1}|A| = (-1)^{i+j}|A|,$$

whence $|A| = (-1)^{i+j}|B|$. But the sum of the terms of $|A|$ containing $a_{ij}$ is $a_{ij}A_{ij}$. It follows that $A_{ij} = (-1)^{i+j}|A_{(i)(j)}|$.

Combining Theorems 3.4.5 and 3.4.2 we have:

**3.4.6** COROLLARY. *If* A *is a square matrix of order* n, *then*

(1)
$$|A| = \sum_{j=1}^{n} (-1)^{i+j}a_{ij}|A_{(i)(j)}|,$$

(2)
$$|A| = \sum_{i=1}^{n} (-1)^{i+j}a_{ij}|A_{(i)(j)}|.$$

**3.4.7** REMARKS. Corollary 3.4.6 (1) furnishes an expression for $|A|$ in terms of determinants of orders one lower than that of $|A|$. This yields an *inductive definition* of the determinant of a square matrix $A$. The definition may be formulated as follows: If matrix $A$ is $1 \times 1$, then $|A| = a_{11}$; if matrix $A$ is $n \times n$, then $|A| = \sum_{j=1}^{n}(-1)^{1+j}a_{1j}|A_{(1)(j)}|$.

The expression (1) of Corollary 3.4.6, or the equivalent expression (1) of Theorem 3.4.2, is called "the expansion of $|A|$ by the elements and cofactors of the $i$th row of $A$," and the expression (2) of Corollary 3.4.6, or the equivalent expression (2) of Theorem 3.4.2, is called "the expansion of $|A|$ by the elements and cofactors of the $j$th column of $A$."

**3.4.8** THEOREM (alternative proof of Theorem 3.3.6). *If two rows or two columns of a square matrix* $A_{(n)}$ *are identical, then* $|A| = 0$.

The theorem is certainly true for $n = 2$. Assume it to be true for $n = k$ and let $A$ be a $(k + 1) \times (k + 1)$ matrix with its $i_1$th and $i_2$th rows identical. Expanding $|A|$ by the elements and cofactors of row $i \neq i_1$, $i_2$, we have

$$|A| = \sum_{j=1}^{k+1} a_{ij}A_{ij} = \sum_{j=1}^{k+1} (-1)^{i+j}a_{ij}|A_{(i)(j)}| = 0$$

by our inductive hypothesis, since each $|A_{(i)(j)}|$ is of order $k$ with two identical rows. The part of the theorem pertaining to two identical rows now follows by the principle of mathematical induction. The other part of the theorem may be similarly established.

*Note.* This proof circumvents the difficulty noted in connection with the proof of Theorem 3.3.6.

**3.4.9** THEOREM. *If* A *is a square matrix of order* n, *then*

(1) $$\sum_{j=1}^{n} a_{ij}A_{kj} = 0 \qquad \text{if } i \neq k,$$

(2) $$\sum_{i=1}^{n} a_{ij}A_{ik} = 0 \qquad \text{if } j \neq k.$$

Let $B$ be obtained from $A$ by replacing the $k$th row of $A$ by its $i$th row. Then, by Theorem 3.4.8, $|B| = 0$. But, expanding $|B|$ by the elements and cofactors of its $k$th row, we obtain

$$|B| = \sum_{j=1}^{n} a_{ij}A_{kj},$$

and (1) follows. A similar proof can be given for (2).

*Note.* The expressions (1) and (2) are often referred to as the *zero formulas.*

**3.4.10** THEOREM (alternative proof of Theorem 3.3.10). *If matrix* B *is obtained from square matrix* A *by replacing the* $k$th *row* $A_k$ *of* A *by* $A_k + cA_i$, *or the* $k$th *column* $A'_k$ *of* A *by* $A'_k + cA'_i$, *where* i $\neq$ k *and* c *is a scalar, then* $|B| = |A|$.

Suppose $B$ is obtained from $A_{(n)}$ by replacing the $k$th row $A_k$ of $A$ by $A_k + cA_i$. Then, expanding $|B|$ by the elements and cofactors of its $k$th row, we have

$$|B| = \sum_{j=1}^{n} (a_{kj} + ca_{ij})A_{kj}$$

$$= \sum_{j=1}^{n} a_{kj}A_{kj} + c \sum_{j=1}^{n} a_{ij}A_{kj}$$

$$= |A| + c(0) \qquad \text{[by Theorems 3.4.2 (1) and 3.4.8 (1)]}$$

$$= |A|.$$

A similar proof can be given of the other part of the theorem.

**3.4.11** DEFINITION. A *triangular matrix* is a square matrix, all of whose elements above (or below) the principal diagonal are zero elements. (See Definition 1.4.15.)

The proofs of the following two theorems will be left to the reader.

**3.4.12** THEOREM. *The determinant of a triangular matrix is equal to the product of the elements along the principal diagonal.*

**3.4.13** JACOBI'S THEOREM.† *The determinant of a skew-symmetric 𝒞 matrix of odd order equals zero.*

## PROBLEMS

**3.4–1** Give a proof, by mathematical induction, of Theorem 3.4.12.

**3.4–2** Prove Theorem 3.4.13.

**3.4–3** Show that each of the following determinants has value 0:

$$\textbf{(a)} \quad \begin{vmatrix} 0 & x & y \\ -x & 0 & -z \\ -y & z & 0 \end{vmatrix}. \qquad \textbf{(b)} \quad \begin{vmatrix} (a-a)^3 & (a-b)^3 & (a-c)^3 \\ (b-a)^3 & (b-b)^3 & (b-c)^3 \\ (c-a)^3 & (c-b)^3 & (c-c)^3 \end{vmatrix}.$$

**3.4–4** Expand the following determinants:

$$\textbf{(a)} \quad \begin{vmatrix} 1 & 2 & 3 & 4 \\ 0 & 0 & 0 & 2 \\ 2 & 4 & 6 & 7 \\ 5 & 6 & 7 & 8 \end{vmatrix}. \qquad \textbf{(b)} \quad \begin{vmatrix} 1 & 0 & 2 & 0 \\ 0 & 3 & 0 & 4 \\ 5 & 0 & 6 & 0 \\ 0 & 7 & 0 & 8 \end{vmatrix}.$$

**3.4–5** In matrix $A_{(n)}$, $a_{ii} = i$, $i = 1, \cdots, n$, and $a_{ij} = 1$, $i \neq j$. Prove that $|A| = (n-1)!$.

**3.4–6** Prove by mathematical induction that it is impossible for all $n!$ terms of the determinant of an $n \times n$ 𝓡 matrix to be positive if $n > 2$. (See Problem 3.2.10.)

**3.4–7** (a) Each element of matrix $B_{(n)}$ is 0 or 2. Prove that $|B|$ is equal to an integer divisible by $2^n$.

---

† Given by Carl Gustav Jacob Jacobi (1804–1851) in 1827. Next to Cauchy, Jacobi was perhaps the most prolific contributor to determinant theory. It was with him that the word "determinant" received its final acceptance. He early used the functional determinant that Sylvester later called the *Jacobian*, and which is met by all students of function theory.

**(b)** Each element of matrix $A_{(n)}$ is 1 or $-1$. Prove that $|A|$ is equal to an integer divisible by $2^{n-1}$.

**3.4–8** Let $(i_1, i_2, \cdots, i_n)$ be a permutation of $1, 2, \cdots, n$ and let a matrix $B_{(n)}$ be formed from a matrix $A_{(n)}$ by choosing as the $i_j$th column of $B$ the $j$th column of $A$, $j = 1, \cdots, n$. Prove that $|B| = (-1)^k|A|$, where $k$ is the number of inversions in $(i_1, i_2, \cdots, i_n)$.

**3.4–9** Show that

$$
\begin{vmatrix}
x & 0 & 0 & \cdots & 0 & a_0 \\
-1 & x & 0 & \cdots & 0 & a_1 \\
0 & -1 & x & \cdots & 0 & a_2 \\
\cdots & \cdots & \cdots & \cdots & \cdots & \cdots \\
0 & 0 & 0 & \cdots & x & a_{n-1} \\
0 & 0 & 0 & \cdots & -1 & a_n
\end{vmatrix}
$$

$$
= a_n x^n + a_{n-1} x^{n-1} + \cdots + a_2 x^2 + a_1 x + a_0.
$$

(This problem gives a simple way of expressing a polynomial of degree $n$ as an $n \times n$ determinant.)

**3.4–10** Let $A$ be an $n \times n$ matrix for which every $k \times k$ subdeterminant is 0 for some fixed $k < n$. Prove that $|A| = 0$.

**3.4–11** Show that

$$
\begin{vmatrix}
0 & 1 & 1 & 1 \\
1 & 0 & z^2 & y^2 \\
1 & z^2 & 0 & x^2 \\
1 & y^2 & x^2 & 0
\end{vmatrix}
=
\begin{vmatrix}
0 & x & y & z \\
x & 0 & z & y \\
y & z & 0 & x \\
z & y & x & 0
\end{vmatrix}.
$$

**3.4–12** Show that

$$
\begin{vmatrix}
t & x & y & z \\
x & t & z & y \\
y & z & t & x \\
z & y & x & t
\end{vmatrix}
= (t + x + y + z)(t + x - y - z)(t - x + y - z) \times
$$
$$
(t - x - y + z).
$$

**3.4–13** Show that

$$
\begin{vmatrix}
a & x & x & x & x \\
x & b & x & x & x \\
x & x & c & x & x \\
x & x & x & d & x \\
x & x & x & x & e
\end{vmatrix}
= \phi(x) - x\phi'(x),
$$

where $\phi(x) = (a - x)(b - x)(c - x)(d - x)(e - x)$.

### 3.5 Cyclic determinants and Vandermonde determinants

Following the historical completion of a general theory of determinants, mathematicians interested in this field of work undertook the study of many special forms of determinants. The number of these special forms that have been examined is simply legionary, and a great many of them can be shown to arise naturally, or to find valuable application, in some area of mathematical study. In this section we shall informally consider two of these special forms, the so-called *cyclic determinants* and the useful *Vandermonde determinants*. In considering these two special forms of determinants, we shall illustrate some of the tricks and procedures that have been found of value in the expansion of special determinants; the reader is invited to check the details. Further expansion procedures of both a practical and a theoretical interest will be considered in following sections.

*Example* 1.   Evaluate the *cyclic determinant*, or (special) *circulant*,†

$$A = \begin{vmatrix} 1 & 2 & 3 & \cdots & n-1 & n \\ 2 & 3 & 4 & \cdots & n & 1 \\ 3 & 4 & 5 & \cdots & 1 & 2 \\ \cdots & \cdots & \cdots & \cdots & \cdots & \cdots \\ n & 1 & 2 & \cdots & n-2 & n-1 \end{vmatrix}.$$

Subtracting the $(n-1)$th row from the $n$th, the $(n-2)$th from the $(n-1)$th, $\cdots$, the first from the second, and then subtracting the first column from each of the other columns, we find

$$A = \begin{vmatrix} 1 & 1 & 2 & \cdots & n-2 & n-1 \\ 1 & 0 & 0 & \cdots & 0 & -n \\ 1 & 0 & 0 & \cdots & -n & 0 \\ \cdots & \cdots & \cdots & \cdots & \cdots & \cdots \\ 1 & -n & 0 & \cdots & 0 & 0 \end{vmatrix}.$$

Now multiplying the first column by $n$ and compensating by putting the factor $1/n$ outside the determinant, and then adding all columns after the first to the first, we find

† Circulants seem to have been first considered by Eugène Charles Catalan (1814–1894) in 1846. They were then further studied by William Spottiswoode, J. W. L. Glaisher, R. F. Scott, and others.

$$A = (1/n) \begin{vmatrix} n + n(n-1)/2 & 1 & 2 & \cdots & n-2 & n-1 \\ 0 & 0 & 0 & \cdots & 0 & -n \\ 0 & 0 & 0 & \cdots & -n & 0 \\ \cdots & \cdots & \cdots & \cdots & \cdots & \cdots \\ 0 & -n & 0 & \cdots & 0 & 0 \end{vmatrix}$$

$$= (n+1)/2 \begin{vmatrix} 0 & \cdots & 0 & -n \\ 0 & \cdots & -n & 0 \\ \cdots & \cdots & \cdots & \cdots \\ -n & \cdots & 0 & 0 \end{vmatrix}.$$

Finally, using Problem 3.2–9, we obtain

$$A = (-1)^{(n-1)(n-2)/2}(-n)^{n-1}(n+1)/2 = (-1)^{n(n-1)/2}(n+1)n^{n-1}/2.$$

*Example 2.* Evaluate the *Vandermonde determinant*,† or *alternant*,

$$V_n = \begin{vmatrix} 1 & a_1 & a_1{}^2 & \cdots & a_1{}^{n-1} \\ 1 & a_2 & a_2{}^2 & \cdots & a_2{}^{n-1} \\ \cdots & \cdots & \cdots & \cdots & \cdots \\ 1 & a_n & a_n{}^2 & \cdots & a_n{}^{n-1} \end{vmatrix}.$$

Subtracting $a_n$ times the $(n-1)$th column from the $n$th column, $a_n$ times the $(n-2)$th column from the $(n-1)$th column, $\cdots$, $a_n$ times the first column from the second column, we obtain a determinant whose $n$th row is

$$1, 0, 0, \cdots, 0$$

and whose $i$th row, $i < n$, is

$$1, a_i - a_n, a_i(a_i - a_n), \cdots, a_i{}^{n-2}(a_i - a_n).$$

Therefore, expanding the transformed determinant by the elements and cofactors of the $n$th row, we obtain

$$V_n = (-1)^{n+1} \begin{vmatrix} a_1 - a_n & a_1(a_1 - a_n) & \cdots & a_1{}^{n-2}(a_1 - a_n) \\ a_2 - a_n & a_2(a_2 - a_n) & \cdots & a_2{}^{n-2}(a_2 - a_n) \\ \cdots & \cdots & \cdots & \cdots \\ a_{n-1} - a_n & a_{n-1}(a_{n-1} - a_n) & \cdots & a_{n-1}{}^{n-2}(a_{n-1} - a_n) \end{vmatrix}.$$

Taking outside the determinant the common factor $a_1 - a_n$ of the first row, the common factor $a_2 - a_n$ of the second row, $\cdots$, the common factor

---

† Named for Alexandre-Théophile Vandermonde (1735–1796), who in 1771 gave the earliest systematic treatment of the theory of determinants.

$a_{n-1} - a_n$ of the last row, we obtain

$$V_n = (-1)^{n-1}(a_1 - a_n)(a_2 - a_n) \cdots (a_{n-1} - a_n)V_{n-1}$$

$$= (a_n - a_1)(a_n - a_2) \cdots (a_n - a_{n-1})V_{n-1}.$$

Since

$$V_2 = \begin{vmatrix} 1 & a_1 \\ 1 & a_2 \end{vmatrix} = a_2 - a_1,$$

we finally have

$$V_n = (a_n - a_{n-1})(a_n - a_{n-2}) \cdots (a_n - a_2)(a_n - a_1)$$

$$(a_{n-1} - a_{n-2}) \cdots (a_{n-1} - a_2)(a_{n-1} - a_1)$$

$$\cdots\cdots\cdots\cdots\cdots\cdots\cdots\cdots\cdots$$

$$(a_3 - a_2)(a_3 - a_1)$$

$$(a_2 - a_1).$$

*Alternative.* If $a_2 = a_1$, two rows of the Vandermonde determinant are equal, and consequently $a_2 - a_1$ is a factor of $V_n$. Similarly, $a_r - a_s$, where $r > s$, is a factor of $V_n$. Hence

(1)  $V_n = k(a_n - a_{n-1}) \cdots (a_n - a_1) \times$

$$(a_{n-1} - a_{n-2}) \cdots (a_{n-1} - a_1) \cdots (a_2 - a_1).$$

Since $V_n$ is of degree $n(n - 1)/2$ in the $a_i$'s, $k$ is independent of these quantities. Now the term $a_2 a_3^2 \cdots a_n^{n-1}$ occurs in $V_n$, and on the right-hand side of (1) the coefficient of this term is $k$. Hence $k = 1$.

*Note.* The polynomial expansion of $V_n$ is known as the *alternating function* of $a_1, \cdots, a_n$, since the transposition of any two of the $a_i$'s merely alters the sign of the function. It is this connection of $V_n$ with an alternating function that led $V_n$ to be called an *alternant*. Alternating functions are said by Cauchy to be discernible in certain work of Vandermonde, though it is almost certain that Vandermonde did not see their connection with determinants. Nevertheless, this explains why alternants are frequently called Vandermonde determinants.

## PROBLEMS

**3.5–1**  Check all details of the solutions of Examples 1 and 2 of Section 3.5.

**3.5–2** Prove that

$$
\begin{vmatrix}
1 & a_1 & a_2 & \cdots & a_n \\
1 & a_1 + b_1 & a_2 & \cdots & a_n \\
1 & a_1 & a_2 + b_2 & \cdots & a_n \\
\cdots & \cdots & \cdots & \cdots & \cdots \\
1 & a_1 & a_2 & \cdots & a_n + b_n
\end{vmatrix}
= b_1 b_2 \cdots b_n.
$$

**3.5–3** If $A = [a_{ij}]_{(n)}$, where

$$
a_{ij} = \binom{i + j - 2}{j - 1} = (i + j - 2)!/(i - 1)!(j - 1)!,
$$

prove that $|A| = 1$.

**3.5–4** Prove that

$$
\begin{vmatrix}
a & 0 & \cdots & 0 & b \\
0 & a & \cdots & b & 0 \\
\cdots & \cdots & \cdots & \cdots & \cdots \\
0 & b & \cdots & a & 0 \\
b & 0 & \cdots & 0 & a
\end{vmatrix}
= (a^2 - b^2)^k,
$$

where the determinant is of even order, $n = 2k$.

**3.5–5** If $A$ is an $n \times n$ matrix with $a$'s along the principal diagonal and $b$'s everywhere else, prove that $|A| = (a + nb - b)(a - b)^{n-1}$.

**3.5–6** Prove that

$$
\begin{vmatrix}
1 & 1 & 1 & \cdots & 1 \\
b_1 & a_1 & a_1 & \cdots & a_1 \\
b_1 & b_2 & a_2 & \cdots & a_2 \\
\cdots & \cdots & \cdots & \cdots & \cdots \\
b_1 & b_2 & b_3 & \cdots & a_n
\end{vmatrix}
= (a_1 - b_1)(a_2 - b_2) \cdots (a_n - b_n).
$$

**3.5–7** Prove that

$$
\begin{vmatrix}
a_1 & x & x & \cdots & x \\
x & a_2 & x & \cdots & x \\
\cdots & \cdots & \cdots & \cdots & \cdots \\
x & x & x & \cdots & a_n
\end{vmatrix}
= f(x) - x \, df/dx,
$$

where $f(x) \equiv (a_1 - x)(a_2 - x) \cdots (a_n - x)$.

**3.5–8** Let

$$
D_n =
\begin{vmatrix}
1 & 1 & 1^2 & \cdots & 1^{n-1} \\
1 & 2 & 2^2 & \cdots & 2^{n-1} \\
1 & 3 & 3^2 & \cdots & 3^{n-1} \\
\cdots & \cdots & \cdots & \cdots & \cdots \\
1 & n & n^2 & \cdots & n^{n-1}
\end{vmatrix}.
$$

(a) Show that $D_n = (n - 1)! D_{n-1}$.

(b) Show that $D_n = 1! 2! 3! \cdots (n - 1)!$.

**3.5–9** Let $C_n$ denote the *continuant*†

$$
\begin{vmatrix}
a_1 & 1 & 0 & 0 & \cdots & 0 & 0 \\
-1 & a_2 & 1 & 0 & \cdots & 0 & 0 \\
0 & -1 & a_3 & 1 & \cdots & 0 & 0 \\
\cdots & \cdots & \cdots & \cdots & \cdots & \cdots & \cdots \\
0 & 0 & 0 & 0 & \cdots & a_{n-1} & 1 \\
0 & 0 & 0 & 0 & \cdots & -1 & a_n
\end{vmatrix}.
$$

Show that $C_n = a_n C_{n-1} + C_{n-2}$.

**3.5–10** Using the fact that

$$
\begin{vmatrix}
t & x & y & z \\
1 & 1 & 1 & 1 \\
t & x & y & z \\
t^2 & x^2 & y^2 & z^2
\end{vmatrix} \equiv 0,
$$

obtain the identity

$$t(x - z)(x - y)(z - y) + x(y - z)(y - t)(z - t)$$
$$+ y(z - x)(z - t)(x - t) + z(x - y)(x - t)(y - t) \equiv 0.$$

## 3.6 Chio's expansion

The expansion of a determinant of order $n$ by the elements and cofactors of some row or column expresses the determinant as a linear combination of $n$ determinants of order $n - 1$. The evaluation of a determinant of high order by this process requires that the process be successively repeated many times, with the result that the number of terms soon becomes very large and the work involved becomes prohibitively lengthy. Mathematicians have accordingly devised less cumbersome processes wherein the determinant is reduced to a single determinant of lower order and then evaluated by the systematic repetition of this process. Such a method of evaluating a determinant is called a *condensation method*, and most of these are variants of a procedure now known as *Chio's pivotal condensation method*, since a form of this method was given by F. Chio in a paper of 1853, though an earlier trace of the method can be found in an 1849 paper by C. Hermite. Chio's pivotal condensation

---

† Continuants are intimately connected with the study of continued fractions, from which connection they obtained their name. The history of continuants seems to start with an 1853 paper by James Joseph Sylvester (1814–1897).

method goes considerably beyond its practical aspect, for the process leads to some remarkable determinant identities. It is because of this dual importance of the procedure that we briefly consider it here.

**3.6.1** THEOREM (Chio's pivotal condensation process). *Let* $A = [a_{ij}]$ *be an* $n \times n$ *matrix and suppose* $a_{11} \neq 0$. *Let* $D$ *denote the matrix obtained by replacing each element* $a_{ij}$ *in* $A_{(1)(1)}$ *by* $\begin{vmatrix} a_{11} & a_{1j} \\ a_{i1} & a_{ij} \end{vmatrix}$. *Then* $|A| = |D|/a_{11}^{n-2}$. *That is,*

$$|A| = (1/a_{11}^{n-2}) \begin{vmatrix} \begin{vmatrix} a_{11} & a_{12} \\ a_{21} & a_{22} \end{vmatrix} & \begin{vmatrix} a_{11} & a_{13} \\ a_{21} & a_{23} \end{vmatrix} & \cdots & \begin{vmatrix} a_{11} & a_{1n} \\ a_{21} & a_{2n} \end{vmatrix} \\ \begin{vmatrix} a_{11} & a_{12} \\ a_{31} & a_{32} \end{vmatrix} & \begin{vmatrix} a_{11} & a_{13} \\ a_{31} & a_{33} \end{vmatrix} & \cdots & \begin{vmatrix} a_{11} & a_{1n} \\ a_{31} & a_{3n} \end{vmatrix} \\ \cdots & \cdots & \cdots & \cdots \\ \begin{vmatrix} a_{11} & a_{12} \\ a_{n1} & a_{n2} \end{vmatrix} & \begin{vmatrix} a_{11} & a_{13} \\ a_{n1} & a_{n3} \end{vmatrix} & \cdots & \begin{vmatrix} a_{11} & a_{1n} \\ a_{n1} & a_{nn} \end{vmatrix} \end{vmatrix}.$$

Multiply each row of $A$, except the first, by $a_{11}$, and then perform the elementary row operations $o(2 - a_{21} \cdot 1)$, $o(3 - a_{31} \cdot 1)$, $\cdots$, $o(n - a_{n1} \cdot 1)$. We obtain

$$a_{11}^{n-1}|A| = \begin{vmatrix} a_{11} & a_{12} & \cdots & a_{1n} \\ a_{11}a_{21} & a_{11}a_{22} & \cdots & a_{11}a_{2n} \\ \cdots & \cdots & \cdots & \cdots \\ a_{11}a_{n1} & a_{11}a_{n2} & \cdots & a_{11}a_{nn} \end{vmatrix}$$

$$= \begin{vmatrix} a_{11} & a_{12} & a_{13} & \cdots & a_{1n} \\ 0 & \begin{vmatrix} a_{11} & a_{12} \\ a_{21} & a_{22} \end{vmatrix} & \begin{vmatrix} a_{11} & a_{13} \\ a_{21} & a_{23} \end{vmatrix} & \cdots & \begin{vmatrix} a_{11} & a_{1n} \\ a_{21} & a_{2n} \end{vmatrix} \\ \cdots & \cdots & \cdots & \cdots & \cdots \\ 0 & \begin{vmatrix} a_{11} & a_{12} \\ a_{n1} & a_{n2} \end{vmatrix} & \begin{vmatrix} a_{11} & a_{13} \\ a_{n1} & a_{n3} \end{vmatrix} & \cdots & \begin{vmatrix} a_{11} & a_{1n} \\ a_{n1} & a_{nn} \end{vmatrix} \end{vmatrix}$$

$$= a_{11}|D|,$$

and the theorem follows.

*Note.* If $A$ is symmetric, then $D$ also is symmetric, a fact that, in this case, greatly reduces the work in computing $D$.

*Example*

$$\begin{vmatrix} 1 & 2 & 3 & 4 \\ 8 & 7 & 6 & 5 \\ 1 & 8 & 2 & 7 \\ 3 & 6 & 4 & 5 \end{vmatrix} = \begin{vmatrix} \begin{vmatrix} 1 & 2 \\ 8 & 7 \end{vmatrix} & \begin{vmatrix} 1 & 3 \\ 8 & 6 \end{vmatrix} & \begin{vmatrix} 1 & 4 \\ 8 & 5 \end{vmatrix} \\ \begin{vmatrix} 1 & 2 \\ 1 & 8 \end{vmatrix} & \begin{vmatrix} 1 & 3 \\ 1 & 2 \end{vmatrix} & \begin{vmatrix} 1 & 4 \\ 1 & 7 \end{vmatrix} \\ \begin{vmatrix} 1 & 2 \\ 3 & 6 \end{vmatrix} & \begin{vmatrix} 1 & 3 \\ 3 & 4 \end{vmatrix} & \begin{vmatrix} 1 & 4 \\ 3 & 5 \end{vmatrix} \end{vmatrix}$$

$$= \begin{vmatrix} -9 & -18 & -27 \\ 6 & -1 & 3 \\ 0 & -5 & -7 \end{vmatrix} = -9 \begin{vmatrix} 1 & 2 & 3 \\ 6 & -1 & 3 \\ 0 & -5 & -7 \end{vmatrix}$$

$$= -9 \begin{vmatrix} \begin{vmatrix} 1 & 2 \\ 6 & -1 \end{vmatrix} & \begin{vmatrix} 1 & 3 \\ 6 & 3 \end{vmatrix} \\ \begin{vmatrix} 1 & 2 \\ 0 & -5 \end{vmatrix} & \begin{vmatrix} 1 & 3 \\ 0 & -7 \end{vmatrix} \end{vmatrix} = -9 \begin{vmatrix} -13 & -15 \\ -5 & -7 \end{vmatrix}$$

$$= -9(91 - 75) = -9(16) = -144.$$

**3.6.2** NOTATION. If $A = [a_{ij}]_{(n)}$, it is convenient on occasion to denote $|A|$ by $|a_{11} \ a_{22} \cdots a_{nn}|$, and the subdeterminant involving rows $i_1, \cdots, i_s$ and columns $j_1, \cdots, j_s$ by $|a_{i_1 j_1} \ a_{i_2 j_2} \cdots a_{i_s j_s}|$.

Using the preceding notation and Chio's pivotal condensation process, we now establish a remarkable identity.

**3.6.3** THEOREM. *If* $A = [a_{ij}]_{(n)}$ *and* $n \geq 3$, *then*

$$|a_{11} \ a_{22} \cdots a_{n-2,n-2}| \ |a_{11} \ a_{22} \cdots a_{nn}|$$
$$\equiv \begin{vmatrix} |a_{11} \ a_{22} \cdots a_{n-2,n-2} \ a_{n-1,n-1}| & |a_{11} \ a_{22} \cdots a_{n-2,n-2} \ a_{n-1,n}| \\ |a_{11} \ a_{22} \cdots a_{n-2,n-2} \ a_{n,n-1}| & |a_{11} \ a_{22} \cdots a_{n-2,n-2} \ a_{n,n}| \end{vmatrix}.$$

We shall, for the moment, assume that

(1) $\quad a_{11} \neq 0, \qquad |a_{11} \ a_{22}| \neq 0, \quad \cdots, \quad |a_{11} \ a_{22} \cdots a_{n-2,n-2}| \neq 0,$

and employ the technique of mathematical induction.

The case $n = 3$ follows from Chio's process applied to

$$|a_{11} \ a_{22} \ a_{33}|.$$

Assume the relation holds for $n = 3, 4, \cdots, k - 1$. Then

$$|a_{11} \quad a_{22} \quad \cdots \quad a_{kk}|$$

$$= \frac{1}{a_{11}{}^{k-2}} \begin{vmatrix} |a_{11} \quad a_{22}| & |a_{11} \quad a_{23}| & \cdots & |a_{11} \quad a_{2k}| \\ |a_{11} \quad a_{32}| & |a_{11} \quad a_{33}| & \cdots & |a_{11} \quad a_{3k}| \\ \cdots & \cdots & \cdots & \cdots \\ |a_{11} \quad a_{k2}| & |a_{11} \quad a_{k3}| & \cdots & |a_{11} \quad a_{kk}| \end{vmatrix}$$

(by Chio's process)

$$= \frac{1}{a_{11}{}^{k-2}|a_{11} \quad a_{22}|^{k-3}} \begin{vmatrix} a_{11}|a_{11} \quad a_{22} \quad a_{33}| & a_{11}|a_{11} \quad a_{22} \quad a_{34}| & \cdots & a_{11}|a_{11} \quad a_{22} \quad a_{3k}| \\ a_{11}|a_{11} \quad a_{22} \quad a_{43}| & a_{11}|a_{11} \quad a_{22} \quad a_{44}| & \cdots & a_{11}|a_{11} \quad a_{22} \quad a_{4k}| \\ \cdots & \cdots & \cdots & \cdots \\ a_{11}|a_{11} \quad a_{22} \quad a_{k3}| & a_{11}|a_{11} \quad a_{22} \quad a_{k4}| & \cdots & a_{11}|a_{11} \quad a_{22} \quad a_{kk}| \end{vmatrix}$$

(by Chio's process and the inductive assumption)

$$= \frac{1}{|a_{11} \quad a_{22}|^{k-3}} \begin{vmatrix} |a_{11} \quad a_{22} \quad a_{33}| & |a_{11} \quad a_{22} \quad a_{34}| & \cdots & |a_{11} \quad a_{22} \quad a_{3k}| \\ |a_{11} \quad a_{22} \quad a_{43}| & |a_{11} \quad a_{22} \quad a_{44}| & \cdots & |a_{11} \quad a_{22} \quad a_{4k}| \\ \cdots & \cdots & \cdots & \cdots \\ |a_{11} \quad a_{22} \quad a_{k3}| & |a_{11} \quad a_{22} \quad a_{k4}| & \cdots & |a_{11} \quad a_{22} \quad a_{kk}| \end{vmatrix}$$

(canceling)

$$= \frac{1}{|a_{11} \quad a_{22} \quad a_{33}|^{k-4}} \begin{vmatrix} |a_{11} \quad a_{22} \quad a_{33} \quad a_{44}| & |a_{11} \quad a_{22} \quad a_{33} \quad a_{45}| & \cdots & |a_{11} \quad a_{22} \quad a_{33} \quad a_{4k}| \\ |a_{11} \quad a_{22} \quad a_{33} \quad a_{54}| & |a_{11} \quad a_{22} \quad a_{33} \quad a_{55}| & \cdots & |a_{11} \quad a_{22} \quad a_{33} \quad a_{5k}| \\ \cdots & \cdots & \cdots & \cdots \\ |a_{11} \quad a_{22} \quad a_{33} \quad a_{k4}| & |a_{11} \quad a_{22} \quad a_{33} \quad a_{k5}| & \cdots & |a_{11} \quad a_{22} \quad a_{33} \quad a_{kk}| \end{vmatrix}$$

(by repeating the above procedure)

$$= \cdots$$

$$= \frac{\begin{vmatrix} |a_{11} \quad a_{22} \quad \cdots \quad a_{k-2,k-2} \quad a_{k-1,k-1}| & |a_{11} \quad a_{22} \quad \cdots \quad a_{k-2,k-2} \quad a_{k-1,k}| \\ |a_{11} \quad a_{22} \quad \cdots \quad a_{k-2,k-2} \quad a_{k,k-1}| & |a_{11} \quad a_{22} \quad \cdots \quad a_{k-2,k-2} \quad a_{kk}| \end{vmatrix}}{|a_{11} \quad a_{22} \quad \cdots \quad a_{k-2,k-2}|}$$

(by continued repetition of the above process). It follows, then, that the relation also holds for the case $n = k$.

We now conclude, by the principle of mathematical induction, that the concerned relation holds for any $n \geqq 3$, provided conditions (1) also hold. But the concerned relation is of the form $\theta = \phi$, where $\theta$ and $\phi$ are polynomials of degree at most 2 in each of the $n^2$ letters $a_{ij}$. It is not difficult to show that the concerned relation holds, under the restrictions (1), for three sets of numerical values of the $n^2$ letters $a_{ij}$, where the three values taken on by each $a_{ij}$ are distinct. For example, one may choose any three distinct nonzero values for $a_{11}$, then three distinct associated values for each of $a_{12}$, $a_{21}$, $a_{22}$ such that $|a_{11} \quad a_{22}| \neq 0$, then three distinct associated values for each of $a_{13}$, $a_{23}$, $a_{31}$, $a_{32}$, $a_{33}$ such that $|a_{11} \quad a_{22} \quad a_{33}| \neq 0$, etc. It now follows, by the algebraic theorem given in the Appendix to this section, that the concerned relation is an identity and therefore holds whether the restrictions (1) prevail or not.

Theorem 3.6.3 may be stated in more compact symbolism in the form:

*If* M *is a square matrix and* a, b, c, d *are scalars, then*

$$|M| \begin{vmatrix} M & U & V \\ \hline R & a & b \\ \hline S & c & d \end{vmatrix} = \begin{vmatrix} \begin{vmatrix} M & U \\ \hline R & a \end{vmatrix} & \begin{vmatrix} M & V \\ \hline R & b \end{vmatrix} \\ \begin{vmatrix} M & U \\ \hline S & c \end{vmatrix} & \begin{vmatrix} M & V \\ \hline S & d \end{vmatrix} \end{vmatrix}.$$

It is now seen that the identity of Theorem 3.6.3 is essentially the Chio process applied to the partitioned matrix

$$\begin{bmatrix} M & U & V \\ \hline R & a & b \\ \hline S & c & d \end{bmatrix}.$$

Many other remarkable identities involving determinants of partitioned matrices can be derived. The reader can find an application of the above identity to the theory of Pfaffians in an Addendum to the present chapter. There also can be found an application of the Chio pivotal condensation process to the solution of systems of linear equations.

### Appendix

**3.6.4** THEOREM. *Suppose* $\theta(x_1, \cdots, x_n)$ *and* $\phi(x_1, \cdots, x_n)$ *are polynomials of degrees less than* m *in each of the* n *variables* $x_i$ *and suppose*

$$\theta(a_{i1}, \cdots, a_{in}) = \phi(a_{i1}, \cdots, a_{in})$$

*for* m *sets of numbers* $a_{i1}, \cdots, a_{in}$, i = 1, $\cdots$, m, *where* $a_{rj} \neq a_{sj}$ *for each* j = 1, $\cdots$, n *if* r $\neq$ s. *Then* $\theta \equiv \phi$.

The case $n = 1$ is well known.

Suppose the theorem is true for $n = 1, \cdots, k - 1$, and suppose $\theta(x_1, \cdots, x_k)$ and $\phi(x_1, \cdots, x_k)$ are polynomials in $x_1, \cdots, x_k$ of degrees less than $m$ in each of the variables $x_1, \cdots, x_n$. Further suppose that

$$\theta(a_{i1}, \cdots, a_{ik}) = \phi(a_{i1}, \cdots, a_{ik})$$

for $m$ sets of numbers $a_{i1}, \cdots, a_{ik}$, $i = 1, \cdots, m$, where $a_{rj} \neq a_{sj}$ for each $j = 1, \cdots, k$ if $r \neq s$. Order the terms of the polynomials $\theta$ and $\phi$ by increasing powers of $x_k$:

$$\theta = \theta_0 + \theta_1 x_k + \theta_2 x_k^2 + \cdots + \theta_s x_k^s, \qquad s < m,$$

$$\phi = \phi_0 + \phi_1 x_k + \phi_2 x_k^2 + \cdots + \phi_t x_k^t, \qquad t < m.$$

Here the $\theta_j$ and $\phi_j$ are polynomials of degrees less than $m$ in each of the $k - 1$ variables $x_1, \cdots, x_{k-1}$. Then, by hypothesis,

$$\sum_{j=0}^{s} \theta_j(a_{i1}, \cdots, a_{i,k-1}) a_{ik}^j = \sum_{j=0}^{t} \phi_j(a_{i1}, \cdots, a_{i,k-1}) a_{ik}^j$$

for each $i = 1, \cdots, m$. By the initial case,

$$\sum_{j=0}^{s} \theta_j(a_{i1}, \cdots, a_{i,k-1}) x_k^j \equiv \sum_{j=0}^{t} \phi_j(a_{i1}, \cdots, a_{i,k-1}) x_k^j.$$

Therefore $s = t$ and

$$\theta_j(a_{i1}, \cdots, a_{i,k-1}) = \phi_j(a_{i1}, \cdots, a_{i,k-1})$$

for $i = 1, \cdots, m$ and $j = 0, \cdots, s$. By the inductive hypothesis it follows that

$$\theta_j(x_1, \cdots, x_{k-1}) \equiv \phi_j(x_1, \cdots, x_{k-1})$$

for all $j = 0, \cdots, s$. We conclude that $\theta \equiv \phi$.

## PROBLEMS

**3.6–1**   Evaluate the following determinants by Chio's pivotal condensation process:

(a) $\begin{vmatrix} 2 & 3 & 3 & 2 \\ 2 & 2 & 0 & 0 \\ 2 & -1 & 1 & 4 \\ 0 & 0 & 2 & 2 \end{vmatrix}$,   (b) $\begin{vmatrix} 1 & 2 & 3 & 4 \\ 2 & 3 & 4 & 1 \\ 3 & 4 & 1 & 2 \\ 4 & 1 & 2 & 3 \end{vmatrix}$,

(c) $\begin{vmatrix} 1 & 1 & 1 & 1 \\ 1 & 2 & 1 & 1 \\ 1 & 1 & 3 & 1 \\ 1 & 1 & 1 & 4 \end{vmatrix}$,   (d) $\begin{vmatrix} 1 & 1 & 2 & 3 \\ 1 & 2 & 2 & 3 \\ 1 & 1 & 4 & 3 \\ 1 & 1 & 2 & 8 \end{vmatrix}$,

(e) $\begin{vmatrix} 1 & 0 & 0 & 2 \\ 0 & 1 & 2 & 0 \\ 0 & 2 & 1 & 0 \\ 2 & 0 & 0 & 1 \end{vmatrix}$,  (f) $\begin{vmatrix} 1 & 1 & 1 & 1 \\ 1 & 2 & 2 & 2 \\ 1 & 2 & 3 & 3 \\ 1 & 2 & 3 & 4 \end{vmatrix}$.

**3.6–2**  Show that the evaluation of a general $n$th-order determinant by Chio's pivotal condensation process requires the evaluation of

$$n(n + 1)(2n + 1)/6$$

second-order determinants.

**3.6–3**  (a) Prove that if matrix $A$ in Theorem 3.6.1 is symmetric, then so also is matrix $D$.

(b) What can be said about matrix $D$ in Theorem 3.6.1 if $A$ is skew-symmetric?

**3.6–4**  What must one do when evaluating a determinant by Chio's pivotal condensation process if in the course of the reduction a determinant appears in which the top left element is 0?

**3.6–5**  Let $A^{(1)} = [a_{ij}]$ be a nonzero matrix (not necessarily square) and let $B^{(1)} = [b_{ij}]$ be obtained from $A^{(1)}$ by interchange of rows and interchange of columns so that $b_{11} \neq 0$. Let $A^{(2)}$ denote the matrix, if it exists, obtained by replacing each element $b_{ij}$ in $B_{(1)(1)}^{(1)}$ by $\begin{vmatrix} b_{11} & b_{ij} \\ b_{i1} & b_{ij} \end{vmatrix}$. Then $A^{(2)}$ will be called a *condensed matrix* of $A^{(1)}$. A condensed matrix of $A^{(2)}$, if it exists, will be denoted by $A^{(3)}$. In general, a condensed matrix of $A^{(k)}$, if it exists, will be denoted by $A^{(k+1)}$. The sequence of matrices $A^{(1)}, A^{(2)}, \cdots$ will be called a *condensation chain* of $A^{(1)}$. Calculate condensation chains for the following matrices:

(a) $\begin{bmatrix} 2 & 1 & 0 & 3 \\ 1 & 4 & 2 & 1 \\ 2 & -2 & 1 & 0 \end{bmatrix}$,  (b) $\begin{bmatrix} 2 & 1 & 0 & 3 \\ 1 & 4 & 2 & 1 \\ 3 & 5 & 2 & 4 \end{bmatrix}$,

(c) $\begin{bmatrix} 2 & 1 & 0 & 3 \\ 4 & 2 & 0 & 6 \\ 6 & 3 & 0 & 9 \end{bmatrix}$,  (d) $\begin{bmatrix} 0 & 1 & 2 & 3 \\ 1 & 2 & 3 & 0 \\ 2 & 3 & 0 & 1 \end{bmatrix}$.

**3.6–6**  Carry through the details, in full expanded form, of the proof of Theorem 3.6.3 for the matrix $A = [a_{ij}]_{(5)}$.

**3.6–7**  Look up a proof, in an elementary theory of equations text, of the case $n = 1$ of Theorem 3.6.4.

**3.6–8**  (a) Prove the following theorem (by paralleling the proof of Theorem 3.6.4):

Suppose $\theta(x_1, \cdots, x_n)$ and $\phi(x_1, \cdots, x_n)$ are polynomials of degrees less than m in each of the n variables $x_i$ and suppose

$$\theta(a_1, \cdots, a_n) = \phi(a_1, \cdots, a_n)$$

for arbitrary numbers $a_1, \cdots, a_n$ belonging to a set M of m distinct numbers. Then $\theta \equiv \phi$.

**(b)** Show why the theorem in part (a) cannot be used instead of Theorem 3.6.4 toward the end of the proof of Theorem 3.6.3.

### 3.7   Laplace's expansion

In 1772 the great French mathematician Pierre-Simon Laplace (1749–1827) wrote a lengthy paper entitled, "Recherches sur le calcul intégral et sur le système du monde." In the course of this work, Laplace found that he must eliminate a set of variables from a system of linear equations, and in his consideration of this problem, he arrived at a somewhat cloudy statement of the very general expansion theorem in the theory of determinants that now bears his name. Like Chio's expansion, Laplace's expansion is of both practical and theoretical interest. It is a remarkable generalization of the expansion of a determinant by the elements and cofactors of some row or column. We commence our discussion by generalizing the notion of *minor* that was introduced in Definition 3.4.4.

**3.7.1**   DEFINITIONS.   A *minor* of a square $n \times n$ matrix $A$ is the determinant of any square submatrix $M$ of $A$. If $|M|$ is one of the $m$-rowed minors of $A$, then the $(n - m)$-rowed minor $|N|$ obtained by deleting from $A$ the rows and columns represented in $M$ is called the *complement* of $|M|$.

If $|M|$ is the $m$-rowed minor of $A$ in which rows $r_1, \cdots, r_m$ and columns $c_1, \cdots, c_m$ are represented, then the *algebraic complement*, or *cofactor*, of $|M|$ is

$$(-1)^{r_1 + \cdots + r_m + c_1 + \cdots + c_m} |N|.$$

An $(n - m)$-rowed minor of $A$ is called an $m$th *minor* of $A$; the *zeroth minor* of $A$ is just $|A|$.

**3.7.2**   NOTATION.   Though we shall find no use for it, we introduce here a consistent notation for minors of a matrix $A_{(n)}$; the notation generalizes that of Definition and Notation 3.4.4. Let

$$A_{(i_1,\cdots,i_m)(j_1,\cdots,j_m)}$$

denote the $(n - m) \times (n - m)$ submatrix of $A$ obtained by deleting rows $i_1, \cdots, i_m$ and columns $j_1, \cdots, j_m$ from $A$. Then

$$\left| A_{(i_1,\cdots,i_m)(j_1,\cdots,j_m)} \right|$$

is the corresponding $m$th minor of $A$.

**3.7.3** LAPLACE'S EXPANSION THEOREM. *Select any* m *rows* (*columns*) *from matrix* $A_{(n)}$ *and form all the* m-*rowed minors of* A *found in these* m *rows* (*columns*). *Then* $|A|$ *is equal to the sum of the products of each of these minors and its algebraic complement.*

Let

$$|M_1| = \begin{vmatrix} a_{11} & \cdots & a_{1m} \\ \cdots & \cdots & \cdots \\ a_{m1} & \cdots & a_{mm} \end{vmatrix}, \qquad |N_1| = \begin{vmatrix} a_{m+1,m+1} & \cdots & a_{m+1,n} \\ \cdots & \cdots & \cdots \\ a_{n,m+1} & & a_{nn} \end{vmatrix}.$$

Then

$$|M_1| = \sum_p e_p(a_{i_1 1} \cdots a_{i_m m}),$$

where $p = (i_1, \cdots, i_m)$ is a permutation of $1, \cdots, m$, and

$$|N_1| = \sum_q e_q(a_{i_{m+1}m+1} \cdots a_{i_n n}),$$

where $q = (i_{m+1}, \cdots, i_n)$ is a permutation of $m + 1, \cdots, n$. It follows that

$$|M_1| \, |N_1| = \sum_w e_p e_q(a_{i_1 1} \cdots a_{i_m m} a_{i_{m+1}m+1} \cdots a_{i_n n}),$$

where $w$ is the permutation $(i_1, \cdots, i_m; i_{m+1}, \cdots, i_n)$ of $1, \cdots, m; m + 1, \cdots, n$. Now $e_p e_q = e_w$, and we see that the terms in the product $|M_1| \, |N_1|$ are terms of $|A|$.

Now let $|M_k|$ be the $m$-rowed minor of $A$ involving rows $r_1, \cdots, r_m$ and columns $c_1, \cdots, c_m$, both in natural order, of $A$. By making

$$(r_1 - 1) + (r_2 - 2) + \cdots + (r_m - m) = (r_1 + \cdots + r_m) - m(m + 1)/2$$

successive interchanges of adjacent rows, and

$$(c_1 - 1) + (c_2 - 2) + \cdots + (c_m - m) = (c_1 + \cdots + c_m) - m(m + 1)/2$$

successive interchanges of adjacent columns, we may shift $M_k$ into the top left corner, and $N_k$ will be in the bottom right corner, of a matrix $B$. By the opening paragraph of the proof, the terms of the product $|M_k|\,|N_k|$ are terms of $|B|$. But

$$|B| = (-1)^{r_1 + \cdots + r_m + c_1 + \cdots + c_m - m(m+1)}|A|$$

$$= (-1)^{r_1 + \cdots + r_m + c_1 + \cdots + c_m}|A|,$$

since $m(m+1)$ is even. It follows that the terms of the product $|M_k|\,|N_k|$ are $(-1)^{r_1 + \cdots + r_m + c_1 + \cdots + c_m}$ times terms of $|A|$, or the terms of the product

$$(-1)^{r_1 + \cdots + r_m + c_1 + \cdots + c_m}|M_k|\,|N_k|$$

are precisely terms of $|A|$.

Now let $|M_k|$ and $|M_j|$ represent two distinct $m$-rowed minors chosen from the same $m$ rows (columns) of $A$. Then it is clear that the terms of the product $|M_k|\,|N_k|$ are distinct from the terms of the product $|M_j|\,|N_j|$.

Since there are $C(n, m) = n!/[m!(n-m)!]$ $m$-rowed minors lying in the same $m$ rows (columns) of $A$, there are $n!/[m!(n-m)!]$ products of a minor and its algebraic complement in the associated Laplace expansion. Each such product contains $m!(n-m)!$ terms, whence, by the preceding paragraph, the Laplace expansion contains $n!$ distinct terms, each term being a term of $|A|$. It follows that the Laplace expansion is simply the expansion of $|A|$, and the theorem is established.

*Examples*

**1.** Prove that for any real numbers $a$, $b$, $c$,

$$D = \begin{vmatrix} a & 0 & b & 0 & c & 0 \\ 0 & b & 0 & c & 0 & a \\ b & 0 & c & 0 & a & 0 \\ 0 & c & 0 & a & 0 & b \\ c & 0 & a & 0 & b & 0 \\ 0 & a & 0 & b & 0 & c \end{vmatrix} \geqq 0.$$

Using a Laplace expansion involving the first, third, and fifth rows, we find

$$D = \begin{vmatrix} a & b & c \\ b & c & a \\ c & a & b \end{vmatrix} \begin{vmatrix} b & c & a \\ c & a & b \\ a & b & c \end{vmatrix} = \begin{vmatrix} a & b & c \\ b & c & a \\ c & a & b \end{vmatrix}^2 \geqq 0.$$

**2.** Expand

$$D = \begin{vmatrix} a & 1 & 0 & 0 & 0 \\ b & a & 1 & 0 & 0 \\ 0 & b & a & 1 & 0 \\ 0 & 0 & b & a & 1 \\ 0 & 0 & 0 & b & a \end{vmatrix}.$$

Using a Laplace expansion involving the first two columns, we find

$$D = \begin{vmatrix} a & 1 \\ b & a \end{vmatrix} \begin{vmatrix} a & 1 & 0 \\ b & a & 1 \\ 0 & b & a \end{vmatrix} - \begin{vmatrix} a & 1 \\ 0 & b \end{vmatrix} \begin{vmatrix} 1 & 0 & 0 \\ b & a & 1 \\ 0 & b & a \end{vmatrix}$$

$$= (a^2 - b)(a^3 - 2ab) - ab(a^2 - b)$$

$$= a(a^2 - b)(a^2 - 3b).$$

**3.** Expand

$$D = \begin{vmatrix} a & x & y & b \\ x & 0 & 0 & y \\ y & 0 & 0 & x \\ c & y & x & d \end{vmatrix}.$$

Using a Laplace expansion involving the second and third rows, we find

$$D = \begin{vmatrix} x & y \\ y & x \end{vmatrix} \begin{vmatrix} x & y \\ y & x \end{vmatrix} = (x^2 - y^2)^2.$$

## PROBLEMS

**3.7–1** Evaluate the following determinants by appropriate Laplace expansions:

**(a)** $\begin{vmatrix} 1 & 2 & 2 & 1 \\ 0 & 2 & 0 & 1 \\ 1 & 0 & 2 & 2 \\ 0 & 1 & 0 & 1 \end{vmatrix},$    **(b)** $\begin{vmatrix} 1 & 2 & 0 & 0 \\ 2 & 1 & 1 & 0 \\ 0 & 1 & 2 & 1 \\ 0 & 0 & 2 & 1 \end{vmatrix},$

**(c)** $\begin{vmatrix} 1 & 2 & 3 & 4 \\ 2 & 1 & 2 & 1 \\ 0 & 0 & 1 & 1 \\ 3 & 4 & 1 & 2 \end{vmatrix},$    **(d)** $\begin{vmatrix} a & 0 & b & 0 \\ 0 & c & 0 & d \\ e & 0 & f & 0 \\ 0 & g & 0 & h \end{vmatrix}.$

**3.7–2** By an appropriate Laplace expansion of the determinant

$$\begin{vmatrix} x_1 & x_2 & x_3 & x_4 \\ y_1 & y_2 & y_3 & y_4 \\ x_1 & x_2 & x_3 & x_4 \\ y_1 & y_2 & y_3 & y_4 \end{vmatrix} \equiv 0,$$

establish the identity

$$p_{12}p_{34} + p_{13}p_{42} + p_{14}p_{23} \equiv 0,$$

where $p_{ij} = x_i y_j - y_i x_j$.

**3.7–3** If $A$, $B$, $C$ are $n \times n$ matrices, prove that

$$\begin{vmatrix} A & O \\ C & B \end{vmatrix} = |A|\,|B|.$$

**3.7–4** If $A_1, A_2, \cdots, A_s$ are square matrices, show by use of the Laplace expansion theorem that

$$|\mathrm{diag}(A_1, A_2, \cdots, A_s)| = |A_1|\,|A_2| \cdots |A_s|.$$

**3.7–5** Show that the value of the $n$th-order determinant

$$\begin{vmatrix} O_{(k)} & A \\ B & C \end{vmatrix}$$

is 0 if $k > n/2$.

## 3.8   The product theorem

There is an important and useful theorem to the effect that if $A$ and $B$ are both $n \times n$ matrices, then $|AB| = |A|\,|B|$.† In this section we shall give three different proofs of this theorem, picking up other pieces of useful information in the process.

**3.8.1** THEOREM. $|R(j, k)| = |C(j, k)| = -1$, $|R(ck| = |C(ck)| = c$, $|R(j + ak)| = |C(j + ak)| = 1$.

We have $|I| = 1$ (by Theorem 3.4.12). Hence $|R(j, k)| = |C(j, k)| = -1$ (by Theorem 3.3.5), $|R(ck)| = |C(ck)| = c$ (by Theorem 3.3.7), $|R(j + ak)| = |C(j + ak)| = 1$ (by Theorem 3.3.9).

† The theorem appeared in 1812 in two masterful papers, both read on November 30 before the French Academy of Sciences, one by Jacques P. M. Binet (1786–1856) and one by Augustin-Louis Cauchy (1789–1857). The Cauchy memoire is particularly important in the history of the theory of determinants.

**3.8.2** THEOREM. *If* R *is an* nth*-order ERT matrix,* C *an* nth*-order ECT matrix, and* A *any* nth*-order matrix, then* $|RA| = |R|\,|A|$ *and* $|AC| = |A|\,|C|$.

For $|RA| = -|A|$, $c|A|$, or $|A|$, according as $|R| = -1$, $c$, or 1. Similarly, $|AC| = -|A|$, $c|A|$, or $|A|$, according as $|C| = -1$, $c$, or 1.

**3.8.3.** THEOREM. *If* $R_1, \cdots, R_s$ *are ERT matrices, then*

$$|R_1 \cdots R_s A| = |R_1| \cdots |R_s|\,|A| = |R_1 \cdots R_s|\,|A|,$$

*and if* $C_1, \cdots, C_t$ *are ECT matrices, then*

$$|AC_1 \cdots C_t| = |A|\,|C_1| \cdots |C_t| = |A|\,|C_1 \cdots C_t|.$$

For, by Theorem 3.8.2,

$$
\begin{aligned}
|R_1 \cdots R_s A| &= |R_1|\,|R_2 \cdots R_s A| \\
&= |R_1|\,|R_2|\,|R_3 \cdots R_s A| \\
&= \cdots \\
&= |R_1|\,|R_2| \cdots |R_s|\,|A| \\
&= |R_1|\,|R_2| \cdots |R_{s-2}|\,|R_{s-1}R_s|\,|A| \\
&= |R_1|\,|R_2| \cdots |R_{s-3}|\,|R_{s-2}R_{s-1}R_s|\,|A| \\
&= \cdots \\
&= |R_1 \cdots R_s|\,|A|.
\end{aligned}
$$

The other part of the theorem can be similarly established.

**3.8.4** THEOREM. *An* n × n *matrix* A *is nonsingular if and only if* $|A| \neq 0$.

There exist ERT matrices $R_i$ and ECT matrices $C_j$ such that

$$R_1 R_2 \cdots R_s A C_1 C_2 \cdots C_t = M,$$

where $M$ is a canonical matrix. By Theorem 3.8.3,

$$|M| = |R_1 R_2 \cdots R_s A C_1 C_2 \cdots C_t| = |R_1| \cdots |R_s|\,|A|\,|C_1| \cdots |C_t|.$$

Since $|R_i| \neq 0$ and $|C_j| \neq 0$, we see that $|M| \neq 0$ if and only if $|A| \neq 0$. But $|M| \neq 0$ if and only if rank $A = n$, and rank $A = n$ if and only if $A$ is nonsingular.

**3.8.5** THEOREM. *If* A *and* B *are* n × n *matrices, then* $|AB| = |A|\,|B|$.

*Case* 1 ($A$ nonsingular). We have $A = R_1 \cdots R_s$, where the $R_i$ are ERT matrices. Then $|AB| = |R_1 \cdots R_s B| = |R_1 \cdots R_s|\,|B| = |A|\,|B|$, by Theorem 3.8.3.

*Case* 2 ($A$ singular). We have $|A| = 0$, whence $|A|\,|B| = 0$. Now rank $AB \leq$ rank $A$. That is, $AB$ is also singular, whence $|AB| = 0$. It follows that $|AB| = |A|\,|B|$.

**3.8.6** THEOREM. *If* A *and* B *are* n × n *matrices, then* $|AB| = |AB'| = |A'B| = |A'B'| = |A|\,|B|$.

This follows because $|AB'| = |A|\,|B'| = |A|\,|B|$, etc.

*Note.* This theorem yields four ways in which we may express the product of two $n$th-order determinants $|A|$ and $|B|$ as an $n$th-order determinant. In forming the elements of the product determinant, we may multiply rows of $A$ by columns of $B$, rows of $A$ by rows of $B$, columns of $A$ by rows of $B$, or columns of $A$ by columns of $B$.

**3.8.7** THEOREM (alternative proof of Theorem 3.8.5). *If* A *and* B *are* n × n *matrices, then* $|AB| = |A|\,|B|$.

Clearly,

$$\begin{bmatrix} I & A \\ O & I \end{bmatrix} \begin{bmatrix} A & O \\ -I & B \end{bmatrix} = \begin{bmatrix} O & AB \\ -I & B \end{bmatrix}.$$

Now $\begin{bmatrix} I & A \\ O & I \end{bmatrix}$ can be obtained from $I_{(2n)}$ by elementary row operations of the type $o(j + ak)$, and consequently is equal to a product of ERT matrices of the type $R(j + ak)$. It follows that $\begin{bmatrix} O & AB \\ -I & B \end{bmatrix}$ is obtainable from $\begin{bmatrix} A & O \\ -I & B \end{bmatrix}$ by elementary row operations of the type $o(j + ak)$. Since elementary row operations of this type do not alter the value of the determinant of a matrix,

$$\begin{vmatrix} O & AB \\ -I & B \end{vmatrix} = \begin{vmatrix} A & O \\ -I & B \end{vmatrix},$$

or, by Laplace expansions involving the first $n$ rows,

$$(-1)^{1 + \cdots + n + (n+1) + \cdots + (2n)}|AB|(-1)^n = (-1)^{1 + \cdots + n + 1 + \cdots + n}|A|\,|B|,$$

or

$$(-1)^{n(1 + 2n) + n}|AB| = (-1)^{n(1 + n)}|A|\,|B|,$$

or

$$|AB| = |A|\,|B|.$$

**3.8.8**   THEOREM.   *Let* $n \times n$ *matrices* A, B, C *be identical except perhaps for their jth columns* $A'_j$, $B'_j$, $C'_j$, *and suppose* $C'_j = A'_j + B'_j$. *Then* $|C| = |A| + |B|$.

The theorem is easily established by expanding $|C|$ by the elements and cofactors of its *j*th column.

The easy proofs of the following two corollaries will be left to the reader.

**3.8.9**   COROLLARY.   *If* A *and* B *are two* $n \times n$ *matrices, then* $|A + B|$ *can be expressed as the sum of* $2^n$ *nth-order determinants, in each of which the jth column,* $j = 1, \cdots, n$, *is the jth column of* A *or the jth column of* B.

**3.8.10**   COROLLARY.   *If* A, B, $\cdots$, M *are* $m$ $n \times n$ *matrices, then*

$$|A + B + \cdots + M|$$

*can be expressed as the sum of* $m^n$ *nth-order determinants, in each of which the jth column,* $j = 1, \cdots, n$, *is the jth column of some one of* A, B, $\cdots$, M.

**3.8.11**   THEOREM (alternative proof of Theorem 3.8.5).   *If* A *and* B *are* $n \times n$ *matrices, then* $|AB| = |A| \, |B|$.

Let $A = [a_{ij}]$, $B = [b_{ij}]$. Then

$$|AB| = \begin{vmatrix} \sum_{t=1}^{n} a_{1t}b_{t1} & \sum_{t=1}^{n} a_{1t}b_{t2} & \cdots & \sum_{t=1}^{n} a_{1t}b_{tn} \\ \cdots & \cdots & \cdots & \cdots \\ \sum_{t=1}^{n} a_{nt}b_{t1} & \sum_{t=1}^{n} a_{nt}b_{t2} & \cdots & \sum_{t=1}^{n} a_{nt}b_{tn} \end{vmatrix}$$

$$= \left| \sum_{t=1}^{n} \begin{bmatrix} a_{1t}b_{t1} & a_{1t}b_{t2} & \cdots & a_{1t}b_{tn} \\ \cdots & \cdots & \cdots & \cdots \\ a_{nt}b_{t1} & a_{nt}b_{t2} & \cdots & a_{nt}b_{tn} \end{bmatrix} \right|.$$

Applying Corollary 3.8.10, we then have

$$|AB| = \sum \begin{vmatrix} a_{1i_1}b_{i_11} & a_{1i_2}b_{i_22} & \cdots & a_{1i_n}b_{i_nn} \\ \cdots & \cdots & \cdots & \cdots \\ a_{ni_1}b_{i_11} & a_{ni_2}b_{i_22} & \cdots & a_{ni_n}b_{i_nn} \end{vmatrix},$$

where each $i_j$ independently ranges over $1, \cdots, n$, yielding a sum of $n^n$ determinants. Many of these $n^n$ determinants have two columns proportional, and are therefore equal to zero. Eliminating these, we find

$$|AB| = \sum_p \begin{vmatrix} a_{1i_1} & a_{1i_2} & \cdots & a_{1i_n} \\ \cdots & \cdots & \cdots & \cdots \\ a_{ni_1} & a_{ni_2} & \cdots & a_{ni_n} \end{vmatrix} (b_{i_11} b_{i_22} \cdots b_{i_nn})$$

(where $p$ ranges over all permutations $(i_1, \cdots, i_n)$ of $1, \cdots, n$)

$$= \begin{vmatrix} a_{11} & a_{12} & \cdots & a_{1n} \\ \cdots & \cdots & \cdots & \cdots \\ a_{n1} & a_{n2} & \cdots & a_{nn} \end{vmatrix} \sum_p e_p(b_{i_11} b_{i_22} \cdots b_{i_nn})$$

$$= |A|\,|B|.$$

The product theorem can often be used to advantage when evaluating a determinant, as illustrated by the following two examples.

*Example 1.*  Taking all summations over $i = 1, 2, 3, 4$,

$$\begin{vmatrix} 4 & \sum a_i & \sum a_i^2 & \sum a_i^3 \\ \sum a_i & \sum a_i^2 & \sum a_i^3 & \sum a_i^4 \\ \sum a_i^2 & \sum a_i^3 & \sum a_i^4 & \sum a_i^5 \\ \sum a_i^3 & \sum a_i^4 & \sum a_i^5 & \sum a_i^6 \end{vmatrix}$$

$$= \begin{vmatrix} 1 & 1 & 1 & 1 \\ a_1 & a_2 & a_3 & a_4 \\ a_1^2 & a_2^2 & a_3^2 & a_4^2 \\ a_1^3 & a_2^3 & a_3^3 & a_4^3 \end{vmatrix} \begin{vmatrix} 1 & a_1 & a_1^2 & a_1^3 \\ 1 & a_2 & a_2^2 & a_2^3 \\ 1 & a_3 & a_3^2 & a_3^3 \\ 1 & a_4 & a_4^2 & a_4^3 \end{vmatrix}$$

$$= (a_4 - a_3)^2(a_4 - a_2)^2(a_4 - a_1)^2(a_3 - a_2)^2(a_3 - a_1)^2(a_2 - a_1)^2,$$

since each of the factor determinants is a Vandermonde determinant.

*Example 2*

$$\begin{vmatrix} 1 & \cos(\alpha - \beta) & \cos(\alpha - \gamma) \\ \cos(\beta - \alpha) & 1 & \cos(\beta - \gamma) \\ \cos(\gamma - \alpha) & \cos(\gamma - \beta) & 1 \end{vmatrix}$$

$$= \begin{vmatrix} \cos\alpha & \sin\alpha & 0 \\ \cos\beta & \sin\beta & 0 \\ \cos\gamma & \sin\gamma & 0 \end{vmatrix} \begin{vmatrix} \cos\alpha & \cos\beta & \cos\gamma \\ \sin\alpha & \sin\beta & \sin\gamma \\ 0 & 0 & 0 \end{vmatrix} = 0.$$

*Example* 3.   We shall show that if the quadratic equation

$$ax^2 + 2hxy + by^2 + 2gx + 2fy + c = 0$$

represents a pair of straight lines, then

$$\begin{vmatrix} a & h & g \\ h & b & f \\ g & f & c \end{vmatrix} = 0.$$

To this end, suppose

$$ax^2 + 2hxy + by^2 + 2gx + 2fy + c = (\alpha x + \beta y + \gamma)(\alpha' x + \beta' y + \gamma')$$
$$= \alpha\alpha' x^2 + (\alpha\beta' + \beta\alpha')xy + \beta\beta' y^2 + (\alpha\gamma' + \gamma\alpha')x + (\beta\gamma' + \gamma\beta')y + \gamma\gamma'.$$

Then

$$8\begin{vmatrix} a & h & g \\ h & b & f \\ g & f & c \end{vmatrix} = \begin{vmatrix} 2\alpha\alpha' & \alpha\beta' + \beta\alpha' & \alpha\gamma' + \gamma\alpha' \\ \alpha\beta' + \beta\alpha' & 2\beta\beta' & \beta\gamma' + \gamma\beta' \\ \alpha\gamma' + \gamma\alpha' & \beta\gamma' + \gamma\beta' & 2\gamma\gamma' \end{vmatrix}$$

$$= \begin{vmatrix} \alpha & \alpha' & 0 \\ \beta & \beta' & 0 \\ \gamma & \gamma' & 0 \end{vmatrix}\begin{vmatrix} \alpha' & \beta' & \gamma' \\ \alpha & \beta & \gamma \\ 0 & 0 & 0 \end{vmatrix} = 0.$$

## PROBLEMS

**3.8–1**   Complete the proof of Theorem 3.8.3.

**3.8–2**   Supply proofs for Theorem 3.8.8 and Corollaries 3.8.9 and 3.8.10.

**3.8–3**   Show that

$$\begin{vmatrix} 0 & (a-b)^2 & (a-c)^2 \\ (b-a)^2 & 0 & (b-c)^2 \\ (c-a)^2 & (c-b)^2 & 0 \end{vmatrix}$$

$$= \begin{vmatrix} a^2 & -2a & 1 \\ b^2 & -2b & 1 \\ c^2 & -2c & 1 \end{vmatrix}\begin{vmatrix} 1 & 1 & 1 \\ a & b & c \\ a^2 & b^2 & c^2 \end{vmatrix} = 2\begin{vmatrix} 1 & 1 & 1 \\ a & b & c \\ a^2 & b^2 & c^2 \end{vmatrix}^2.$$

**3.8–4**    Taking all summations over $i = 1, 2, 3, 4$, evaluate

$$\begin{vmatrix} \sum a_i & \sum a_i^3 & \sum a_i^5 & \sum a_i^7 \\ \sum a_i^3 & \sum a_i^5 & \sum a_i^7 & \sum a_i^9 \\ \sum a_i^5 & \sum a_i^7 & \sum a_i^9 & \sum a_i^{11} \\ \sum a_i^7 & \sum a_i^9 & \sum a_i^{11} & \sum a_i^{13} \end{vmatrix}.$$

**3.8–5**    Show that

$$\begin{vmatrix} \sin(\alpha + \alpha') & \sin(\alpha + \beta') & \sin(\alpha + \gamma') \\ \sin(\beta + \alpha') & \sin(\beta + \beta') & \sin(\beta + \gamma') \\ \sin(\gamma + \alpha') & \sin(\gamma + \beta') & \sin(\gamma + \gamma') \end{vmatrix} = 0.$$

**3.8–6**    If $A = [a_{ij}]_{(n)}$, $f_j(x) = a_{1j} + a_{2j}x + \cdots + a_{nj}x^{n-1}$, $b_{jk} = f_j(x_k)$ for $k = 1, \cdots, n$, $B = [b_{jk}]_{(n)}$, show that $|B| = V_n|A|$, where $V_n$ is the Vandermonde determinant formed by the numbers $x_1, \cdots, x_n$.

**3.8–7**    Evaluate $|A| = |a_{ij}|_{(n)}$, where $a_{1j} = 1$ and $a_{ij} = x_j^{i-1} + x_j^{i-2}$ if $i \geqq 2$.

**3.8–8**    If $\epsilon$ is the $n$th primitive root of 1, prove that

$$\begin{vmatrix} 1 & 1 & 1 & \cdots & 1 \\ 1 & \epsilon & \epsilon^2 & \cdots & \epsilon^{n-1} \\ 1 & \epsilon^2 & \epsilon^4 & \cdots & \epsilon^{2(n-1)} \\ \cdots & \cdots & \cdots & \cdots & \cdots \\ 1 & \epsilon^{n-1} & \epsilon^{2(n-1)} & \cdots & \epsilon^{(n-1)^2} \end{vmatrix} = n^{n/2}.$$

**3.8–9**    Establish the following result due to Arthur Cayley. If 1, 2, 3, 4 denote four points on a circle, then

$$\begin{vmatrix} 0 & (12)^2 & (13)^2 & (14)^2 \\ (21)^2 & 0 & (23)^2 & (24)^2 \\ (31)^2 & (32)^2 & 0 & (34)^2 \\ (41)^2 & (42)^2 & (43)^2 & 0 \end{vmatrix} = 0.$$

**3.8–10**    If $a, b, c, d$ denote four points on a sphere, show that

$$\begin{vmatrix} 1 & \cos \widehat{ab} & \cos \widehat{ac} & \cos \widehat{ad} \\ \cos \widehat{ba} & 1 & \cos \widehat{bc} & \cos \widehat{bd} \\ \cos \widehat{ca} & \cos \widehat{cb} & 1 & \cos \widehat{cd} \\ \cos \widehat{da} & \cos \widehat{db} & \cos \widehat{dc} & 1 \end{vmatrix} = 0.$$

**3.8–11** By considering the product

$$\begin{vmatrix} a & b & c \\ b & c & a \\ c & a & b \end{vmatrix} \begin{vmatrix} -a & c & b \\ -b & a & c \\ -c & b & a \end{vmatrix}$$

prove that

$$\begin{vmatrix} 2bc - a^2 & c^2 & b^2 \\ c^2 & 2ca - b^2 & a^2 \\ b^2 & a^2 & 2ab - c^2 \end{vmatrix} = (a^3 + b^3 + c^3 - 3abc)^2.$$

**3.8–12** Find the value of

$$|A| = \begin{vmatrix} 1/1! & 1/0! & 0 & 0 \\ 1/2! & 1/1! & 1/0! & 0 \\ 1/3! & 1/2! & 1/1! & 1/0! \\ 1/4! & 1/3! & 1/2! & 1/1! \end{vmatrix}$$

by considering the product $|A| \, |B|$, where $|B|$ is defined by $|B| = |b_{ij}|_{(4)}$,

$$b_{ij} = (-1)^{i+j}(j-1)!/(i-1)!$$

for $i \leq j$ and $b_{ij} = 0$ for $i > j$.

**3.8–13** Evaluate the determinant

$$\begin{vmatrix} 3! & \binom{3}{1} & \binom{3}{2} & \binom{3}{3} \\ 2! & \binom{2}{0} & \binom{2}{1} & \binom{2}{2} \\ 1! & 0 & \binom{1}{0} & \binom{1}{1} \\ 0! & 0 & 0 & \binom{0}{0} \end{vmatrix}$$

by multiplying it by the triangular determinant $|A| = |a_{ij}|_{(4)}$, where

$$a_{1j} = (-1)^{j+1} \binom{3}{j-1}$$

for $j = 1, 2, 3, 4$; $a_{ij} = 1$ for $i = j$; all other $a_{ij} = 0$.

**3.8–14** Prove that

$$\begin{vmatrix} 1 + x_1 y_1 & 1 + x_1 y_2 & 1 + x_1 y_3 \\ 1 + x_2 y_1 & 1 + x_2 y_2 & 1 + x_2 y_3 \\ 1 + x_3 y_1 & 1 + x_3 y_2 & 1 + x_3 y_3 \end{vmatrix} = 0.$$

**3.8–15**   Solve Problem 3.4–10 by use of Theorem 3.8.8.

**3.8–16**   Prove that

$$\begin{vmatrix} a_{11} + x & a_{12} + x & \cdots & a_{1n} + x \\ a_{21} + x & a_{22} + x & \cdots & a_{2n} + x \\ \cdots & \cdots & \cdots & \cdots \\ a_{n1} + x & a_{n2} + x & \cdots & a_{nn} + x \end{vmatrix} = |a_{ij}| + x \sum_{i,j} A_{ij},$$

where $A_{ij}$ is the cofactor of $a_{ij}$.

**3.8–17**   Prove by means of Problem 3.8–16 that if all the elements of a row of $[a_{ij}]_{(n)}$ are equal to 1, then $|a_{ij}| = \sum_{i,j} A_{ij}$, where $A_{ij}$ is the cofactor of $a_{ij}$.

**3.8–18**   Suppose the elements of an $n \times n$ matrix $A$ are differentiable functions of a variable $x$. Denote the $j$th column of $A$ by $C_j$ and let $dC_j/dx$ denote the column vector whose elements are the derivatives with respect to $x$ of the corresponding elements of $C_j$. Prove that

$$d|A|/dx = |dC_1/dx, C_2, \cdots, C_n| + |C_1, dC_2/dx, \cdots, C_n| \\ + \cdots + |C_1, C_2, \cdots, dC_n/dx|.$$

**3.8–19**   Consider matrices of the form $\begin{bmatrix} a & b \\ -b & a \end{bmatrix}$. **(a)** Show that sums and products of matrices of this form are of the same form. **(b)** Prove that the product of two sums of four squares of positive integers can be expressed as the sum of four squares of positive integers.

### 3.9   Determinant rank of a matrix

In a paper of 1877, Georg Frobenius (1849–1917) defined the "rank" of a square matrix to be the order of the highest-ordered, nonvanishing minor (that is, subdeterminant) in the matrix. Since this definition, which is applicable to any rectangular matrix, is not the definition of rank of a matrix as given in Definition 2.7.2, we shall, at least for the time being, distinguish the two approaches by calling the Frobenius concept the *determinant rank* of the matrix. It turns out that the rank of a matrix (as defined in Definition 2.7.2) and the determinant rank of the matrix are equal. It is the purpose of this section to establish this interesting equality and to prove several allied theorems. The Frobenius approach to the rank of a matrix is the older of the two approaches, and some of the later theorems in this section reflect the natural efforts of mathematicians of the time to reduce the amount of work required in finding the determinant rank of a matrix.

**3.9.1**   DEFINITION.   A matrix $A$ is said to have *determinant rank* $d$ if and only if there exists at least one nonvanishing subdeterminant of $A$ of order $d$, but no nonvanishing subdeterminant of order higher than $d$, and $A$ is said to have *determinant rank zero* if and only if $A = O$.

*Example*.   The matrix

$$\begin{bmatrix} 1 & 2 & 3 & 0 \\ 0 & 0 & 0 & 1 \\ 1 & 2 & 3 & 0 \\ 0 & 0 & 0 & 1 \end{bmatrix}$$

has determinant rank 2, since the determinant of the matrix itself and all subdeterminants of order 3 vanish (because they possess pairs of identical rows), but the second-order subdeterminant

$$\begin{vmatrix} 1 & 0 \\ 0 & 1 \end{vmatrix},$$

obtained by deleting the two middle rows and the two middle columns, does not vanish.

**3.9.2**   THEOREM.   *If all subdeterminants of order* q *of a matrix* A *vanish, then all higher-order subdeterminants of* A *also vanish.*

For a higher-ordered subdeterminant may be expanded, by an appropriate Laplace expansion, in terms of $q \times q$ subdeterminants and their algebraic complements.

*Note.*   In view of this theorem, it is sufficient, when examining for determinant rank of a matrix $A$, to find a nonvanishing subdeterminant of $A$ of order $d$ and to show that all subdeterminants of order $d + 1$ vanish.

**3.9.3**   THEOREM.   *Any* d *rows (columns) of an* m × n *matrix* A *that contain a nonvanishing subdeterminant* |D| *of* A *of order* d *are linearly independent.*

The rows (columns) of $D$ are linearly independent, since $D$ is nonsingular. Hence the $d$ rows (columns) of $A$ are linearly independent, for a linear relation among these rows (columns) of $A$ is also a linear relation among the rows (columns) of $D$.

**3.9.4** THEOREM. *The rank and the determinant rank of an* m × n *matrix* A *are equal.*

Let $r = $ rank $A$ and $d = $ determinant rank $A$. Then $A$ has a nonsingular $d \times d$ submatrix $D$ and, by Theorem 3.9.3, the $d$ rows of $A$ in which $D$ lies are linearly independent. Hence $r \geqq d$.

Now consider any $r$ linearly independent rows of $A$ and denote by $R$ the $r \times n$ submatrix of $A$ formed by these rows. Since row rank $R = r$, then also column rank $R = r$. That is, $R$ contains $r$ linearly independent columns. Let $S$ be the $r \times r$ submatrix of $R$ containing these $r$ columns. Then $S$ is nonsingular, whence $|S| \neq 0$. It follows that $d \geqq r$.

It now follows that $r = d$, and the theorem is established.

**3.9.5** THEOREM. *If a subdeterminant* |D| *of order* d *of* A *is nonvanishing, and every subdeterminant of order* d + 1 *of* A *and containing* |D| *vanishes, then* $\rho(A) = d$.

By interchanges of rows and of columns we may convert $A$ into a matrix $B$ having $D$ in the top left corner and having the elements of any other row of $A$ not through $D$ as $(d + 1)$st row. Since rank is preserved under elementary row and column transformations, $\rho(A) = \rho(B)$. By Theorem 3.9.3, the first $d$ rows of $B$ are linearly independent. Let us designate the first $d + 1$ rows of $B$ by $B_1, B_2, \cdots, B_d, B_{d+1}$, and those parts of these rows in the first $d + 1$ columns of $B$ by $R_1, R_2, \cdots, R_d, R_{d+1}$. Since, by hypothesis, the top left subdeterminant $|R|$ of $B$ of order $d + 1$ vanishes, and since the first $d$ rows of $R$ are linearly independent, we must have

$$(1) \qquad R_{d+1} = c_1R_1 + c_2R_2 + \cdots + c_dR_d.$$

Now let $S$ be any $(d + 1) \times (d + 1)$ submatrix of $B$ chosen from the first $d + 1$ rows of $B$ and differing from $R$ only in its last column. Then, by hypothesis, $|S| = 0$. Let us replace the last row of $S$ by

$$S_{d+1} - c_1S_1 - c_2S_2 - \cdots - c_dS_d,$$

converting $S$ into $T$, where $|T| = |S|$. Because of (1), the first $d$ elements of the last row of $T$ are 0's, and $0 = |S| = |T| = s|D|$, where $s$ is the last element of the last row of $T$. Since $|D| \neq 0$, it follows that $s = 0$, and

$$S_{d+1} = c_1S_1 + c_2S_2 + \cdots + c_dS_d.$$

We can now conclude that

$$B_{d+1} = c_1B_1 + c_2B_2 + \cdots + c_dB_d,$$

and the $(d + 1)$st row of $B$ is linearly dependent upon the first $d$ (linearly independent) rows of $B$. That is, any row of $A$ is linearly dependent upon the $d$ linearly independent rows of $A$ passing through $D$. It follows that $\rho(A) = d$.

*Note.* This theorem further reduces the work required to find the determinant rank of a matrix $A$.

**3.9.6** Definition. In a square matrix $A$, those minors obtained from $A$ by deleting the same columns of $A$ as rows of $A$ are called the *principal minors* of $A$.

**3.9.7** Theorem. *If* A *is symmetric or Hermitian and possesses a nonvanishing principal minor* $|D|$ *of order* d, *and all principal minors of orders* d $+$ 1 *and* d $+$ 2 *containing* D *vanish, then* $\rho(A) = d$.

Let $A$ be symmetric. By interchanges of rows and corresponding interchanges of columns, we may convert $A$ into a symmetric matrix $B$ having $D$ in the top left corner and having any nonprincipal $(d + 1)$-rowed minor of $A$ containing $D$ lying in rows $1, 2, \cdots, d, d + 1$ and columns $1, 2, \cdots, d, d + 2$ of $B$. Then, by Theorem 3.6.3,

$$|b_{11} \quad b_{22} \quad \cdots \quad b_{dd}| \, |b_{11} \quad b_{22} \quad \cdots \quad b_{d+2,d+2}|$$
$$= \begin{vmatrix} |b_{11} & \cdots & b_{dd} & b_{d+1,d+1}| \, |b_{11} & \cdots & b_{dd} & b_{d+1,d+2}| \\ |b_{11} & \cdots & b_{dd} & b_{d+2,d+1}| \, |b_{11} & \cdots & b_{dd} & b_{d+2,d+2}| \end{vmatrix},$$

or, since $|b_{11} \cdots b_{d+2,d+2}| = 0$ and $|b_{11} \cdots b_{d+1,d+1}| = 0$, and because $B$ is symmetric,

$$|b_{11} \quad \cdots \quad b_{dd} \quad b_{d+2,d+1}| = |b_{11} \quad \cdots \quad b_{dd} \quad b_{d+1,d+2}|,$$

we have

$$|b_{11} \quad \cdots \quad b_{dd} \quad b_{d+1,d+2}| = 0.$$

That is, any nonprincipal minor of $A$ of order $d + 1$ and containing $D$ vanishes, as well as the principal minors of $A$ of order $d + 1$ and containing $D$. The theorem now follows by Theorem 3.9.5.

The reader can supply a similar argument for the case where $A$ is Hermitian.

**3.9.8** Theorem. *A symmetric or Hermitian matrix* A *of rank* r *contains at least one nonvanishing principal minor of order* r.

Suppose $A$ is symmetric. Then, since $\rho(A) = r$, $A$ contains a set of $r$

linearly independent rows. By elementary row operations we may convert the remaining rows into zero rows. Since $A$ is symmetric, the corresponding elementary column operations will convert the corresponding columns into zero columns. Denote the resulting matrix by $B$. Since rank is preserved under elementary row and column operations, $\rho(B) = \rho(A) = r$. But the only nonvanishing $r$-rowed minor in $B$ is the principal $r$-rowed minor of $A$ lying in the $r$ linearly independent rows of $A$. This proves the theorem.

The reader can supply a similar argument for the case where $A$ is Hermitian.

*Note.*    Theorems 3.9.8 and 3.9.7 considerably reduce the work required to find the determinant rank of a symmetric or Hermitian matrix $A$.

## PROBLEMS

**3.9–1**    Supply a proof of Theorem 3.9.7 for the case where $A$ is Hermitian.

**3.9–2**    Supply a proof of Theorem 3.9.8 for the case where $A$ is Hermitian.

**3.9–3**    If $\rho(A) = r$, prove that not all square submatrices of any order $s < r$ can be singular.

**3.9–4**    Construct an example to show that the rank of a product of two matrices may be less than the rank of either factor.

**3.9–5**    Prove that the three points $(x_1, y_1)$, $(x_2, y_2)$, $(x_3, y_3)$ of the Cartesian plane are collinear if and only if the rank of the matrix

$$\begin{bmatrix} x_1 & y_1 & 1 \\ x_2 & y_2 & 1 \\ x_3 & y_3 & 1 \end{bmatrix}$$

is less than 3.

**3.9–6**    Find the rank of matrix $A = [a_{ij}]_{(n)}$ if $a_{ij} = n - 1$ for $i = j$ and $a_{ij} = 1$ for $i \neq j$.

**3.9–7**    Suppose $A$ is a square matrix of rank $r$ and suppose the leading principal minor of order $r$ is not 0. Show that if the first $r$ elements in any row or column with index $s > r$ are all 0, then so are all the remaining elements in that row or column.

**3.9–8**    Prove that if $A_{(n)}$ and $(A + A')/2$ both have rank 1, then $A$ is symmetric.

**3.9-9** Prove that the rank of a skew-symmetric $\mathscr{C}$ matrix is even. (This result was given by Jacobi in 1827.)

**3.9-10** (a) Let $A^{(1)}$, $A^{(2)}$, $\cdots$ be a condensation chain of matrix $A^{(1)}$ (see Problem 3.6–5 for definition). Show that if $A^{(r)}$ is the last nonzero matrix in the chain, then $\rho(A) = r$. (This is considered to be an expeditious way of determining the rank of a given matrix.)

(b) Find, by the method of part (a), the ranks of the four $3 \times 4$ matrices given in Problem 3.6–5.

**3.9-11** Show that Theorem 3.9.8 also holds for skew-symmetric and skew-Hermitian matrices.

**3.9-12** Prove that if a symmetric, skew-symmetric, Hermitian, or skew-Hermitian matrix $A$ of rank $r$ has a nonsingular $r$-rowed submatrix that is not a principal submatrix, then $A$ contains at least two nonsingular $r$-rowed principal submatrices.

## 3.10 Adjoint of a square matrix

From a given matrix $A$, various associated matrices can be formed whose elements are subdeterminants, or signed subdeterminants, of $A$. One of the most important of these is the so-called *adjoint* of a square matrix.

**3.10.1** DEFINITION AND NOTATION: Let $A = [a_{ij}]$ be an $n \times n$ matrix, $n > 1$, and let $A_{ij}$ denote the cofactor of $a_{ij}$. Set $M = [A_{ij}]$. Then we define $M'$, the transpose of $M$, to be the *adjoint* of $A$, and we denote it by adj $A$. In other words, if we set adj $A = [\alpha_{ij}]$, then $\alpha_{ij} = A_{ji}$.

Some writers use the term *adjugate* for the adjoint, and some denote the adjoint of $A$ by $A^{\mathbf{A}}$.

*Example.* Find the adjoint of

$$A = \begin{bmatrix} 2 & -2 & 4 \\ 2 & 3 & 2 \\ -1 & 1 & -1 \end{bmatrix}.$$

We have

$$A_{11} = (-1)^2 \begin{vmatrix} 3 & 2 \\ 1 & -1 \end{vmatrix} = -5, \qquad A_{21} = (-1)^3 \begin{vmatrix} -2 & 4 \\ 1 & -1 \end{vmatrix} = 2,$$

$$A_{12} = (-1)^3 \begin{vmatrix} 2 & 2 \\ -1 & -1 \end{vmatrix} = 0, \qquad A_{22} = (-1)^4 \begin{vmatrix} 2 & 4 \\ -1 & -1 \end{vmatrix} = 2,$$

$$A_{13} = (-1)^4 \begin{vmatrix} 2 & 3 \\ -1 & 1 \end{vmatrix} = 5, \qquad A_{23} = (-1)^5 \begin{vmatrix} 2 & -2 \\ -1 & 1 \end{vmatrix} = 0,$$

$$A_{31} = (-1)^4 \begin{vmatrix} -2 & 4 \\ 3 & 2 \end{vmatrix} = -16,$$

$$A_{32} = (-1)^5 \begin{vmatrix} 2 & 4 \\ 2 & 2 \end{vmatrix} = 4,$$

$$A_{33} = (-1)^6 \begin{vmatrix} 2 & -2 \\ 2 & 3 \end{vmatrix} = 10.$$

Therefore

$$\text{adj } A = \begin{bmatrix} -5 & 2 & -16 \\ 0 & 2 & 4 \\ 5 & 0 & 10 \end{bmatrix}.$$

**3.10.2** THEOREM. *If* A *is a square matrix of order* n > 1, *then* A($adj$ A) = ($adj$ A)A = |A|$I_{(n)}$.

The $(i, j)$th element of $A$(adj $A$) is

$$\sum_{k=1}^{n} a_{ik}\alpha_{kj} = \sum_{k=1}^{n} a_{ik}A_{jk} = \begin{cases} |A| & \text{if } i = j \\ 0 & \text{if } i \neq j. \end{cases}$$

Therefore $A$(adj $A$) is a scalar matrix with diagonal elements all equal to |A|. It follows that $A$(adj $A$) = |A|$I_{(n)}$. One may similarly show that (adj $A$)$A$ = |A|$I_{(n)}$.

**3.10.3** THEOREM. *If* A *is a nonsingular matrix of order* n > 1, *then* A$^{-1}$ = ($adj$ A)/|A|.

For if $A$ is nonsingular, then |A| $\neq$ 0, and the theorem follows from Theorem 3.10.2.

*Note.* The relation $A^{-1}$ = (adj $A$)/|A| is perhaps the oldest way of computing the inverse of a nonsingular matrix $A$. The method soon becomes prohibitively laborious as the order of $A$ increases.

*Example.* Find the inverse of

$$A = \begin{bmatrix} 2 & -2 & 4 \\ 2 & 3 & 2 \\ -1 & 1 & -1 \end{bmatrix}.$$

We have already computed adj $A$ in the example following Definition 3.10.1. One also finds

$$|A| = 2A_{11} - 2A_{12} + 4A_{13} = 2(-5) - 2(0) + 4(5) = 10.$$

Therefore

$$A^{-1} = (adj\ A)/|A| = (1/10) \begin{bmatrix} -5 & 2 & -16 \\ 0 & 2 & 4 \\ 5 & 0 & 10 \end{bmatrix} = \begin{bmatrix} -1/2 & 1/5 & -8/5 \\ 0 & 1/5 & 2/5 \\ 1/2 & 0 & 1 \end{bmatrix}.$$

The proofs of the next two theorems are so easy that they are left to the reader.

**3.10.4** THEOREM. *If* A *is a square matrix of order* n > 1, *then* $(adj\ A)' = adj\ (A')$.

**3.10.5** THEOREM. *If* A *is a square matrix of order* n > 1 *and* k *is a scalar, then* $adj\ (kA) = k^{n-1}(adj\ A)$.

**3.10.6** THEOREM. *If* A *is a square matrix of order* n > 1 *and if* $\rho(A) = r$, *then* $\rho(adj\ A) = $ n, 1, *or* 0, *according as* r = n, n − 1, *or* < n − 1.

If $r = n$, the corresponding part of the theorem follows by Theorem 3.10.3.

If $r < n$, there exist nonsingular matrices $P$ and $Q$ such that

$$PAQ = \begin{bmatrix} I_{(r)} & O \\ O & O \end{bmatrix}.$$

By Theorem 3.10.2,

$$PAQ(Q^{-1}adj\ A) = PA(adj\ A) = P|A| = O.$$

Thus the first $r$ rows of $(Q^{-1}\ adj\ A)$ must be zero rows and $\rho(Q^{-1}\ adj\ A)$ is at most $n - r$. Hence (by Theorem 2.7.6), $\rho(adj\ A)$ is at most $n - r$. If $r = n - 1$, then $\rho(adj\ A)$ is at most 1; that it actually is 1 follows from the fact that $A$ contains a nonvanishing subdeterminant of order $n - 1$, whence adj $A$ contains at least one nonzero element. If $r < n - 1$, then $\rho(adj\ A) = 0$, since $A$ contains no nonvanishing subdeterminant of order $n - 1$; in this case, adj $A = O$.

**3.10.7**   THEOREM.   *If* A *is a square matrix of order* n > 1 *and if* $|A| = 0$, *then* $|adj\ A| = 0$.

If $|A| = 0$, then $\rho(A) < n$ and (by Theorem 3.10.6) $\rho(\text{adj } A)$ is 0 or 1. It follows that $|\text{adj } A| = 0$.

**3.10.8**   CAUCHY'S THEOREM.   *If* A *is a square matrix of order* n > 1, *then* $|adj\ A| = |A|^{n-1}$.

By Theorem 3.10.2, we have $|A||\text{adj } A| = |A|^n$. If $|A| \neq 0$, then $|\text{adj } A| = |A|^{n-1}$. If $|A| = 0$, then $|\text{adj } A| = 0$ (by Theorem 3.10.7), and again $|\text{adj } A| = |A|^{n-1}$.

**3.10.9**   THEOREM.   *If* A *is a square matrix of order* n > 2, *then* $adj(adj\ A) = A|A|^{n-2}$.

We have, by Theorems 3.10.2 and 3.10.8,

$$(\text{adj } A)[\text{adj}(\text{adj } A)] = |\text{adj } A|I_{(n)} = |A|^{n-1}I_{(n)}.$$

Therefore

$$[A(\text{adj } A)][\text{adj}(\text{adj } A)] = A|A|^{n-1},$$

or (again by Theorem 3.10.2)

$$|A|[\text{adj}(\text{adj } A)] = A|A|^{n-1}.$$

Hence, if $|A| \neq 0$,

$$\text{adj}(\text{adj } A) = A|A|^{n-2}.$$

If $|A| = 0$, then $\rho(\text{adj } A)$ is at most 1. If $n > 2$, it follows (by Theorem 3.10.6) that adj(adj $A$) = $O$, and the theorem is trivially true in this case. If $n = 2$, it can be easily shown that adj(adj $A$) = $A$, whether $|A| = 0$ or not.

**3.10.10**   DEFINITION AND NOTATION.   We shall denote adj(adj $A$) by $\text{adj}_2 A$ and, in general, $\text{adj}(\text{adj}_{k-1} A)$ by $\text{adj}_k A$ when $k > 2$; $\text{adj}_1 A$ will mean adj $A$. We shall call $\text{adj}_k A$ the $k$th *adjoint* of $A$.

**3.10.11**   THEOREM.   *If* A *is a square matrix of order* n > 1, *then* $|adj_k A| = |A|^{(n-1)^k}$.

We leave the proof, by mathematical induction, to the reader.

**3.10.12**   THEOREM.   *If* A *and* B *are nonsingular matrices of order* n > 1, *then* $adj(AB) = (adj\ B)(adj\ A)$.

Since $|A| \neq 0$ and $|B| \neq 0$, we have $|AB| \neq 0$ and $(AB)^{-1}$ exists. Then, by Theorem 3.10.3, $\text{adj}(AB) = |AB|(AB)^{-1} = |A||B|B^{-1}A^{-1} = (|B|B^{-1})(|A|A^{-1}) = (\text{adj } B)(\text{adj } A)$.

*Note.* This theorem, but not the proof given here, still holds when either $A$ or $B$ is singular. A proof of this case will be given later.

We conclude the section with a proof of the celebrated *Cramer's rule* for solving a nonsingular system of $n > 1$ linear equations in $n$ unknowns. The rule was given by Gabriel Cramer (1704–1752) in an appendix to his famous treatise, *Introduction à l'analyse des lignes courbes algebraiques*, published in Geneva in 1750.

**3.10.13** CRAMER'S RULE. *A nonsingular system of* $n > 1$ *linear equations in* $n$ *unknowns, expressed in matrix form as* $AX = B$, $|A| \neq 0$, *has the unique solution*

$$x_i = |A(i)|/|A|, \qquad i = 1, \cdots, n,$$

*where* $|A(i)|$ *denotes* $|A|$ *with its* $i$th *column replaced by the column vector* **B**.

We have $X = A^{-1}B = (\text{adj } A)B/|A|$, whence, setting $\text{adj } A = [\alpha_{ij}]$,

$$x_i = \sum_{k=1}^{n} \alpha_{ik}b_k/|A| = \sum_{k=1}^{n} b_k A_{ki}/|A| = |A(i)|/|A|.$$

## PROBLEMS

**3.10–1** Prove Theorem 3.10.4.

**3.10–2** Prove Theorem 3.10.5.

**3.10–3** Prove Theorem 3.10.11.

**3.10–4** Find the adjoints and the inverses of:

(a) $\begin{bmatrix} 3 & -1 & 2 \\ 1 & 0 & 3 \\ 4 & 0 & 2 \end{bmatrix}$, (b) $\begin{bmatrix} 2 & 1 & 2 \\ 0 & 1 & -1 \\ 2 & 3 & -1 \end{bmatrix}$, (c) $\begin{bmatrix} -4 & -3 & -3 \\ 1 & 0 & 1 \\ 4 & 4 & 3 \end{bmatrix}$.

**3.10–5** Solve, by Cramer's rule:

(a) $\quad 2x_1 - 2x_2 + 4x_3 = 1,$
$\quad\;\; 2x_1 + 3x_2 + 2x_3 = 2,$
$\quad -x_1 + x_2 - x_3 = 3.$

(b) $\quad 3x_1 + x_2 + x_3 + x_4 = 0,$
$\quad\;\; 2x_1 - x_2 + 2x_3 - x_4 = 4,$
$\quad\;\; 2x_1 \qquad\;\; + 3x_3 + x_4 = 1,$
$\quad\;\;\; x_1 \qquad\;\; + 2x_3 + x_4 = 3.$

**3.10–6**   If $A$ is a $2 \times 2$ matrix, show that adj(adj $A$) = $A$.

**3.10–7**   **(a)** If $A$ is a diagonal matrix, prove that adj $A$ is also a diagonal matrix.

**(b)** If $A$ is triangular, prove that adj $A$ is also triangular.

**(c)** If $A$ is symmetric, prove that adj $A$ is also symmetric.

**(d)** If $A$ is Hermitian, prove that adj $A$ is also Hermitian.

**(e)** If $A$ is skew-symmetric of order $n$, prove that adj $A$ is symmetric or skew-symmetric, according as $n$ is odd or even.

**3.10–8**   Let $P$, $Q$, $A$, $B$ be $n \times n$ matrices such that $|P| = |Q| = 1$, $A = $ adj $B$. Show that $PAQ = $ adj$(Q^{-1}BP^{-1})$.

**3.10–9**   Let matrix $B$ be obtained from square matrix $A$ by deleting the $r$th and $s$th rows and the $p$th and $q$th columns. Show that

$$\begin{vmatrix} A_{rp} & A_{sp} \\ A_{rq} & A_{sq} \end{vmatrix} = (-1)^{r+s+p+q}|B||A|.$$

**3.10–10**   Let $A$ and $B$ be two $m \times n$ matrices and define the *inner product* $A|B$ of $A$ and $B$ to be the scalar

$$A|B = \sum_{i,j} a_{ij}b_{ij}.$$

Establish the following relations:

**(a)** If $A$ and $B$ are $m \times n$, then $A|B = B|A$.

**(b)** If $A$ and $B$ are $m \times n$, then $A|B = A'|B'$.

**(c)** If $A$, $B$, $C$ are $m \times n$, then $A|(B + C) = (A|B) + (A|C)$.

**(d)** If $A$, $B$, $M$ are $n \times n$, then $A|(BM) = (AM')|B$.

**(e)** If $A$, $B$, $M$ are $n \times n$, then $A|(MB) = (M'A)|B$.

**(f)** If $A$ and $B$ are $n \times n$, then $A|B = I_{(n)}|(BA') = I_{(n)}|(AB') = I_{(n)}|(A'B) = I_{(n)}|(B'A)$.

**(g)** If $A$ is $n \times n$, then $A|$adj $A' = n|A|$.

**(h)** If $A$ is $n \times n$, $U = [u_1, \cdots, u_n]$, $V = \{v_1, \cdots, v_n\}$, then $UAV = A|(U'V')$.

**(i)** If $A$ and $B$ are $3 \times 3$ and $k$ is a scalar, then
$|kA + B| = k^3|A| + k^2(B|$adj $A') + k(A|$adj $B') + |B|$.

*Note.*   The inner product of two matrices finds application in the geometry of a pair of conics. For example, suppose that $A$ and $B$ are the matrices associated with two conics (as in Example 4 of Section 0.1). Then a necessary and sufficient condition that it be possible to inscribe in conic $A$ a triangle that is self-polar to conic $B$ is that $A|$adj $B = 0$. Again, a necessary and sufficient condition that it be possible to inscribe in conic

$A$ a triangle that is circumscribed about conic $B$ is that

$$(A|\text{adj } B)^2 - 4|B|(B|\text{adj } A) = 0.$$

**3.10–11**   Let $A$ and $B$ be two $n \times n$ matrices. Then we define the *cojoint* of $A$ and $B$, denoted by $\text{coj}(A, B)$, to be the matrix

$$\text{coj}(A, B) = \text{adj}(A + B) - \text{adj } A - \text{adj } B.$$

(This name and notation were first suggested by Alexander D. Wallace in 1934.) Establish the following relations:

(a) If $A$ and $B$ are $n \times n$, then $\text{coj}(A, B) = \text{coj}(B, A)$.

(b) If $A$ and $B$ are $n \times n$, then $[\text{coj}(A, B)]' = \text{coj}(A', B')$.

(c) If $A$ and $B$ are $n \times n$ and $k$ is a scalar, then

$$\text{coj}(kA, kB) = k^{n-1}\text{coj}(A, B).$$

(d) If $A$, $B$, $M$ are $n \times n$ and $M$ is nonsingular, then

$$\text{coj}(MA, B)M = |M|\text{coj}(A, M^{-1}B).$$

(e) If $A$, $B$, $M$ are $n \times n$ and $M$ is nonsingular, then

$$M \text{coj}(AM, B) = |M|\text{coj}(A, BM^{-1}).$$

(f) If $A$ is $n \times n$, then $\text{coj}(A, A) = 2(2^{n-2} - 1)\text{adj } A$.

(g) If $A$, $B$, $M$ are $n \times n$ and $M$ is nonsingular, then

$$M' \text{coj}(MAM', MBM')M = |M|^2 \text{coj}(A, B).$$

(h) If $A$ and $B$ are $3 \times 3$ and $p$ and $q$ are scalars, then

$$\text{adj}(pA + qB) = p^2 \text{adj } A + pq \text{coj}(A, B) + q^2 \text{adj } B$$

$$= p(p - q)\text{adj } A + q(q - p)\text{adj } B + pq \text{adj}(A + B).$$

(i) If $A$ and $B$ are $3 \times 3$, then

$$(B|\text{adj } A)\text{adj } A - (\text{adj } A)B(\text{adj } A) = |A|\text{coj}(A, B).$$

(j) If $A$ and $B$ are $3 \times 3$, then

$$(A|\text{adj } B)A - A(\text{adj } B)A = \text{coj}(\text{adj } A, \text{adj } B).$$

(k) If $A$ and $B$ are $3 \times 3$, then

$$(A|\text{adj } B)A - A(\text{adj } B)A = (B|\text{adj } A)B - B(\text{adj } A)B.$$

(l) If $A$ and $B$ are $3 \times 3$ and $A$ is nonsingular, then

$$\text{adj}(A + B) = \text{adj } A + \text{adj } B + (B|\text{adj } A)A^{-1} - (\text{adj } A)BA^{-1}.$$

*Note.* The cojoint of two matrices finds application in the geometry of a pair of conics. For example, suppose that $A$ and $B$ are the matrices associated with two conics (as in Example 4 of Section 0.1). Then the eight points of contact with the two conics $A$ and $B$ of their four common tangents all lie on a third conic whose matrix is coj(adj $A$, adj $B$).

## ADDENDA TO CHAPTER 3

Each of the following items is associated with material of Chapter 3 and may serve as a topic for a "junior" research project. The items are of various lengths and degrees of difficulty.

### 3.1A    A Geometric Study of Permutations

We may associate with each permutation $p$ of 1, 2, $\cdots$, $n$, two geometrical figures as follows:

Let the $n$ integers 1, 2, $\cdots$, $n$ be written uniformly spaced out and in natural order in a horizontal row. Directly beneath and a short distance below write out in the same way the successive integers in the permutation $p$. Now join by straight-line segments each integer in the upper row with the same integer in the lower row. The resulting figure may be called the *lower diagram* of $p$.

Similarly, by placing the row of successive integers in the permutation $p$ directly over and a short distance above the row of integers 1, 2, $\cdots$, $n$, and then joining by straight-line segments each integer in the lower row with the same integer in the upper row, we obtain a figure that may be called the *upper diagram* of $p$.

As an illustration, the lower and upper diagrams of the permutation $p = (4, 3, 5, 2, 1)$ of 1, 2, 3, 4, 5 are shown in the accompanying diagram.

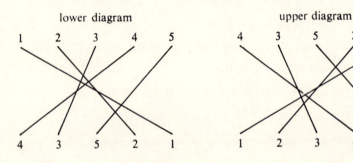

The theory of permutations of 1, 2, $\cdots$, $n$ can be developed by a geometrical study of lower and upper diagrams. As a start, the interested reader might prove the following theorems.

**I** THEOREM. *The number of inversions in a permutation* p *of* 1, 2, $\cdots$, n *is equal to the number of simple intersections occurring in either the lower or the upper diagram of* p, *where a multiple intersection of order* m *is to be counted as* m *simple intersections.*

**II** DEFINITION. Two permutations $p$ and $q$ of 1, 2, $\cdots$, $n$ are said to be *conjugate* if the element and place occupied in one are the place and element occupied in the other.

Thus the two permutations (3, 2, 4, 1) and (4, 2, 1, 3) of 1, 2, 3, 4 are conjugate.

**III** THEOREM. *A necessary and sufficient condition for two permutations* p *and* q *of* 1, 2, $\cdots$, n *to be conjugate is that the upper diagram of one be the lower diagram of the other.*

**IV** THEOREM. *Two conjugate permutations* p *and* q *of* 1, 2, $\cdots$, n, *possess the same number of inversions.*

*Note.* This is essentially the important Lemma 3.3.1 of the text.

**V** DEFINITION. A permutation $p$ of 1, 2, $\cdots$, $n$ is said to be *self-conjugate* if and only if it is its own conjugate.

**VI** DEFINITION. The horizontal line midway between the upper and lower rows of integers of a permutation diagram is called the h-*line* of the diagram.

**VII** THEOREM. *A necessary and sufficient condition for a permutation* p *of* 1, 2, $\cdots$, n *to be self-conjugate is that its diagrams be symmetrical with respect to their* h-*lines.*

**VIII** THEOREM. *If* $P_k$ *denotes the number of self-conjugate permutations of* 1, 2, $\cdots$, k, *then* $P_n = P_{n-1} + (n-1)P_{n-2}$.

**IX** DEFINITION.   By the *product pq* of two permutations *p* and *q* of 1, 2, $\cdots$, *n*, we mean the permutation *w* of 1, 2, $\cdots$, *n* whose lower diagram is obtained by placing the lower diagram of *q* directly below the lower diagram of *p* so that the initial points of joins in the lower diagram of *q* fall upon the terminal points of joins in the lower diagram of *p*, and then straightening all broken-line joins in the resulting figure.

As an illustration, let *p* = (3, 2, 4, 1) and *q* = (3, 1, 4, 2) be permutations of 1, 2, 3, 4. Then we have the arrangements shown in the illustration here. Thus *w* = *pq* = (4, 3, 1, 2).

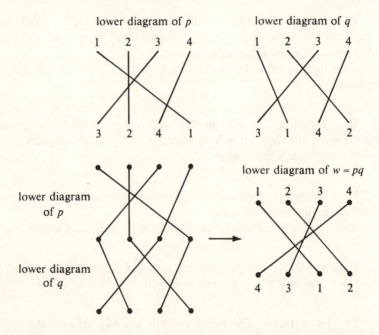

The reader may now prove that permutation multiplication is associative but not, in general, commutative. By defining the *identity permutation i* of 1, 2, $\cdots$, *n* to be the permutation (1, 2, $\cdots$, *n*)—that is, the permutation whose upper and lower diagrams consist only of vertical line segments—and by defining an *inverse permutation* of permutation *p* of 1, 2, $\cdots$, *n* to be a permutation $p^{-1}$ of 1, 2, $\cdots$, *n* such that $pp^{-1} = i$, the reader may easily show by the use of permutation diagrams that: (1) *p* and $p^{-1}$ are conjugate, (2) $p^{-1}$ is unique, (3) $p^{-1}p = i$.

Such concepts of permutation theory as transposition, cycle, degree and order of a permutation, circular permutation, cyclic permutation, transform of one permutation by another, and so on, can be looked up by the reader and then their geometrical counterparts examined and studied.

### 3.2A    Permutation Matrices

An elegant way to study the theory of permutations $(i_1, i_2, \cdots, i_n)$ of $1, 2, \cdots, n$ is to associate with each such permutation $p$ the $n$th-order matrix $P$ obtained from $I_{(n)}$ by subjecting the rows of $I_{(n)}$ to the permutation $p$. Such a matrix is called a *permutation matrix*. The value of the association lies in the fact that if $P$ and $Q$ are the permutation matrices associated with the permutations $p$ and $q$ of $1, 2, \cdots, n$, then the matrix $PQ$ is the permutation matrix associated with the permutation $pq$. It follows that $P^{-1}$ is the permutation matrix associated with permutation $p^{-1}$. A permutation $p$ is even or odd according as $|P| = +1$ or $-1$. The important Lemma 3.3.1 of the text is then taken care of by showing that if $P$ is a permutation matrix, so is $P'$. It is easily shown that $P' = P^{-1}$. A permutation matrix $P$ is called a *transposition* if $P^2 = I$, that is, if $P$ is involutoric. Such concepts of permutation theory as degree and order of a permutation, circular permutation, cyclic permutation, transform of one permutation by another, and so on, are neatly studied by elementary matrix theory when these concepts are assigned to the associated permutation matrices. The details are left to the reader.

### 3.3A    Permanents

The determinant of an $n \times n$ matrix $A = [a_{ij}]$ is a scalar-valued function of $A$ defined as

$$\sum_q e_q a_{i_1 1} a_{i_2 2} \cdots a_{i_n n}$$

and denoted by $|A|$ or $d(A)$; here the summation is taken over all permutations $q = (i_1, i_2, \cdots, i_n)$ of $1, 2, \cdots, n$. Another, and related, scalar-valued function of $A$ is the *permanent*, or *plus determinant*, of $A$, defined as

$$\sum_q a_{i_1 1} a_{i_2 2} \cdots a_{i_n n}$$

and denoted by $\overset{+}{|}A\overset{+}{|}$ or $p(A)$.

The following properties of $p(A)$ are easily established:

(1) If $P$ and $Q$ are $n$th-order permutation matrices (see Addenda 3.2A for definition), then $p(PAQ) = p(A)$.

(2) If $D$ and $E$ are $n$th-order diagonal matrices, then $p(DAE) = |D|\,|E|\,p(A)$.

(3) $p(A') = p(A)$.

(4) $p(A^*) = \overline{p(A)}$.

(5) If $k$ is a scalar, then $p(kA) = k^n p(A)$.

(6) If $P$ is a permutation matrix, then $p(P) = 1$.

(7) $p(O_{(n)}) = 0$.

(8) Expansion of a permanent by a row or a column:

$$p(A) = \sum_{j=1}^{n} a_{ij} p(A_{(i)(j)}) = \sum_{i=1}^{n} a_{ij} p(A_{(i)(j)}).$$

(9) If $M$ is any square submatrix of $A$, then $p(M)$ is called a *subpermanent* of $A$. If $N$ is the square submatrix of $A$ obtained by deleting from $A$ the rows and columns through $M$, then $p(N)$ is called the *complementary subpermanent* of subpermanent $M$.

(10) Laplace's expansion theorem for permanents: Select any $m$ rows (columns) from matrix $A_{(n)}$ and form all the $m$-rowed subpermanents of $A$ found in these $m$ rows (columns). Then $p(A)$ is equal to the sum of the products of each of these subpermanents and its complementary subpermanent.

Permanents are not nearly so rich in application as are determinants, but the reader may care to verify the following:

(a) Let $S = \{s_1, \cdots, s_n\}$ be a set of $n$ elements and let $S_1, \cdots, S_n$ denote $n$ subsets of $S$. An ordering $s_{i_1}, \cdots, s_{i_n}$ of the elements of $S$ such that $s_{i_j} \in S_j$ is called a *system of distinct representatives* of the given situation. The problem of determining the total number $m$ of systems of distinct representatives naturally arises. To find $m$, one first forms the so-called *incidence matrix* $A = [a_{ij}]_{(n)}$ for the situation, where $a_{ij} = 1$ or $0$, according as $s_i \in S_j$ or not. Then $m = p(A)$.

(b) Let each of $n$ distinct points $c_1, \cdots, c_n$ be connected by $n$ paths with itself and with each of the other $n - 1$ points. Suppose there is a particle $q_i$ at point $c_i$ $(i = 1, \cdots, n)$, and that at time zero, each particle $q_i$ begins to move uniformly to some point $c_j$ along one of the $n$ paths radiating from point $c_i$. Let $p_{ij}$ denote the probability that particle $q_i$ will move along the path from point $c_i$ to point $c_j$ $(i = 1, \cdots, n; j = 1, \cdots, n)$. What is the probability that, in the ultimate arrangement of the particles, there will be precisely one particle at each of the points $c_i$? The answer is $p(P)$, where $P = [p_{ij}]_{(n)}$.

### 3.4A    Postulational Definitions of Determinant

The determinant of a square matrix $A_{(n)}$ is a function (actually a polynomial function) of the elements $a_{ij}$ of $A$. A number of properties of this function are given in Section 3.3 and in some of the sections following Section 3.3. It is natural to wonder if a certain basic set of these properties completely characterizes the concerned function and thus implies all other properties. If this should be the case, we could define the determinant of $A_{(n)}$ to be the function of the elements $a_{ij}$ of $A$ which satisfies the basic set of properties. Such a definition of the determinant of a square matrix is called a *postulational definition*, and as might be expected, it turns out that it is possible to formulate a number of postulational definitions. We give here three of these postulational definitions and invite the interested reader to show that they are equivalent to one another and to either the explicit definition of Section 3.3 or the inductive definition of Section 3.4. The great German mathematician Karl Weierstrass (1815–1897) was perhaps the first to conceive this way of defining the determinant of a square matrix.

(1) Weierstrass's definition. The determinant of $A_{(n)}$ is a polynomial in the elements $a_{ij}$ of $A$, which is homogeneous and linear in the elements of each row of $A$, which merely changes sign when two rows of $A$ are permuted, and which reduces to 1 when $A_{(n)} = I_{(n)}$. (See C. C. MacDuffee, *The Theory of Matrices*, Chelsea Publishing Company, 1946, p. 7.)

(2) The determinant of $A_{(n)}$ is a scalar function of $A$ such that: (a) multiplying a row of $A$ by a scalar $c$ multiplies the determinant by $c$, (b) adding one row of $A$ to another row of $A$ leaves the determinant unaltered, (c) the determinant of $A$ reduces to 1 when $A_{(n)} = I_{(n)}$. (See R. R. Stoll, *Linear Algebra and Matrix Theory*, McGraw-Hill Book Co., Inc., 1952, p. 88.)

(3) The determinant $d(A)$ of $A_{(n)}$ is a polynomial in the elements $a_{ij}$ of $A$ of lowest possible positive degree such that $d(AB) = d(A)d(B)$ for any two $n$th-order matrices $A$ and $B$. (See C. C. MacDuffee, *Vectors and Matrices*, Carus Mathematical Monograph No. 7, The Mathematical Association of America, 1943, p. 50.)

### 3.5A    The Sweep-out Process for Evaluating Determinants

If a square matrix $A$ can be reduced to a triangular matrix by a succession of operations of the type where two rows are interchanged or where a multiple of some row is added to another row, then (see Theorems 3.3.5, 3.3.10, 3.4.12) the value of $|A|$ is merely plus or minus the product of the elements along the principal diagonal of the reduced triangular matrix. But matrix $A$ can be reduced to a triangular matrix by the types of operation described above as follows:

(1) If all the elements in the first column of $A$ are zeros, then $|A| = 0$. Otherwise, by an operation of the first type, if necessary, obtain a nonzero element (preferably a 1) in the (1, 1) position.

(2) Add appropriate multiples of the first row to the other rows to produce zeros in the remainder of the first column.

(3) Repeat steps (1) and (2), starting with the element in the (2, 2) position.

(4) Continue in this fashion down the principal diagonal until either the end of the diagonal is reached or all remaining elements in the matrix are zeros.

This method of computing the value of a determinant is called the *sweep-out process* and is nicely adapted to a desk calculator. One should have on hand calculating forms for the sweep-out calculation of determinants of various orders. Such a form for third-order determinants might look like the one in the accompanying table.

After filling in the top nine boxes with the values of the elements of the

| Third-order determinant | | | Instructions |
|---|---|---|---|
| $a_{11} =$ | $a_{12} =$ | $a_{13} =$ | row (1) |
| $a_{21} =$ | $a_{22} =$ | $a_{23} =$ | row (2) |
| $a_{31} =$ | $a_{32} =$ | $a_{33} =$ | row (3) |
| 1 | | | $(1)' = (1) \div a_{11}$ |
| $a_{21} =$ | | | $(1)'' = (1)' \times a_{21}$ |
| $a_{31} =$ | | | $(1)''' = (1)' \times a_{31}$ |
| $b_{21} = 0$ | $b_{22} =$ | $b_{23} =$ | $(2)' = (2) - (1)''$ |
| $b_{31} = 0$ | $b_{32} =$ | $b_{33} =$ | $(3)' = (3) - (1)'''$ |
| 0 | 1 | | $(2)'' = (2)' \div b_{22}$ |
| 0 | $b_{32} =$ | | $(2)''' = (2)'' \times b_{32}$ |
| $c_{31} = 0$ | $c_{32} = 0$ | $c_{33} =$ | $(3)'' = (3)' - (2)'''$ |
| $D =$ | | | $(4)\ D = a_{11}b_{22}c_{33}$ |

given matrix, there are only 14 more boxes to be filled in with computed values, the last one being the sought value of the determinant of the matrix.

The reader may care to construct similar forms for fourth-order and fifth-order determinants and to consider the alterations needed if accidental zeros should appear as diagonal elements. These forms can also be made to serve as computing forms for finding the rank of the concerned matrix. The forms are essentially self-explanatory.

### 3.6A  Pfaffians

It is easy to show that $|a_{ij}|_{(n)}$, considered as a polynomial $P$ in the $n^2$ independent elements $a_{ij}$, is irreducible—that is, cannot be factored into a product of two nonconstant polynomials. For suppose $P = UV$, where $U$ and $V$ are polynomials in the elements $a_{ij}$, and suppose that element $a_{11}$ occurs in $U$. Since $P$ is linear in every element, $a_{11}$ cannot occur in $V$. Now no term of $P$ contains $a_{11}a_{r1}$; hence $V$ is of degree 0 in every $a_{r1}$. It follows that all $a_{r1}$ must occur in $U$. Since no term of $P$ contains $a_{r1}a_{rs}$, $V$ is of degree 0 in every $a_{rs}$. It follows that $V$ is a constant.

With a slight modification of the preceding proof, it can be similarly shown that if $[a_{ij}]_{(n)}$ is symmetric, then $|a_{ij}|$, considered as a polynomial in its $n(n + 1)/2$ independent elements $a_{ij}$, $i \leq j$, is irreducible. We leave this to the reader.

The case where $[a_{ij}]_{(n)}$ is skew-symmetric and the $a_{ij}$ are considered as variables over the complex number field is particularly interesting. If $n$ is odd, we have, by Jacobi's Theorem 3.4.13, $|a_{ij}| \equiv 0$. If $n$ is even, it turns out that $|a_{ij}|$ is the square of a polynomial in the $n(n - 1)/2$ elements $a_{ij}$, $i < j$. We here give a proof of this, using the identity of Theorem 3.6.3. We first establish a lemma.

**I   LEMMA.**   *If* $|a_{11} a_{22} \cdots a_{nn}|$ *is skew-symmetric of even order* n, *then*

$$|a_{11} a_{22} \cdots a_{n-2,n-2} a_{n-1,n}| = -|a_{11} a_{22} \cdots a_{n-2,n-2} a_{n,n-1}|.$$

Set

$$M = [a_{11} a_{22} \cdots a_{n-2,n-2}], \qquad R = [a_{11} a_{22} \cdots a_{n-2,n-2} a_{n-1,n}],$$
$$S = [a_{11} a_{22} \cdots a_{n-2,n-2} a_{n,n-1}].$$

Then $M$ is skew-symmetric and

$$R = \left[ \begin{array}{ccc|c} & & & a_{1n} \\ & & & a_{2n} \\ & M & & \cdots \\ & & & a_{n-2,n} \\ \hline a_{n-1,1} & a_{n-1,2} & \cdots \; a_{n-1,n-2} & a_{n-1,n} \end{array} \right],$$

whence

$$R' = \left[ \begin{array}{ccc|c} & & & a_{n-1,1} \\ & & & a_{n-1,2} \\ & -M & & \cdots \\ & & & a_{n-1,n-2} \\ \hline a_{1n} & a_{2n} & \cdots \; a_{n-2,n} & a_{n-1,n} \end{array} \right]$$

$$= \left[ \begin{array}{ccc|c} & & & -a_{1,n-1} \\ & & & -a_{2,n-1} \\ & -M & & \cdots \\ & & & -a_{n-2,n-1} \\ \hline -a_{n1} & -a_{n2} & \cdots \; -a_{n,n-2} & -a_{n,n-1} \end{array} \right] = -S.$$

Therefore $|R| = |R'| = |-S| = -|S|$.

**II**    THEOREM.    *If* $[a_{ij}]_{(n)}$ *is an even-ordered skew-symmetric* $\mathscr{C}$ *matrix, then* $|a_{ij}|$, *expressed as a polynomial in the* $n(n-1)/2$ *elements* $a_{ij}$, $i < j$, *is the square of a polynomial in those elements.*

The theorem is certainly true for such matrices of order two, for we have

$$\left| \begin{array}{cc} 0 & a_{12} \\ -a_{12} & 0 \end{array} \right| = (a_{12})^2.$$

Assume the theorem is true for even-ordered skew-symmetric $\mathscr{C}$ matrices of order $k - 2$, where $k$ is an even number. Now, by Theorem 3.6.3,

$$|a_{11}\, a_{22} \cdots a_{k-2,k-2}| \; |a_{11}\, a_{22} \cdots a_{kk}|$$

$$= \left| \begin{array}{cc} |a_{11} \cdots a_{k-2,k-2}\, a_{k-1,k-1}| & |a_{11} \cdots a_{k-2,k-2}\, a_{k-1,k}| \\ |a_{11} \cdots a_{k-2,k-2}\, a_{k,k-1}| & |a_{11} \cdots a_{k-2,k-2}\, a_{k,k}| \end{array} \right|$$

$$= |D|, \text{ say.}$$

It follows, by Lemma I and by Jacobi's Theorem 3.4.13, that $D$ is skew-symmetric of order two, whence

$$|D| = |a_{11} a_{22} \cdots a_{k-2,k-2} a_{k-1,k}|^2 = p^2,$$

where $p$ is some polynomial. Also, $|a_{11} a_{22} \cdots a_{k-2,k-2}|$ is skew-symmetric of order $k - 2$, whence, by the inductive assumption,

$$|a_{11} a_{22} \cdots a_{k-2,k-2}| = q^2,$$

where $q$ is some polynomial. It follows that

$$q^2 |a_{11} a_{22} \cdots a_{kk}| = p^2,$$

whence we must have

$$|a_{11} a_{22} \cdots a_{kk}| = r^2,$$

where $r$ is some polynomial. The theorem now follows by the principle of mathematical induction.

We have shown that if $[a_{ij}]_{(n)}$ is an even-ordered skew-symmetric $\mathscr{C}$ matrix, then $|a_{ij}| = r^2$, where $r$ is a polynomial in the $n(n - 1)/2$ elements $a_{ij}, i < j$. The polynomial $r$ is called the *Pfaffian of order* $m = n/2$, the adjective "Pfaffian" being attached to the polynomial because Johann Friedrich Pfaff (1765–1825) encountered such polynomials in an 1815 research paper on partial differential equations. We leave to the reader the proof of the following theorem given by Pfaff in his paper.

**III** Theorem. *The Pfaffian of order* m *involves* m(2m − 1) *variables, is of degree* m, *and contains* 1 · 3 ⋯ (2m − 1) *terms.*

### 3.7A  Solution of Systems of Linear Equations

The solution of systems of linear equations can be a burdensome task; accordingly, mathematicians have devised many procedures to facilitate the computation. We partially describe and illustrate here a procedure that essentially consists of nothing but the evaluation of a sequence of second-order determinants.

**I** *Description of the procedure for* n *equations in* n *unknowns.* Consider the system of linear equations $AX = K$, where

$$A = [a_{ij}]_{(n)}, \qquad X = \{x_1, \cdots, x_n\}, \qquad K = \{k_1, \cdots, k_n\}.$$

First set up the $n \times (n + 1)$ augmented matrix of the system,

$$\begin{bmatrix} a_{11} & a_{12} & a_{13} & \cdots & a_{1n} & k_1 \\ a_{21} & a_{22} & a_{23} & \cdots & a_{2n} & k_2 \\ \cdots & \cdots & \cdots & \cdots & \cdots & \cdots \\ a_{n1} & a_{n2} & a_{n3} & \cdots & a_{nn} & k_n \end{bmatrix}.$$

Then form the $(n - 1) \times n$ *reduced matrix*

$$(1) \qquad \begin{bmatrix} b_{11} & b_{12} & b_{13} & \cdots & b_{1,n-1} & k'_1 \\ b_{21} & b_{22} & b_{23} & \cdots & b_{2,n-1} & k'_2 \\ \cdots & \cdots & \cdots & \cdots & \cdots & \cdots \\ b_{n-1,1} & b_{n-1,2} & b_{n-1,3} & \cdots & b_{n-1,n-1} & k'_{n-1} \end{bmatrix},$$

in which

$$b_{ij} = \begin{vmatrix} a_{11} & a_{1,j+1} \\ a_{i+1,1} & a_{i+1,j+1} \end{vmatrix}, \qquad k'_i = \begin{vmatrix} a_{11} & k_1 \\ a_{i+1,1} & k_{i+1} \end{vmatrix}.$$

(Note that the process by which the reduced matrix is obtained from the first matrix is similar to that used in the evaluation of a determinant by the Chio pivotal condensation process.) Now reduce the matrix (1) in the same way, and repeat the procedure until the reduction has been carried out $n - 1$ times. We shall assume that no one of the matrices formed has a zero element in the top left corner. Denote the final matrix, resulting from the $(n - 1)$th reduction, by

$$[g_{11} \quad k_1^{(n-1)}].$$

Then $x_n$ can be found by solving the equation

$$g_{11}x_n = k_1^{(n-1)}.$$

Let the next-to-the-last reduced matrix be

$$\begin{bmatrix} f_{11} & f_{12} & k_1^{(n-2)} \\ f_{21} & f_{22} & k_2^{(n-2)} \end{bmatrix}.$$

Then, $x_n$ being known, the value of $x_{n-1}$ can be determined by means of either the equation

$$(2) \qquad f_{11}x_{n-1} + f_{12}x_n = k_1^{(n-2)}$$

or the equation

$$(3) \qquad f_{21}x_{n-1} + f_{22}x_n = k_2^{(n-2)},$$

where one naturally chooses whichever seems the simpler to use. Now that $x_n$ and $x_{n-1}$ are known, we can determine $x_{n-2}$ by means of an equation analogous to (2) or (3), obtained from the $(n-3)$th reduced matrix. By continuing in this way, all the unknowns can be determined.

II  *An illustration.*  Consider the system of linear equations

$$x_1 + x_2 - 2x_3 + x_4 = 2,$$
$$3x_1 - x_2 - x_3 + x_4 = 0,$$
$$9x_1 + 3x_2 + 3x_3 - x_4 = 0,$$
$$2x_1 - 3x_2 - 3x_3 + 2x_4 = -2.$$

The initial matrix is

$$\begin{bmatrix} 1 & 1 & -2 & 1 & 2 \\ 3 & -1 & -1 & 1 & 0 \\ 9 & 3 & 3 & -1 & 0 \\ 2 & -3 & -3 & 2 & -2 \end{bmatrix},$$

and the successive reduced matrices are

$$\begin{bmatrix} -4 & 5 & -2 & -6 \\ -6 & 21 & -10 & -18 \\ -5 & 1 & 0 & -6 \end{bmatrix}, \qquad \begin{bmatrix} -54 & 28 & 36 \\ 21 & -10 & -6 \end{bmatrix}, \qquad [-48 \quad -432].$$

Then

$$-48x_4 = -432, \qquad x_4 = 9,$$
$$21x_3 - 10x_4 = -6, \qquad x_3 = 4,$$
$$-5x_2 + x_3 = -6, \qquad x_2 = 2,$$
$$x_1 + x_2 - 2x_3 + x_4 = 2, \qquad x_1 = -1.$$

III  *Suggestions for "junior" research.*  It is suggested that the reader justify the procedure described above; show that the assumption made in part I, concerning the nonzero character of all top left elements in the matrices, assures that the original system has a unique solution; generalize the method for systems of $m$ equations in $n$ unknowns; and adjust the process for solutions where a top left element in one of the matrices turns out to be a zero element. The reader may care to consult

N. B. Conkwright and J. D. Heide, "The solution of simultaneous linear equations," *The Mathematics Teacher*, April 1945, pp. 177–180.

### 3.8A   Continuants

The determinant of a square matrix in which all elements are zero except perhaps those on the principal diagonal and the two bordering minor diagonals is called a *continuant*. One can produce a short paper on certain special continuants by expanding the following suggested development:

(1) Show that

$$
\begin{vmatrix}
a & 1 & 0 & 0 & 0 \\
-1 & b & 1 & 0 & 0 \\
0 & -1 & c & 1 & 0 \\
0 & 0 & -1 & d & 1 \\
0 & 0 & 0 & -1 & e
\end{vmatrix}
= \begin{aligned}
& abcde + abc + abe + ade + cde \\
& \qquad\qquad + a + c + e.
\end{aligned}
$$

(2) Establish the following rule for evaluating continuants of the form appearing in (1): For the terms of the determinant take the product of the consecutive elements along the principal diagonal and the products obtained from this by omitting, in every possible way, pairs of consecutive literal factors.

(3) Denote the determinant in (1) by the symbol $(a, b, c, d, e)$.

(4) Show that the successive approximants to the value of the simple continued fraction

$$
a + \cfrac{1}{b + \cfrac{1}{c + \cfrac{1}{d + \cdots}}}
$$

are

$$
(a), \quad \frac{(a, b)}{(b)}, \quad \frac{(a, b, c)}{(b, c)}, \quad \frac{(a, b, c, d)}{(b, c, d)}, \quad \cdots .
$$

(5) Show that $(a_1, a_2, \cdots, a_n) = a_1(a_2, \cdots, a_n) + (a_3, \cdots, a_n)$.

(6) Let $\phi(n)$ denote the number of terms in the expansion of the continuant $(a_1, a_2, \cdots, a_n)$. Show that $\phi(n) = \phi(n-1) + \phi(n-2)$, and thus show that the sequence $\{\phi(n)\}$ is the Fibonacci sequence 1, 2, 3, 5, 8, 13, $\cdots$, $x$, $y$, $x + y$, $\cdots$.

### 3.9A    An Application of Determinants to Triangles and Tetrahedra

Determinants can be made to play a very useful role in analytic geometry. We give here an interesting example, letting the reader supply certain details.

Take the origin of a rectangular Cartesian coordinate system at the circumcenter of a triangle $(x_1, y_1)$, $(x_2, y_2)$, $(x_3, y_3)$, where the successive vertices appear in counterclockwise order. Denote the area of the triangle by $K$, the circumradius by $R$, and the sides opposite $(x_1, y_1)$, $(x_2, y_2)$, $(x_3, y_3)$ by $a$, $b$, $c$. Then

$$2KR = \begin{vmatrix} x_1 & y_1 & R \\ x_2 & y_2 & R \\ x_3 & y_3 & R \end{vmatrix}, \qquad -2KR = \begin{vmatrix} x_1 & x_2 & x_3 \\ y_1 & y_2 & y_3 \\ -R & -R & -R \end{vmatrix}.$$

Using the product theorem of determinants and using such facts as

$$2(x_1 x_2 + y_1 y_2) - 2R^2 = 2x_1 x_2 + 2y_1 y_2 - (x_1{}^2 + y_1{}^2) \\ - (x_2{}^2 + y_2{}^2)$$

$$= -[(x_1 - x_2)^2 + (y_1 - y_2)^2] = -c^2,$$

one obtains

$$-4K^2 R^2 = -(1/8)\begin{vmatrix} 0 & c^2 & b^2 \\ c^2 & 0 & a^2 \\ b^2 & a^2 & 0 \end{vmatrix} = -(1/4)a^2 b^2 c^2$$

or

$$4RK = abc.$$

That is, *the volume of the right prism having the triangle as base and an altitude equal to four times the circumradius of the triangle is exactly equal to the volume of the rectangular parallelepiped having the sides of the triangle for its three dimensions.*

Stepping up a dimension (see Figure 8), let $(x_1, y_1, z_1)$, $(x_2, y_2, z_2)$, $(x_3, y_3, z_3)$, $(x_4, y_4, z_4)$ be the rectangular Cartesian coordinates of the vertices $P_1$, $P_2$, $P_3$, $P_4$ of a tetrahedron, referred to a coordinate system having its origin at the circumcenter of the tetrahedron. Denote the volume of the tetrahedron by $V$, the circumradius by $R$, and the edges $P_2 P_3$, $P_3 P_1$, $P_1 P_2$, $P_4 P_1$, $P_4 P_2$, $P_4 P_3$ by $a$, $b$, $c$, $d$, $e$, $f$, respectively. By paralleling the preceding treatment for a triangle, the reader may show that

$$-36R^2V^2 = (1/16)\begin{vmatrix} 0 & c^2 & b^2 & d^2 \\ c^2 & 0 & a^2 & e^2 \\ b^2 & a^2 & 0 & f^2 \\ d^2 & e^2 & f^2 & 0 \end{vmatrix},$$

whence

$$36R^2V^2 = S(S - ad)(S - be)(S - cf),$$

where $2S = ad + be + cf$.

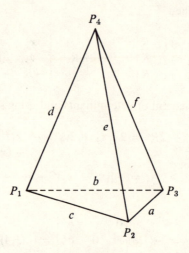

FIG. 8

### 3.10A   Quantitative Aspect of Linear Independence of Vectors

A set of $m$ $n$-dimensional row vectors $V_1, \cdots, V_m$ over a field $F$ is said to be *linearly dependent* in $F$ if and only if there exist scalars $c_1, \cdots, c_m$ of $F$, not all zero, such that

$$c_1V_1 + \cdots + c_mV_m = O;$$

otherwise the set of vectors is said to be *linearly independent* in $F$. Thus the concept of linear dependence and independence of a set of vectors is purely qualitative. In this note we shall supply a quantitative measure of the linear independence of a set of linearly independent real or complex vectors, so that of two such sets of these vectors, we can tell if one set is more, or less, independent than the other. We shall restrict the treatment in the first part of the note to the field $\mathscr{R}$ of real numbers, and then in the second part of the note, generalize the results of the first part to the field $\mathscr{C}$ of complex numbers. Several allied and incidental results will be incorporated. All proofs (some of which admittedly are not easy) will be left to the reader. Details can be found in John W. Royal, "Quantitative aspect of linear independence of vectors," Master of Arts Thesis, University of Maine, June, 1958.

### Real Number Field—Part 1

In this first part, all matrices, vectors, and scalars will be understood as taken over the real number field $\mathscr{R}$.

**I** DEFINITION. Let $A$ and $B$ be two $m \times n$ $(m \leq n)$ matrices. Then we define the *scalar product* of $A$ and $B$, here written as $A \circ B$, to be the sum of all products of corresponding $m$-rowed determinants in $A$ and $B$. Thus

$$A \circ B \equiv \sum_{s_1, \cdots, s_m} d_{s_1 \cdots s_m}(A) \, d_{s_1 \cdots s_m}(B),$$

where $d_{s_1 \cdots s_m}(A)$ represents the determinant formed by the columns $s_1, \cdots, s_m$ $(s_1 < \cdots < s_m)$ of matrix $A$, and the summation is extended over all possible selections of $s_1, \cdots, s_m$.

**II** THEOREM (the generalized Lagrange identity). *If we denote the row vectors of matrix* $A_{(m,n)}$ *by* $A_1, \cdots, A_m$, *and those of matrix* $B_{(m,n)}$ *by* $B_1, \cdots, B_m$, *then*

(1)
$$A \circ B \equiv \begin{vmatrix} A_1 \cdot B_1 & \cdots & A_1 \cdot B_m \\ \cdots & \cdots & \cdots \\ A_m \cdot B_1 & \cdots & A_m \cdot B_m \end{vmatrix},$$

*where* $A_i \cdot B_j$ *denotes the ordinary scalar product of the two row vectors* $A_i$ *and* $B_j$.

**III** REMARKS. The two extreme cases of Theorem II, where $m = 1$ and $m = n$, are of interest. If $m = 1$, then $A$ and $B$ are row vectors, and we have $A \circ B = A \cdot B$. Thus the scalar product of two $m \times n$ matrices is, in this sense, a generalization of the scalar product of two $n$-dimensional vectors.

If $m = n$, then $A \circ B = |A| \, |B|$, and Theorem II says that $|A| \, |B'| = |A| \, |B| = |AB'|$, which is the product theorem for determinants.

If $m = 2$ and $n = 3$, Theorem II reduces to the familiar Lagrange identity of vector analysis, namely,

$$(A_1 \times A_2) \cdot (B_1 \times B_2) \equiv \begin{vmatrix} A_1 \cdot B_1 & A_1 \cdot B_2 \\ A_2 \cdot B_1 & A_2 \cdot B_2 \end{vmatrix},$$

where $A_1 \times A_2$ represents the vector product of the two vectors $A_1$ and $A_2$, etc.

**IV   DEFINITION.**   If $A_1, \cdots, A_m$ are $m$ $n$-dimensional row vectors $(m \leqq n)$, and if $A$ is the matrix having $A_1, \cdots, A_m$ as its successive rows, then we define the *independence* of the vectors $A_1, \cdots, A_m$ to be the scalar

$$I(A_1, \cdots, A_m) \equiv (A \circ A)^{1/2}.$$

**V   REMARK.**   There is a geometrical motivation for Definition IV. The $m$ $n$-dimensional row vectors $A_1, \cdots, A_m$ determine an $m$-dimensional parallelepiped in rectangular Cartesian $n$ space, whose $m$-volume, $P_{mn}$, is defined by

$$(P_{mn})^2 = \begin{vmatrix} A_1 \cdot A_1 & \cdots & A_1 \cdot A_m \\ \cdots & \cdots & \cdots \\ A_m \cdot A_1 & \cdots & A_m \cdot A_m \end{vmatrix}.$$

It follows that $I(A_1, \cdots, A_m)$ is the absolute value of the $m$-volume of the $m$-dimensional parallelepiped determined by the vectors $A_1, \cdots, A_m$ in $n$-space. It is natural to assume that the greater this volume becomes, the more independent we should regard the vectors $A_1, \cdots, A_m$.

**VI   THEOREM.**   $I(A_1, \cdots, A_m) > 0$ *if and only if* $A_1, \cdots, A_m$ *are linearly independent.*

**VII   THEOREM.**   *Let* $A_1, \cdots, A_m$ *be* m *linearly independent* n-*dimensional row vectors* (m $\leqq$ n). *There exists a row vector* $B_1$ *such that* $B_1 \cdot A_i = 0$, i $= 2, \cdots,$ m, *and* $B_1 \cdot B_1 = A_1 \cdot B_1$.

**VIII   THEOREM.**   *If* $A_1, A_2, \cdots, A_m$ *are* m *linearly independent* n-*dimensional row vectors* (m $\leqq$ n), *and if* $B_1$ *is the row vector of Theorem* VII *then*

$$I(B_1, A_2, \cdots, A_m) = I(A_1, A_2, \cdots, A_m).$$

**IX** THEOREM. *If* $A_1, A_2, \cdots, A_m$ *are* m n-*dimensional row vectors* (m $\leq$ n), *then*

$$I(A_1, A_2, \cdots, A_m) \leq (A_1 \cdot A_1)^{1/2} I(A_2, \cdots, A_m).$$

**X** THEOREM (the generalized Hadamard theorem). *If* $A_1, A_2, \cdots, A_m$ *are* m n-*dimensional row vectors* (m $\leq$ n), *then*

$$I(A_1, A_2, \cdots, A_m) \leq (A_1 \cdot A_1)^{1/2} (A_2 \cdot A_2)^{1/2} \cdots (A_m \cdot A_m)^{1/2}.$$

**XI** COROLLARY (Hadamard's theorem). *Suppose that in* $|a_{ij}|_{(n)}$ *we have*

$$absolute\ value\ a_{ij} \leq a,$$

*then*

$$absolute\ value\ |a_{ij}| \leq n^{n/2} a^n.$$

### Complex Number Field—Part 2

In this second part, all matrices, vectors, and scalars will be understood as taken over the complex number field $\mathscr{C}$.

**XII** NOTATION. If $s$ is a scalar, we shall denote its complex conjugate by $\bar{s}$. If $A$ is a matrix or a vector, then by $\bar{A}$ we mean the matrix or vector whose elements are the complex conjugates of the elements of $A$.

**XIII** DEFINITION. Let $X = [x_1, \cdots, x_n]$, $Y = [y_1, \cdots, y_n]$ be two $n$th-order row vectors. We define the *scalar product* of $X$ and $Y$, written $X \cdot Y$, to be

$$X \cdot Y = x_1 \bar{y}_1 + x_2 \bar{y}_2 + \cdots + x_n \bar{y}_n.$$

That is, in matrix notation, $X \cdot Y = X \bar{Y}'$.

It is to be noted that when $X$ and $Y$ are real row vectors, the foregoing definition of the scalar product of $X$ and $Y$ coincides with the definition customarily given for the scalar product of two real row vectors.

**XIV** DEFINITION. Let $A$ and $B$ be two $m \times n$ ($m \leq n$) matrices. We define the *scalar product* of $A$ and $B$, here written as $A \circ B$, to be the sum of all products of corresponding $m$-rowed determinants in $A$ and $\bar{B}$. We note that this definition is a generalization of Definition I.

**XV**  THEOREM.   *If* A *is an* m × n (m ≦ n) *matrix, then* A ∘ A *is a nonnegative real number.*

**XVI**  DEFINITION.   If $A_1, \cdots, A_m$ $(m \leqq n)$ are $m$ $n$-dimensional row vectors and $A$ is the matrix having $A_1, \cdots, A_m$ as its successive rows, then we define the *independence* of the vectors $A_1, \cdots, A_m$ to be the real scalar

$$I(A_1, \cdots, A_m) = (A \circ A)^{1/2}.$$

**XVII**  THEOREM.   *Theorems VI, VII, VIII, IX, X, XI continue to hold in the complex number field* $\mathscr{C}$.

### 3.11A   Sylvester's Dialytic Method of Elimination

Let

$$f(x) \equiv a_0 x^m + a_1 x^{m-1} + \cdots + a_m = 0, \qquad a_0 \neq 0,$$

$$g(x) \equiv b_0 x^n + b_1 x^{n-1} + \cdots + b_n = 0, \qquad b_0 \neq 0,$$

be two polynomial equations in $x$. An important question that can arise is: Do the equations $f(x) = 0$ and $g(x) = 0$ have a common root? In 1840, J. J. Sylvester established the following attractive theorem.

THEOREM.   *The two polynomial equations* f(x) = 0 *and* g(x) = 0 *have a common root if and only if the* (m + n) × (m + n) *determinant*

$$
\left|
\begin{array}{ccccccccccccc}
a_0 & a_1 & a_2 & \cdots & & \cdots & & \cdots & a_{m-1} & a_m & 0 & \cdots & 0 \\
0 & a_0 & a_1 & \cdots & & \cdots & & \cdots & a_{m-2} & a_{m-1} & a_m & \cdots & 0 \\
\cdots & \cdots & \cdots & \cdots & & \cdots & & \cdots & \cdots & \cdots & \cdots & \cdots & \cdots \\
0 & 0 & 0 & \cdots & a_0 & a_1 & a_2 & \cdots & & & & \cdots & a_m \\
b_0 & b_1 & b_2 & \cdots & b_{n-1} & b_n & 0 & \cdots & 0 & 0 & 0 & \cdots & 0 \\
0 & b_0 & b_1 & \cdots & b_{n-2} & b_{n-1} & b_n & \cdots & 0 & 0 & 0 & \cdots & 0 \\
\cdots & \cdots & \cdots & \cdots & \cdots & \cdots & \cdots & \cdots & \cdots & \cdots & \cdots & \cdots & \cdots \\
0 & 0 & 0 & \cdots & 0 & 0 & 0 & \cdots & b_0 & b_1 & b_2 & \cdots & b_n
\end{array}
\right|
\begin{array}{l}
\left.\rule{0pt}{30pt}\right\} \begin{array}{l} n \\ rows \end{array} \\
\left.\rule{0pt}{30pt}\right\} \begin{array}{l} m \\ rows \end{array}
\end{array}
$$

*is equal to zero.*

The establishment of this theorem makes a good "junior" research project, and the interested student may consult an appropriate textbook in the theory of equations.

# 4. MATRICES WITH POLYNOMIAL ELEMENTS

*4.1. Review of some polynomial theory. Problems.   4.2. Lambda matrices. Problems.   4.3. The Smith normal form. Problems.   4.4 Invariant factors and elementary divisors. Problems.   4.5. The characteristic function of a square matrix. Problems.   4.6. Some results related to the characteristic function of a square matrix. Problems. 4.7. Characteristic vectors of a square matrix. Problems.   4.8. The minimum function of a square matrix. Problems.   4.9. Finding the minimum function of a square matrix. Problems. ADDENDA.   4.1A. Elementary λ matrices. 4.2A. Systems of linear differential equations with constant coefficients.   4.3A. Equivalence of pairs of matrices.   4.4A. kth roots of nonsingular matrices.   4.5A. The coefficients in the characteristic function.   4.6A. Computation of $A^{-1}$ by the Hamilton-Cayley equation.   4.7A. Frame's recursion formula for inverting a matrix.   4.8A. Characteristic roots of a polynomial function of a matrix A.*

## 4.1  Review of some polynomial theory

In this section we recall some theory of polynomials in one variable. This theory, which the student has probably encountered in earlier mathematics courses, will be useful to us in subsequent sections of the present chapter. Since the material is largely of a review nature, the proofs of some of the theorems will be omitted.

*179*

**4.1.1** DEFINITIONS AND NOTATION.   An expression of the form

$$f(\lambda) = a_n\lambda^n + a_{n-1}\lambda^{n-1} + \cdots + a_1\lambda + a_0,$$

where the coefficients $a_i$ are complex numbers, is called a *complex polynomial in $\lambda$*. If every $a_i = 0$, the polynomial is called the *zero polynomial* and we write $f(\lambda) \equiv 0$.

Henceforth in this chapter, unless explicitly stated otherwise, we shall assume all polynomials in $\lambda$ to be *complex* polynomials in $\lambda$. In Chapter 6 we shall consider an extension of much of the theory of the present chapter to polynomials in $\lambda$ where the coefficients of the polynomials are elements chosen from a more general appropriate set $S$.

**4.1.2** DEFINITION AND NOTATION.   Let

$$f(\lambda) = a_n\lambda^n + a_{n-1}\lambda^{n-1} + \cdots + a_1\lambda + a_0$$

and

$$g(\lambda) = b_m\lambda^m + b_{m-1}\lambda^{m-1} + \cdots + b_1\lambda + b_0$$

be two polynomials. Then we say $f(\lambda)$ and $g(\lambda)$ are *equal,* and we write $f(\lambda) = g(\lambda)$, if and only if $m = n$ and $b_i = a_i$ for $i = 0, \cdots, n$.

**4.1.3** DEFINITIONS.   Let

$$f(\lambda) = a_n\lambda^n + a_{n-1}\lambda^{n-1} + \cdots + a_1\lambda + a_0$$

be a polynomial. The integer $n$ is called the *virtual degree* of $f(\lambda)$. If $a_n \neq 0$, $n$ is called the *degree* of $f(\lambda)$. It follows that any polynomial of degree $n$ can be written as a polynomial of virtual degree any integer $m \geq n$, and we may speak of any such $m$ as *a* virtual degree of $f(\lambda)$. In particular, the polynomial $f(\lambda) = a_0 \neq 0$ is said to be of *degree zero*. The zero polynomial will be said to be of *degree minus infinity* $(-\infty)$. The coefficient $a_n$ is called the *virtual leading coefficient* of $f(\lambda)$ and is called the *leading coefficient* of $f(\lambda)$ if and only if it is not zero. If $a_n = 1$, $f(\lambda)$ is called a *monic polynomial.*

We assume the reader knows how to add and to multiply two polynomials in $\lambda$. The following four theorems then present no difficulty.

**4.1.4** THEOREM.   *If* $f(\lambda)$ *and* $g(\lambda)$ *are two polynomials, the degree of* $f(\lambda) + g(\lambda)$ *is at most the larger of the two degrees of* $f(\lambda)$ *and* $g(\lambda)$.

**4.1.5** THEOREM. *If* $f(\lambda)$ *and* $g(\lambda)$ *are two polynomials, the degree of* $f(\lambda)g(\lambda)$ *is the sum of the degrees of* $f(\lambda)$ *and* $g(\lambda)$. *The leading coefficient of* $f(\lambda)g(\lambda)$ *is the product of the leading coefficients of* $f(\lambda)$ *and* $g(\lambda)$, *and thus, if* $f(\lambda)$ *and* $g(\lambda)$ *are monic, so is* $f(\lambda)g(\lambda)$.

**4.1.6** THEOREM. *The product of two nonzero polynomials is a nonzero polynomial and is a constant if and only if both polynomials are constants.*

**4.1.7** THEOREM. *If* $f(\lambda)$, $g(\lambda)$, $h(\lambda)$ *are three polynomials, if* $f(\lambda) \not\equiv 0$, *and if* $f(\lambda)g(\lambda) = f(\lambda)h(\lambda)$, *then* $g(\lambda) = h(\lambda)$.

**4.1.8** THEOREM (the division algorithm). *Let* $f(\lambda)$ *and* $g(\lambda)$ *be two polynomials of degrees* n *and* m, *respectively, and suppose* $g(\lambda) \not\equiv 0$. *Then there exist unique polynomials* $q(\lambda)$ *and* $r(\lambda)$ *such that* $r(\lambda)$ *has virtual degree* m − 1, $q(\lambda)$ *is either the zero polynomial or has degree* n − m, *and*

$$f(\lambda) = q(\lambda)g(\lambda) + r(\lambda).$$

Let

$$f(\lambda) = a_n\lambda^n + a_{n-1}\lambda^{n-1} + \cdots + a_1\lambda + a_0, \qquad a_n \neq 0,$$

and

$$g(\lambda) = b_m\lambda^m + b_{m-1}\lambda^{m-1} + \cdots + b_1\lambda + b_0, \qquad b_m \neq 0.$$

If $n < m$, we have

$$f(\lambda) = 0 \cdot g(\lambda) + f(\lambda),$$

which is a representation of the desired form. Suppose, then, that $n \geq m$ and form the difference

$$f_1(\lambda) = f(\lambda) - (a_n/b_m)\lambda^{n-m}g(\lambda).$$

Now $f_1(\lambda)$ is a polynomial of degree less than $n$. We construct a proof by mathematical induction and assume the division algorithm is true for all polynomials of degree less than $n$. Since $f_1(\lambda)$ is such a polynomial, we have

$$f_1(\lambda) = q_1(\lambda)g(\lambda) + r(\lambda),$$

where $r(\lambda)$ has virtual degree $m - 1$ and $q_1(\lambda)$ has degree less than $n - m$. Then

$$f(\lambda) - (a_n/b_m)\lambda^{n-m}g(\lambda) = q_1(\lambda)g(\lambda) + r(\lambda)$$

or

$$f(\lambda) = [(a_n/b_m)\lambda^{n-m} + q_1(\lambda)]g(\lambda) + r(\lambda)$$
$$= q(\lambda)g(\lambda) + r(\lambda),$$

which is a representation of $f(\lambda)$ of the desired form.

All that remains to be proved is that $q(\lambda)$ and $r(\lambda)$ are unique. Suppose there is a second representation,

$$f(\lambda) = q'(\lambda)g(\lambda) + r'(\lambda),$$

where $r'(\lambda)$ has virtual degree $m - 1$ and $q'(\lambda)$ is either the zero polynomial or has degree $n - m$. Then

$$q'(\lambda)g(\lambda) + r'(\lambda) = q(\lambda)g(\lambda) + r(\lambda)$$

or

$$g(\lambda)[q'(\lambda) - q(\lambda)] = r(\lambda) - r'(\lambda).$$

The right side of the last equation has virtual degree $m - 1$. Hence, unless $q'(\lambda) - q(\lambda) = 0$, we have a contradiction. It follows that $q'(\lambda) = q(\lambda)$ and $r(\lambda) = r'(\lambda)$.

**4.1.9**   DEFINITIONS.   The unique polynomials $q(\lambda)$ and $r(\lambda)$ of the division algorithm are called the *quotient* and the *remainder*, respectively, when polynomial $f(\lambda)$ is divided by polynomial $g(\lambda)$.

**4.1.10**   COROLLARY (the remainder theorem).   *The remainder when a polynomial* $f(\lambda)$ *is divided by* $\lambda - a$, *where* $a$ *is a number, is the number* $f(a)$.

If we replace $g(\lambda)$ in the division algorithm by $\lambda - a$, the remainder becomes a number $r$ and we have

$$f(\lambda) = q(\lambda)(\lambda - a) + r,$$

whence

$$f(a) = q(a)(a - a) + r = r.$$

**4.1.11**   COROLLARY (the factor theorem).   *A polynomial* $f(\lambda)$ *is divisible by* $\lambda - a$ *if and only if* $f(a) = 0$.

By the remainder theorem we have

$$f(\lambda) = q(\lambda)(\lambda - a) + f(a),$$

whence $f(\lambda)$ has $\lambda - a$ as a factor if and only if $f(a) = 0$.

We conclude the section with a definition of a concept that will be needed later in the chapter.

**4.1.12** DEFINITION. The *greatest common divisor* of polynomials $f_1(\lambda), \cdots, f_t(\lambda)$, not all zero, is a monic polynomial $d(\lambda)$, which divides all the $f_i(\lambda)$ and is such that if polynomial $g(\lambda)$ divides every $f_i(\lambda)$, then $g(\lambda)$ divides $d(\lambda)$.

We leave it to the reader to assure himself that a set of polynomials, not all zero, has a unique greatest common divisor.

## PROBLEMS

**4.1–1** Show, by examining the statements of Theorems 4.1.4, 4.1.5, 4.1.7, that the reason for defining the degree of the zero polynomial to be minus infinity is to ensure that certain theorems on polynomials shall hold without exception.

**4.1–2** Supply proofs for Theorems 4.1.4, 4.1.5, and 4.1.6.

**4.1–3** Establish Theorem 4.1.7 by an indirect argument.

**4.1–4** Let $f(\lambda)$ and $g(\lambda)$ be two polynomials in $\lambda$.

(a) Under what conditions will the degree of $f(\lambda) + g(\lambda)$ be less than the degree of either $f(\lambda)$ or $g(\lambda)$?

(b) What can one say about the degree of $f(\lambda) + g(\lambda)$ if $f(\lambda)$ and $g(\lambda)$ have either both positive or both negative leading coefficients?

(c) What can one say about the degree of $[f(\lambda)]^k$, where $k$ is a positive integer?

**4.1–5** (a) State a result about the degree and the leading coefficient of the polynomial

$$h(\lambda) = [f_1(\lambda)]^2 + \cdots + [f_k(\lambda)]^2$$

if $k$ is a positive integer and each $f_i(\lambda)$ is a *real* polynomial.

(b) State a result about the degree and the leading coefficient of the polynomial $g(\lambda)h(\lambda)$, where $g(\lambda)$ is a *real* polynomial of odd degree and $h(\lambda)$ is the polynomial of part (a).

**4.1–6** Let $f(\lambda)$ and $g(\lambda)$ be two polynomials in $\lambda$. Show that if any one of the following relations holds, then both $f(\lambda)$ and $g(\lambda)$ are zero polynomials:

(a) $[f(\lambda)]^2 + x[g(\lambda)]^2 = 0$.

(b) $[f(\lambda)]^3 - x^2[g(\lambda)]^3 = 0$.

(c) $[f(\lambda)]^4 + 2x[f(\lambda)]^2[g(\lambda)]^2 - x(x - 1)[g(\lambda)]^4 = 0.$

(d) $[f(\lambda)]^2 + 2xf(\lambda)g(\lambda) - x[g(\lambda)]^2 = 0.$

**4.1-7** Show that if $f(\lambda)$, $g(\lambda)$, $h(\lambda)$ are *real* polynomials such that

$$[f(\lambda)]^2 - x[g(\lambda)]^2 + [h(\lambda)]^2 = 0,$$

then $f(\lambda)$, $g(\lambda)$, and $h(\lambda)$ are all zero polynomials.

**4.1-8** Let $f(\lambda)$ be a polynomial of degree $n$ with integral coefficients and let $c$ be a root of the equation $f(\lambda) = 0$. Show that any polynomial $g(c)$ with rational coefficients can be expressed as a polynomial in $c$ with rational coefficients and of degree no higher than $n - 1$.

**4.1-9** Show that a necessary and sufficient condition for the number $c$ to be a root of multiplicity $m$ of the equation $f(\lambda) = 0$, where $f(\lambda)$ is a polynomial, is that $c$ be a root of multiplicity $m - 1$ of the equation $f'(\lambda) = 0$, where $f'(\lambda)$ is the derivative with respect to $\lambda$ of $f(\lambda)$.

**4.1-10** Show that all the theorems and proofs of Section 4.1 continue to hold if (a) all numbers are real numbers and all polynomials have real coefficients; (b) all numbers are rational numbers and all polynomials have rational coefficients.

**4.1-11** (a) Prove that a set of polynomials, not all zero, has a unique greatest common divisor.

(b) Show that the greatest common divisor of a set of polynomials, not all zero, is the monic common polynomial divisor of the set of largest possible degree.

**4.1-12** If $d_k(\lambda)$ is the greatest common divisor of the polynomials $f_1(\lambda) \cdots, f_k(\lambda)$, not all zero, and if $d_0(\lambda)$ is the greatest common divisor of $d_k(\lambda)$ and polynomial $f_{k+1}(\lambda)$, prove that $d_0(\lambda)$ is the greatest common divisor of the polynomials $f_1(\lambda), \cdots, f_{k+1}(\lambda)$. (This result reduces the problems of the existence and the calculation of the greatest common divisor of any number of polynomials in $\lambda$, not all zero, to the case of two nonzero polynomials.)

## 4.2    Lambda matrices

Much of the theory of polynomials in one variable can be extended to polynomials in one variable having $n \times n$ matrices as coefficients. Some of this extended theory will be considered in this section.

**4.2.1** DEFINITION.    An $m \times n$ matrix $A(\lambda) = [a_{ij}(\lambda)]_{(m,n)}$, whose elements are polynomials in $\lambda$, will be called a $\lambda$ *matrix*.

We now state an obvious theorem.

**4.2.2**   THEOREM.   *An* m × n *λ matrix* A(λ) *can be expressed as a matric polynomial in the scalar variable λ,*

$$A_p\lambda^p + A_{p-1}\lambda^{p-1} + \cdots + A_1\lambda + A_0,$$

*where* p *is the maximum degree in λ of the polynomials* $a_{ij}(\lambda)$, *the* $A_i$ *are all* m × n *matrices, and* $A_p \neq O_{(m,n)}$.

*Example*

$$A(\lambda) = \begin{bmatrix} \lambda^3 - 4 & 2\lambda - 1 \\ \lambda^2 + \lambda + 1 & \lambda^4 + \lambda^2 + 5\lambda \end{bmatrix}$$

$$= \begin{bmatrix} 0 & 0 \\ 0 & 1 \end{bmatrix}\lambda^4 + \begin{bmatrix} 1 & 0 \\ 0 & 0 \end{bmatrix}\lambda^3 + \begin{bmatrix} 0 & 0 \\ 1 & 1 \end{bmatrix}\lambda^2 + \begin{bmatrix} 0 & 2 \\ 1 & 5 \end{bmatrix}\lambda + \begin{bmatrix} -4 & -1 \\ 1 & 0 \end{bmatrix}.$$

**4.2.3**   DEFINITIONS.   Let $A(\lambda)$ be an $n \times n$ *λ* matrix, where (as in Theorem 4.2.2)

$$A(\lambda) = A_p\lambda^p + A_{p-1}\lambda^{p-1} + \cdots + A_1\lambda + A_0, \qquad A_p \neq O_{(n)}.$$

Then the polynomial on the right is called the *matric-polynomial representation* of $A(\lambda)$; $A(\lambda)$ is said to be *singular* or *nonsingular*, according as $|A(\lambda)|$ is or is not the zero polynomial; $A(\lambda)$ is said to be *proper* or *improper*, according as the matrix $A_p$ is or is not singular; $p$ is called the *degree* of $A(\lambda)$; $A_p$ is called the *leading coefficient* of $A(\lambda)$.

*Example.*   In the example following Theorem 4.2.2, $A(\lambda)$ is non-singular, improper, of degree 4, and with leading coefficient $\begin{bmatrix} 0 & 0 \\ 0 & 1 \end{bmatrix}$.

The proofs of the next two theorems present no difficulties.

**4.2.4**   THEOREM.   *If* A(λ) *and* B(λ) *are two* m × n *λ matrices, then* A(λ) + B(λ) *is a λ matrix of degree at most the larger of the two degrees of* A(λ) *and* B(λ).

**4.2.5**   THEOREM.   *If* A(λ) *and* B(λ) *are two* n × n *λ matrices of degrees* p *and* q, *respectively, then* A(λ)B(λ) *is a λ matrix of degree at most* p + q. *If the leading coefficients of* $A(\lambda)$ *and* $B(\lambda)$ *are* $A_p$ *and* $B_q$, *respectively, and if* $A_pB_q \neq O_{(n)}$, *then the degree of* $A(\lambda)B(\lambda)$ *is exactly* p + q *and the leading coefficient of* $A(\lambda)B(\lambda)$ *is* $A_pB_q$.

**4.2.6**   THEOREM (the division algorithm).   *Let* $A(\lambda)$ *and* $B(\lambda)$ *be two* n × n λ *matrices of degrees* p *and* q, *respectively, and suppose* $B(\lambda)$ *is proper. Then there exist unique* n × n λ *matrices* $Q_1(\lambda)$, $R_1(\lambda)$, $Q_2(\lambda)$, $R_2(\lambda)$ *such that* $R_1(\lambda)$ *and* $R_2(\lambda)$ *have degrees at most* q − 1 *and*

$$A(\lambda) = Q_1(\lambda)B(\lambda) + R_1(\lambda), \qquad A(\lambda) = B(\lambda)Q_2(\lambda) + R_2(\lambda).$$

*Moreover, if* p < q, *then* $Q_1(\lambda) = Q_2(\lambda) = O_{(n)}$, *while if* p ≧ q, *then* $Q_1(\lambda)$ *and* $Q_2(\lambda)$ *each has degree* p − q.

Let

$$A(\lambda) = A_p\lambda^p + A_{p-1}\lambda^{p-1} + \cdots + A_1\lambda + A_0, \qquad A_p \neq O_{(n)},$$

and

$$B(\lambda) = B_q\lambda^q + B_{q-1}\lambda^{q-1} + \cdots + B_1\lambda + B_0, \qquad |B_q| \neq 0.$$

We first show that there exist representations of the desired form and then we show that the representations are unique.

If $p < q$ we have

$$A(\lambda) = O_{(n)}B(\lambda) + A(\lambda), \qquad A(\lambda) = B(\lambda)O_{(n)} + A(\lambda),$$

which are representations of the desired form.

Suppose that $p \geq q$ and form the difference

$$A_1(\lambda) \equiv A(\lambda) - A_pB_q^{-1}B(\lambda)\lambda^{p-q}.$$

Now $A_1(\lambda)$ is an $n \times n$ λ matrix of degree less than $p$. We construct a proof by mathematical induction and assume the division algorithm is true for all $n \times n$ λ matrices of degree less than $p$. Since $A_1(\lambda)$ is such a λ matrix, we have

$$A_1(\lambda) = S_1(\lambda)B(\lambda) + R_1(\lambda),$$

where $S_1(\lambda)$ and $R_1(\lambda)$ are $n \times n$ λ matrices such that $R_1(\lambda)$ has degree at most $q - 1$ and $S_1(\lambda)$ has degree less than $p - q$. Then

$$A(\lambda) - A_pB_q^{-1}B(\lambda)\lambda^{p-q} = S_1(\lambda)B(\lambda) + R_1(\lambda)$$

or

$$A(\lambda) = [A_pB_q^{-1}\lambda^{p-q} + S_1(\lambda)]B(\lambda) + R_1(\lambda)$$

$$= Q_1(\lambda)B(\lambda) + R_1(\lambda),$$

which is a representation of $A(\lambda)$ of the first desired form. Similarly, by considering the difference

$$A_2(\lambda) \equiv A(\lambda) - B(\lambda)B_q^{-1}A_p\lambda^{p-q},$$

we can show that $A(\lambda)$ has a representation of the second desired form.

To prove that the representations are unique, suppose first that there is a second representation

$$A(\lambda) = Q'_1(\lambda)B(\lambda) + R'_1(\lambda),$$

where $Q'_1(\lambda)$ and $R'_1(\lambda)$ are $n \times n$ $\lambda$ matrices such that $R'_1(\lambda)$ has degree at most $q - 1$ and $Q'_1(\lambda)$ is either zero or has degree $p - q$. Then

$$Q'_1(\lambda)B(\lambda) + R'_1(\lambda) = Q_1(\lambda)B(\lambda) + R_1(\lambda)$$

or

$$[Q'_1(\lambda) - Q_1(\lambda)]B(\lambda) = R_1(\lambda) - R'_1(\lambda).$$

The right side of the last equation has degree at most $q - 1$. Hence, unless $Q'_1(\lambda) - Q_1(\lambda) = O_{(n)}$, we have a contradiction. It follows that $Q'_1(\lambda) = Q_1(\lambda)$ and $R_1(\lambda) = R'_1(\lambda)$, and the representation of the first form is unique. One can similarly show that the representation of the second form is unique.

**4.2.7** DEFINITIONS.   The first representation of $A(\lambda)$ in the statement of Theorem 4.2.6 is called the *right division* of $A(\lambda)$ by $B(\lambda)$; the second representation is called the *left division* of $A(\lambda)$ by $B(\lambda)$. We then speak of $Q_1(\lambda)$ and $R_1(\lambda)$ as the *right quotient* and *right remainder*, and of $Q_2(\lambda)$ and $R_2(\lambda)$ as the *left quotient* and *left remainder*. If $R_1(\lambda) = O_{(n)}$, we call $B(\lambda)$ a *right divisor* of $A(\lambda)$; if $R_2(\lambda) = O_{(n)}$, we call $B(\lambda)$ a *left divisor* of $A(\lambda)$.

**4.2.8** DEFINITIONS AND NOTATION.   Let $A(\lambda)$ be an $n \times n$ $\lambda$ matrix and let

$$A_p\lambda^p + A_{p-1}\lambda^{p-1} + \cdots + A_1\lambda + A_0$$

be its matric polynomial representation. Let $C$ be an $n \times n$ matrix. Then we define $A_R(C)$, called the *right functional value of* $A(\lambda)$ *for* C, and $A_L(C)$, called the *left functional value of* $A(\lambda)$ *for* C, by

$$A_R(C) = A_pC^p + A_{p-1}C^{p-1} + \cdots + A_1C + A_0$$

and

$$A_L(C) = C^p A_p + C^{p-1} A_{p-1} + \cdots + CA_1 + A_0.$$

**4.2.9**  THEOREM (the remainder theorem).  *Let* A($\lambda$) *be an* n × n $\lambda$ *matrix and let* C *be an* n × n *matrix. Then the right and left remainders on division of* A($\lambda$) *by* $\lambda I_{(n)} - C$ *are* $A_R(C)$ *and* $A_L(C)$, *respectively.*

In Theorem 4.2.6, replace $B(\lambda)$ by $\lambda I - C$, which is a proper $n \times n$ $\lambda$ matrix. The right division then becomes

$$A(\lambda) = Q_1(\lambda)(\lambda I - C) + R_1,$$

where $R_1$ is a matrix and $Q_1(\lambda)$ is a $\lambda$ matrix of degree $p - 1$. Suppose

$$Q_1(\lambda) = S_{p-1}\lambda^{p-1} + S_{p-2}\lambda^{p-2} + \cdots + S_1\lambda + S_0.$$

Then

$$\begin{aligned}
H(\lambda) &\equiv Q_1(\lambda)(\lambda I - C) \\
&= (S_{p-1}\lambda^p + S_{p-2}\lambda^{p-1} + \cdots + S_1\lambda^2 + S_0\lambda) \\
&\quad - (S_{p-1}C\lambda^{p-1} + S_{p-2}C\lambda^{p-2} + \cdots + S_1C\lambda + S_0C) \\
&= S_{p-1}\lambda^p + (S_{p-2} - S_{p-1}C)\lambda^{p-1} + \cdots + (S_0 - S_1C)\lambda - S_0C.
\end{aligned}$$

If $D$ is any $n \times n$ matrix, we have

$$H_R(D) = S_{p-1}D^p + (S_{p-2} - S_{p-1}C)D^{p-1} + \cdots + (S_0 - S_1C)D - S_0C,$$

while

$$\begin{aligned}
Q_{1R}(D)(DI - C) &= (S_{p-1}D^{p-1} + S_{p-2}D^{p-2} + \cdots + S_1D + S_0)(D-C) \\
&= (S_{p-1}D^p + S_{p-2}D^{p-1} + \cdots + S_1D^2 + S_0D) \\
&\quad - (S_{p-1}D^{p-1}C + S_{p-2}D^{p-2}C \\
&\quad\quad\quad\quad\quad + \cdots + S_1DC + S_0C) \\
&= S_{p-1}D^p + (S_{p-2}D^{p-1} - S_{p-1}D^{p-1}C) \\
&\quad\quad\quad\quad\quad + \cdots + (S_0D - S_1DC) - S_0C.
\end{aligned}$$

Now matrices $H_R(D)$ and $Q_{1R}(D)(D - C)$ are equal in general if and only if $D$ and $C$ commute. They are equal if $D = C$, and thus

$$A(\lambda) = H(\lambda) + R_1$$

implies that

$$A_R(C) = H_R(C) + R_1 = Q_{1R}(C)(C - C) + R_1 = R_1,$$

and it follows that the right remainder, when $A(\lambda)$ is divided by $\lambda I - C$, is

$A_R(C)$. It can be shown similarly that the left remainder, when $A(\lambda)$ is divided by $\lambda I - C$, is $A_L(C)$.

**4.2.10** THEOREM (the factor theorem). *The* n × n $\lambda$ *matrix* A($\lambda$) *has* $\lambda I_{(n)} - C$ *as a right divisor if and only if* $A_R(C) = O_{(n)}$; *it has* $\lambda I_{(n)} - C$ *as a left divisor if and only if* $A_L(C) = O_{(n)}$.

This is an immediate consequence of Theorem 4.2.9.

## PROBLEMS

**4.2–1** (a) Find the matric representation, degree, and leading coefficient of

$$A(\lambda) = \begin{bmatrix} \lambda^2 & \lambda + 1 \\ \lambda - 2 & \lambda^2 + 2 \end{bmatrix}.$$

(b) Is $A(\lambda)$ singular or nonsingular?
(c) Is $A(\lambda)$ proper or improper?
(d) Find both right and left functional values of $A(\lambda)$ for

$$C = \begin{bmatrix} 1 & 2 \\ 3 & 4 \end{bmatrix}.$$

**4.2–2** Prove Theorem 4.2.4.
**4.2–3** Prove Theorem 4.2.5.
**4.2–4** Carry out the details in the proof of Theorem 4.2.6 for the left division of $A(\lambda)$ by $B(\lambda)$.
**4.2–5** Carry out the details in the proof of Theorem 4.2.7 for the left remainder when $A(\lambda)$ is divided by $\lambda I - C$.
**4.2–6** Supply a proof for Theorem 4.2.10.
**4.2–7** Let $A(\lambda)$ and $B(\lambda)$ be two $n \times n$ $\lambda$ matrices and let $C$ be an $n \times n$ matrix. Set

$$N(\lambda) \equiv A(\lambda) + B(\lambda) \qquad \text{and} \qquad M(\lambda) \equiv A(\lambda)B(\lambda).$$

(a) Show that $N_R(C) = A_R(C) + B_R(C)$, $N_L(C) = A_L(C) + B_L(C)$.
(b) Show that we do not necessarily have $M_R(C) = A_R(C)B_R(C)$ or $M_L(C) = A_L(C)B_L(C)$.
**4.2–8** Express

$$A(\lambda) = \begin{bmatrix} \lambda^3 + 5\lambda + 1 & 3\lambda^3 + \lambda - 1 \\ 2\lambda^3 + \lambda^2 + 2 & 4\lambda^3 + 2\lambda + 2 \end{bmatrix}$$

and

$$B(\lambda) = \begin{bmatrix} 2\lambda^2 - 1 & \lambda^2 \\ 3\lambda^2 & 2\lambda^2 \end{bmatrix}$$

as matric polynomials, and then (by following the familiar pattern for division of scalar polynomials) perform the right and the left divisions of $A(\lambda)$ by $B(\lambda)$, and thus find $Q_1(\lambda)$, $R_1(\lambda)$, $Q_2(\lambda)$, $R_2(\lambda)$ of Theorem 4.2.6.

**4.2–9**    By taking

$$A(\lambda) = \begin{bmatrix} 2\lambda^2 + 2 & \lambda^2 + 2 \\ -3\lambda & -\lambda \end{bmatrix}, \qquad B(\lambda) = \begin{bmatrix} \lambda & 2 \\ -2 & \lambda \end{bmatrix},$$

show that division may be exact on one side while not on the other.

**4.2–10**    Taking

$$A(\lambda) = \begin{bmatrix} 2\lambda^4 - \lambda^2 + 2 & -\lambda^3 + \lambda - 1 & 1 - \lambda^2 \\ \lambda^3 - \lambda + 1 & -\lambda^4 + \lambda^2 - 2 & 1 + \lambda^2 \\ \lambda^2 - 1 & -1 - \lambda & \lambda^4 + \lambda^2 - 1 \end{bmatrix},$$

find $A_R(C)$ and $A_L(C)$ by both substitution and synthetic division if

$$C = \begin{bmatrix} 0 & 1 & 0 \\ 0 & 0 & 1 \\ 2 & 0 & 0 \end{bmatrix}.$$

**4.2–11**    An $n \times n$ $\lambda$ matrix is said to be *scalar* if and only if all elements along the principal diagonal are the same polynomial in $\lambda$ and all other elements are zero polynomials.

(**a**) If $A(\lambda)$ is an $n \times n$ scalar $\lambda$ matrix, show that the right functional value of $A(\lambda)$ for an $n \times n$ matrix $C$ is equal to the left functional value of $A(\lambda)$ for $C$.

(**b**) Show that an $n \times n$ $\lambda$ matrix $A(\lambda)$ is divisible by a nonzero scalar $\lambda$ matrix $f(\lambda)I_{(n)}$ if and only if each element $a_{ij}(\lambda)$ of $A(\lambda)$ is divisible by $f(\lambda)$.

**4.2–12**    If $A(\lambda)$ and $B(\lambda)$ are $n \times n$ proper $\lambda$ matrices of degrees $p$ and $q$, respectively, and if $D(\lambda)$ is any nonzero $n \times n$ $\lambda$ matrix, show that the $\lambda$ matrix $A(\lambda)D(\lambda)B(\lambda)$ has degree at least $p + q$.

## 4.3    The Smith normal form

In this section we shall obtain a unique normal form to which any given $\lambda$ matrix may be reduced by analogs of the elementary operations of

Chapter 2. Since the form appears in an 1861–1862 paper of Henry John Stephen Smith (1826–1883), the form has come to be known as *the Smith normal form*. We commence by extending to $\lambda$ matrices some concepts introduced earlier in connection with numerical matrices.

**4.3.1** DEFINITION. By an *elementary $\lambda$ operation* on a $\lambda$ matrix we understand any one of the following:

(1) The interchange of any two rows or of any two columns.

(2) The multiplication of a row, or of a column, by a nonzero number.

(3) The addition to a row (column) of the product of another row (column) by a polynomial in $\lambda$.

**4.3.2** DEFINITION AND NOTATION. Let $A(\lambda)$ and $B(\lambda)$ be two $m \times n$ $\lambda$ matrices. We then say $A(\lambda)$ is *$\lambda$-equivalent* to $B(\lambda)$, and we write

$$A(\lambda) \overset{\lambda E}{=} B(\lambda),$$

if it is possible to pass from $A(\lambda)$ to $B(\lambda)$ by a finite sequence of elementary $\lambda$ operations.

The reader can easily show that the inverse of an elementary $\lambda$ operation is an elementary $\lambda$ operation, and that $\lambda$ equivalence is reflexive, symmetric, and transitive.

**4.3.3** DEFINITION. A nonzero $\lambda$ matrix $A(\lambda)$ is said to be of *rank r* if $r$ is the largest integer such that not all minors of $A(\lambda)$ of order $r$ are identically zero. A zero $\lambda$ matrix is said to be of *rank zero*.

**4.3.4** THEOREM. *If* $A(\lambda) \overset{\lambda E}{=} B(\lambda)$, *then rank* $A(\lambda) = $ *rank* $B(\lambda)$.

Every $t$-rowed minor of $B(\lambda)$ is either a $t$-rowed minor of $A(\lambda)$, the product of a $t$-rowed minor of $A(\lambda)$ by a nonzero number, or a sum

$$M_1(\lambda) + f(\lambda)M_2(\lambda),$$

where $M_1(\lambda)$ and $M_2(\lambda)$ are $t$-rowed minors of $A(\lambda)$ and $f(\lambda)$ is a polynomial in $\lambda$. It follows that rank $B(\lambda)$ cannot exceed rank $A(\lambda)$. But we also have $B(\lambda) \overset{\lambda E}{=} A(\lambda)$, whence rank $A(\lambda)$ cannot exceed rank $B(\lambda)$. We conclude that rank $A(\lambda) = $ rank $B(\lambda)$.

**4.3.5**   THEOREM.   *A nonzero $\lambda$ matrix $A(\lambda)$ of rank r is $\lambda$-equivalent to a $\lambda$ matrix of one of the following forms:*

$$G, \qquad \begin{bmatrix} G & O \\ O & O \end{bmatrix}, \qquad \begin{bmatrix} G \\ O \end{bmatrix}, \qquad [G \quad O],$$

*where* $G = diag\{f_1(\lambda), f_2(\lambda), \cdots, f_r(\lambda)\}$, *the polynomials* $f_i(\lambda)$ *are monic, and* $f_i(\lambda)$ *divides* $f_{i+1}(\lambda)$ *for* $i = 1, 2, \cdots, r - 1$.

The elements of all matrices $\lambda$-equivalent to $A(\lambda)$ are polynomials in $\lambda$, and in the set of all such polynomials there is a nonzero polynomial $f_1(\lambda)$ of lowest degree. By applying elementary $\lambda$ operations of types 2 and 1 in Definition 4.3.1, we may assume that $f_1(\lambda)$ is monic and is in the top left corner of a matrix $A_1(\lambda)$, $\lambda$-equivalent to $A(\lambda)$. Let $f(\lambda)$ be any other element in the first row or the first column of $A_1(\lambda)$. Then, by the division algorithm (Theorem 4.1.8),

$$f(\lambda) = q(\lambda)f_1(\lambda) + r(\lambda),$$

where $r(\lambda)$ is a polynomial of lower degree than $f_1(\lambda)$. Applying an elementary $\lambda$ operation of type 3 to $A_1(\lambda)$, we may obtain a $\lambda$ matrix containing $r(\lambda)$ as an element. By the definition of $f_1(\lambda)$, this is possible only if $r(\lambda) \equiv 0$. It follows that $f_1(\lambda)$ is a factor of every element in the first row of $A_1(\lambda)$ and of every element in the first column of $A_1(\lambda)$, whence, by applying elementary $\lambda$ operations of type 3, we may convert $A_1(\lambda)$ into a matrix of the form

(1) $$\begin{bmatrix} f_1(\lambda) & O \\ O & A_2(\lambda) \end{bmatrix}.$$

Either $A_2(\lambda) = O$ or we may apply the same process to $A_2(\lambda)$. But if in (1) we apply an elementary $\lambda$ operation to only the rows or the columns containing $A_2(\lambda)$, the first row and the first column are not disturbed. It follows that $A(\lambda)$ is $\lambda$-equivalent to a $\lambda$ matrix of the form

$$\begin{bmatrix} f_1(\lambda) & 0 & O \\ 0 & f_2(\lambda) & O \\ O & 0 & A_3(\lambda) \end{bmatrix},$$

where $f_1(\lambda)$ and $f_2(\lambda)$ are monic. If $A_3(\lambda) \neq O$, the process may be continued. Thus $A(\lambda)$ is $\lambda$-equivalent to a $\lambda$ matrix of one of the following forms:

(2) $$G, \qquad \begin{bmatrix} G & O \\ O & O \end{bmatrix}, \qquad \begin{bmatrix} G \\ O \end{bmatrix}, \quad [G \quad O],$$

where, since $\lambda$ matrix (2) must be of rank $r$ (by Theorem 4.3.4),

$$G = \mathrm{diag}(f_1(\lambda), f_2(\lambda), \cdots, f_r(\lambda))$$

and each $f_i(\lambda)$ is monic. Moreover, each $f_i(\lambda)$ is of least degree in the set of all elements of all $\lambda$ matrices $\lambda$-equivalent to $A_i(\lambda)$. Now

$$A_i(\lambda) \overset{\lambda E}{=} \begin{bmatrix} f_i(\lambda) & 0 & O \\ 0 & f_{i+1}(\lambda) & O \\ O & O & A_{i+2}(\lambda) \end{bmatrix}$$

$$\overset{\lambda E}{=} \begin{bmatrix} f_i(\lambda) & 0 & O \\ f_i(\lambda) & f_{i+1}(\lambda) & O \\ O & O & A_{i+2}(\lambda) \end{bmatrix}.$$

But, by the division algorithm (Theorem 4.1.8),

$$f_{i+1}(\lambda) = s_i(\lambda)f_i(\lambda) + t_i(\lambda),$$

where the degree of $t_i(\lambda)$ is less than the degree of $f_i(\lambda)$. If, in the last matrix, we add $-s_i(\lambda)$ times the first column to the second column, we find that

$$A_i(\lambda) \overset{\lambda E}{=} \begin{bmatrix} f_i(\lambda) & -s_i(\lambda)f_i(\lambda) & O \\ f_i(\lambda) & t_i(\lambda) & O \\ O & O & A_{i+2}(\lambda) \end{bmatrix}.$$

By the definition of $f_i(\lambda)$, this implies that $t_i(\lambda) \equiv 0$ and $f_i(\lambda)$ divides $f_{i+1}(\lambda)$.

**4.3.6** THEOREM. *Let* $A(\lambda)$ *be a $\lambda$ matrix, let* $A(\lambda) \overset{\lambda E}{=} B(\lambda)$, *and let* $d_t(\lambda)$ *be the greatest common divisor of all $t$-rowed minors of* $A(\lambda)$. *Then* $d_t(\lambda)$ *is also the greatest common divisor of all $t$-rowed minors of* $B(\lambda)$.

As pointed out in the proof of Theorem 4.3.4, every $t$-rowed minor of $B(\lambda)$ is a $t$-rowed minor of $A(\lambda)$, or the product of a $t$-rowed minor of $A(\lambda)$ by a nonzero number, or a sum

$$M_1(\lambda) + f(\lambda)M_2(\lambda),$$

where $M_1(\lambda)$ and $M_2(\lambda)$ are $t$-rowed minors of $A(\lambda)$, and $f(\lambda)$ is a polynomial in $\lambda$. It follows that any divisor of all the $t$-rowed minors of $A(\lambda)$ is a divisor of all the $t$-rowed minors of $B(\lambda)$. But we also have $B(\lambda) \overset{\lambda E}{=} A(\lambda)$, and any divisor of all the $t$-rowed minors of $B(\lambda)$ is a divisor of all the $t$-rowed minors of $A(\lambda)$. The theorem now follows.

**4.3.7** COROLLARY. *The greatest common divisor* $d_k(\lambda)$ *of the* k-*rowed minors of* $\lambda$ *matrix* $A(\lambda)$ *of rank* r, *where* $k \leqq r$, *is*

$$d_k(\lambda) \equiv f_1(\lambda)f_2(\lambda) \cdots f_k(\lambda),$$

*where the* $f_i(\lambda)$ *are the* $f_i(\lambda)$ *of Theorem* 4.3.5.

**4.3.8** THEOREM. *A necessary and sufficient condition for the* $\lambda$ *equivalence of two* m × n $\lambda$ *matrices is that they have the same rank* r, *and that, for every value of* k *from* 1 *to* r *inclusive, the* k-*rowed minors of one* $\lambda$ *matrix have the same greatest common divisor as the* k-*rowed minors of the other.*

The necessity is covered by Theorem 4.3.6. To prove the sufficiency, suppose each matrix is reduced by elementary $\lambda$ operations to the appropriate form described in Theorem 4.3.5. Using primes to distinguish the form of the second matrix from that of the first matrix, we have (by Corollary 4.3.7)

$$f'_1(\lambda) = f_1(\lambda),$$

$$f'_1(\lambda)f'_2(\lambda) = f_1(\lambda)f_2(\lambda),$$

$$\cdots \cdots \cdots \cdots \cdots \cdots$$

$$f'_1(\lambda) \cdots f'_r(\lambda) = f_1(\lambda) \cdots f_r(\lambda).$$

Since no $f_k(\lambda)$ or $f'_k(\lambda)$ is identically zero, it follows that $f'_k(\lambda) = f_k(\lambda)$ for $k = 1, \cdots, r$. Thus the forms to which the two $\lambda$ matrices can be reduced are identical, and hence the two $\lambda$ matrices are $\lambda$-equivalent, since two $\lambda$ matrices $\lambda$-equivalent to a third are $\lambda$-equivalent to each other.

**4.3.9** COROLLARY. *The matrix of Theorem* 4.3.5 *to which a given* $\lambda$ *matrix* $A(\lambda)$ *is* $\lambda$-equivalent is unique.

**4.3.10** DEFINITION. The unique matrix of Theorem 4.3.5 to which a given $\lambda$ matrix $A(\lambda)$ is $\lambda$-equivalent is called *the Smith normal form* of $A(\lambda)$.

## PROBLEMS

**4.3–1** (a) Show that the inverse of an elementary $\lambda$ operation is an elementary $\lambda$ operation.

(b) Show that an elementary $\lambda$ operation of type 2 may be described as: "the multiplication of a row, or a column, by a polynomial in $\lambda$ whose inverse is a polynomial in $\lambda$."

**4.3–2**    Prove that $\lambda$ equivalence of $\lambda$ matrices is reflexive, symmetric, and transitive.

**4.3**    Show that the greatest common divisor of all $k$-rowed minors ($k = 1, \cdots, r - 1$) of a $\lambda$ matrix $A(\lambda)$ of rank $r$ divides the greatest common divisor of all $(k + 1)$-rowed minors of $A(\lambda)$.

**4.3–4**    Prove Corollary 4.3.7.

**4.3–5**    Prove Corollary 4.3.9.

**4.3–6**    Prove that a square $\lambda$ matrix is $\lambda$-equivalent to its transpose.

**4.3–7**    Carry each of the following matrices into its Smith normal form by a finite sequence of elementary $\lambda$ operations.

(a) $\begin{bmatrix} \lambda & \lambda + 1 \\ \lambda + 2 & \lambda + 3 \end{bmatrix}$,    (b) $\begin{bmatrix} \lambda^2 & \lambda + 1 \\ \lambda - 1 & \lambda^2 \end{bmatrix}$,

(c) $\begin{bmatrix} \lambda^2 + \lambda - 2 & 0 \\ 0 & \lambda^2 + 2\lambda - 3 \end{bmatrix}$.

**4.3–8**    Reduce each of the $\lambda$ matrices of Problem 4.3–7 to its Smith normal form by the use of Corollary 4.3.7.

**4.3–9**    Reduce

$$\begin{bmatrix} 1 + 2\lambda & \lambda^3 + 4\lambda^2 + \lambda + 2 & \lambda^3 + 4\lambda + 2 \\ 0 & \lambda^2 + \lambda & \lambda^2 \\ 1 - 2\lambda & \lambda^3 + 3\lambda^2 - 3\lambda - 1 & \lambda^3 - \lambda^2 + 4\lambda - 2 \end{bmatrix}$$

to Smith normal form.

**4.3–10**    Reduce

$$\begin{bmatrix} \lambda^2 & \lambda & \lambda^2 - 2 \\ \lambda^2 - 1 & \lambda + 1 & \lambda^2 - 2\lambda - 3 \\ \lambda^3 + 2\lambda^2 - 2 & \lambda^2 + 2\lambda + 2 & \lambda^3 - 4\lambda - 6 \end{bmatrix}$$

to Smith normal form.

## 4.4    Invariant factors and elementary divisors

The theory of elementary divisors was created by J. J. Sylvester, H. J. S. Smith, and (especially) Karl Weierstrass, and then later perfected in certain respects by Leopold Kronecker, Georg Frobenius, and others.

We do little more here than introduce this important concept, which has significant applications in both algebra and geometry.

**4.4.1** DEFINITIONS.    The nonzero polynomials $f_1(\lambda), f_2(\lambda), \cdots, f_r(\lambda)$ in the diagonal of the Smith normal form of a $\lambda$ matrix $A(\lambda)$ of rank $r$ are called the *invariant factors* of $A(\lambda)$. An invariant factor that is equal to 1 is called a *trivial invariant factor*.

As an immediate consequence of Corollary 4.3.9, we have

**4.4.2**    THEOREM.    *Two* m × n *$\lambda$ matrices are $\lambda$-equivalent if and only if they have the same invariant factors.*

As a consequence of Corollary 4.3.7, we have

**4.4.3**    THEOREM.    *Let* $f_1(\lambda), \cdots, f_r(\lambda)$ *be the invariant factors, in order of increasing degree, of a $\lambda$ matrix* $A(\lambda)$ *of rank* r, *and let* $d_k(\lambda)$ *denote the greatest common divisor of all the* k-*rowed minors of* $A(\lambda)$. *Then* $f_1(\lambda) = d_1(\lambda)$ *and*

$$f_{i+1}(\lambda) = d_{i+1}(\lambda)/f_1(\lambda) \cdots f_i(\lambda)$$

*for* i = 1, ⋯, r − 1.

**4.4.4**    DEFINITIONS.    Let $A(\lambda)$ be an $m \times n$ $\lambda$ matrix of rank $r$ and let $f_i(\lambda)$ $(i = 1, \cdots, r)$ denote its invariant factors in order of increasing degree. Let

$$f_r(\lambda) = (\lambda - c_1)^{e_{r1}} \cdots (\lambda - c_s)^{e_{rs}},$$

where $c_1, \cdots, c_s$ are the distinct complex roots of the equation $f_r(\lambda) = 0$. Since $f_i(\lambda)$ divides $f_{i+1}(\lambda)$ $(i = 1, \cdots, r - 1)$, it follows that

$$f_i(\lambda) = (\lambda - c_1)^{e_{i1}} \cdots (\lambda - c_s)^{e_{is}},$$

where $0 \leqq e_{ij} \leqq e_{i+1,j}$ $(i = 1, \cdots, r - 1; j = 1, \cdots, s)$. The $rs$ polynomials

$$g_{ij}(\lambda) \equiv (\lambda - c_j)^{e_{ij}}$$

are called the *elementary divisors* of $A(\lambda)$. The elementary divisors of $A(\lambda)$ for which $e_{ij} = 0$ (that is, the elementary divisors that are equal to 1) are called the *trivial elementary divisors* of $A(\lambda)$.

*Example.*    Suppose the invariant factors of a $\lambda$ matrix $A(\lambda)$ are

$$1, \quad 1, \quad \lambda, \quad \lambda^2(\lambda - 1), \quad \lambda^2(\lambda - 1)^2(\lambda + 1).$$

Here $r = 5$, $s = 3$. The $(5)(3) = 15$ elementary divisors of $A(\lambda)$ are

$$\lambda^2, \ (\lambda - 1)^2, \ \lambda + 1; \quad \lambda^2, \ \lambda - 1, \ 1;$$

$$\lambda, \ 1, \ 1; \quad 1, \ 1, \ 1; \quad 1, \ 1, \ 1,$$

The reader should find no difficulty in proving the following theorem.

**4.4.5** THEOREM.    (1) *The invariant factors of a $\lambda$ matrix* $\mathrm{A}(\lambda)$ *uniquely determine the elementary divisors of* $\mathrm{A}(\lambda)$; *conversely, the elementary divisors of* $\mathrm{A}(\lambda)$ *uniquely determine the invariant factors of* $\mathrm{A}(\lambda)$. (2) *The rank and nontrivial invariant factors of a $\lambda$ matrix* $\mathrm{A}(\lambda)$ *uniquely determine the elementary divisors of* $\mathrm{A}(\lambda)$; *the rank and the nontrivial elementary divisors of* $\mathrm{A}(\lambda)$ *uniquely determine the invariant factors of* $\mathrm{A}(\lambda)$.

As a consequence of Theorems 4.4.2 and 4.4.5 we have

**4.4.6** THEOREM.    *Two* m × n *$\lambda$ matrices are $\lambda$-equivalent if and only if they have the same elementary divisors.*

**4.4.7** THEOREM.    *Let* $\mathrm{g}_1(\lambda), \cdots, \mathrm{g}_s(\lambda)$ *be monic polynomials that are relatively prime in pairs. Then the only nontrivial invariant factor of the matrix*

$$\mathrm{A}(\lambda) = diag(\mathrm{g}_1(\lambda), \cdots, \mathrm{g}_s(\lambda))$$

*is the product* $\mathrm{g}_1(\lambda) \cdots \mathrm{g}_s(\lambda)$.
This theorem is a consequence of Theorem 4.4.3.

**4.4.8** THEOREM.    *Let* $\mathrm{c}_1, \cdots, \mathrm{c}_s$ *be distinct complex numbers and set* $\mathrm{g}_i(\lambda) = (\lambda - \mathrm{c}_i)^{n_i}$ *for integers* $\mathrm{n}_i > 0$. *Then the matrix*

$$\mathrm{A}(\lambda) = diag(\mathrm{g}_1(\lambda), \cdots, \mathrm{g}_s(\lambda))$$

*has the* $\mathrm{g}_i(\lambda)$ *as its nontrivial elementary divisors.*
The polynomials $g_i(\lambda)$ are monic and relatively prime in pairs. Therefore, by Theorem 4.4.7, the only nontrivial invariant factor of $A(\lambda)$ is the product $g_1(\lambda) \cdots g_s(\lambda)$. It follows that the nontrivial elementary divisors of $A(\lambda)$ are $g_1(\lambda), \cdots, g_s(\lambda)$.

## PROBLEMS

**4.4–1**   Describe a process for finding the invariant factors of a $\lambda$ matrix $A(\lambda)$ when given the elementary divisors of $A(\lambda)$.

**4.4–2**   Prove Theorem 4.4.5.

**4.4–3**   Prove Theorem 4.4.7.

**4.4–4**   The following polynomials are the invariant factors of a $\lambda$ matrix; find the elementary divisors:

(a) $1, 1, (\lambda - 1)(\lambda + 1), \lambda(\lambda - 1)(\lambda + 1)^2,$
    $\lambda^2(\lambda - 1)^2(\lambda + 1)^2(\lambda - 3);$

(b) $\lambda^2 + \lambda, \lambda^3 + \lambda^2, \lambda^5 + 2\lambda^4 + \lambda^3.$

**4.4–5**   The following polynomials are the elementary divisors of a $\lambda$ matrix; find the invariant factors:

(a) $1, 1, 1, 1, 1, 1, 1, 1, 1, 1, 1, \lambda, \lambda^2, \lambda - 1, \lambda - 1, (\lambda - 1)^2, \lambda + 1,$
    $(\lambda + 1)^2, (\lambda + 1)^2, \lambda - 3;$

(b) $\lambda, \lambda^2, \lambda^3, \lambda + 1, \lambda + 1, (\lambda + 1)^2.$

**4.4–6**   The following polynomials are the nontrivial invariant factors of a $\lambda$ matrix of rank 5; find the elementary divisors:

(a) $\lambda^2 + 1, (\lambda^2 + 1)^3, (\lambda^2 + 1)^4;$

(b) $\lambda, \lambda^3 + \lambda, \lambda^5 + 2\lambda^4 + \lambda, \lambda^6 + \lambda^5 + 2\lambda^4 + 2\lambda^3 + \lambda^2 + \lambda.$

**4.4–7**   The following polynomials are the nontrivial elementary divisors of a $\lambda$ matrix of rank 5; find the invariant factors:

(a) $\lambda - 1, (\lambda - 1)^2, (\lambda - 1)^3, \lambda + 1, (\lambda + 1)^2;$

(b) $\lambda, \lambda^2, \lambda^2, \lambda^3, \lambda - 1, (\lambda - 1)^2, (\lambda + 1)^4.$

**4.4–8**   Formulate a definition of the elementary divisors of a $\lambda$ matrix that is independent of the concept of invariant factors.

**4.4–9**   Use Theorem 4.4.3 to find the invariant factors of

$$A(\lambda) = \begin{bmatrix} a - \lambda & c_1 & 0 & 0 \\ 0 & a - \lambda & c_2 & 0 \\ 0 & 0 & a - \lambda & c_3 \\ 0 & 0 & 0 & a - \lambda \end{bmatrix},$$

where $c_1 c_2 c_3 \neq 0$.

**4.4–10**   Let $A(\lambda)$ be a scalar $\lambda$ matrix and let each element of the principal diagonal be factored into the product of a constant and powers of linear factors of the form $\lambda - a$. Prove that these powers of linear factors are precisely the nontrivial elementary divisors of $A(\lambda)$. (This is an extension of Theorem 4.4.8.)

### 4.5   The characteristic function of a square matrix

Of the $\lambda$ matrices associated with a given numerical matrix $A_{(n)}$, perhaps the most important is the so-called *characteristic matrix* of $A$, whose determinant leads to the *characteristic function* of $A$.

**4.5.1**   DEFINITIONS AND NOTATION.   If $A$ is a square matrix of order $n$, then the $\lambda$ matrix $[A - \lambda I_{(n)}]$ is called the *characteristic matrix* of $A$. The determinant $|A - \lambda I_{(n)}|$ is called the *characteristic determinant* of $A$, and the expansion of this determinant expressed as a polynomial of degree $n$ in $\lambda$,

$$f(\lambda) \equiv (-1)^n[\lambda^n - p_1\lambda^{n-1} + p_2\lambda^{n-2} + \cdots + (-1)^n p_n],$$

is called the *characteristic function* of $A$. The equation $f(\lambda) = 0$ is called the *characteristic equation* of $A$, and the roots of this equation are called the *characteristic roots* of $A$.

It was A. L. Cauchy who, in 1840, introduced the word "characteristic" by calling the equation $|A - \lambda I| = 0$ the characteristic equation of matrix $A$. Other names employed by some writers for the characteristic roots are: *latent roots, hidden roots, proper roots, secular values*, and *eigenvalues*.

*Examples*

1. Take $A = \begin{bmatrix} 5 & -2 \\ -2 & 2 \end{bmatrix}$. Then

$$|A - \lambda I| = \begin{vmatrix} 5 - \lambda & -2 \\ -2 & 2 - \lambda \end{vmatrix} = (5 - \lambda)(2 - \lambda) - 4$$

$$= \lambda^2 - 7\lambda + 6 = (\lambda - 6)(\lambda - 1).$$

Therefore 1 and 6 are the characteristic roots of $A$.

2. Take $A = \begin{bmatrix} a_{11} & a_{12} \\ a_{21} & a_{22} \end{bmatrix}$. Then

$$|A - \lambda I| = \begin{vmatrix} a_{11} - \lambda & a_{12} \\ a_{21} & a_{22} - \lambda \end{vmatrix}$$

$$= \lambda^2 - (a_{11} + a_{22})\lambda + (a_{11}a_{22} - a_{21}a_{12}).$$

Here $p_1 = a_{11} + a_{22}$, $p_2 = a_{11}a_{22} - a_{21}a_{12}$.

The values of $p_1$ and $p_2$ in the preceding Example 2 illustrate the following theorem.

**4.5.2** THEOREM. *In the characteristic function* $f(\lambda)$ *of matrix* A, $p_n = |A|$ *and* $p_1 = \operatorname{tr} A$.

We have $f(0) = (-1)^n(-1)^n p_n = p_n$; also, $f(0) = |A - 0I| = |A|$. Therefore $p_n = |A|$.

Expanding $|A - \lambda I|$ by the elements and cofactors of the first row, one sees that the only term of $|A - \lambda I|$ containing $\lambda^n$ and $\lambda^{n-1}$ is the term

$$(a_{11} - \lambda)(a_{22} - \lambda) \cdots (a_{nn} - \lambda).$$

It follows that when $|A - \lambda I|$ is expressed as a polynomial in $\lambda$, the coefficient of $\lambda^{n-1}$ is

$$-(-1)^n(a_{11} + a_{22} + \cdots + a_{nn}) = -(-1)^n \operatorname{tr} A,$$

and we have $p_1 = \operatorname{tr} A$.

It is a remarkable and important fact that if

$$f(\lambda) \equiv (-1)^n[\lambda^n - p_1\lambda^{n-1} + p_2\lambda^{n-2} + \cdots + (-1)^n p_n]$$

is the characteristic function of an $n \times n$ matrix $A$, then

$$A^n - p_1 A^{n-1} + p_2 A^{n-2} + \cdots + (-1)^n p_n I_{(n)} = O_{(n)}.$$

This fact is sometimes compactly and freely stated by saying that: "A square matrix satisfies its own characteristic equation." This is the famous Hamilton-Cayley Theorem, established for quaternions by W. R. Hamilton in 1853 and stated for the general case by A. Cayley in 1858, who illustrated a general proof with proofs for the cases $n = 2$ and $3$. We now give two proofs of the Hamilton-Cayley Theorem, the first proof depending upon material of Section 4.2 and the second proof being more direct and self-contained.

**4.5.3** THEOREM (the Hamilton-Cayley Theorem). *Let* A *be an* $n \times n$ *matrix and let*

$$f(\lambda) \equiv (-1)^n[\lambda^n - p_1\lambda^{n-1} + p_2\lambda^{n-2} + \cdots + (-1)^n p_n]$$

*be the characteristic function of* A. *Then*

(1)        $A^n - p_1 A^{n-1} + p_2 A^{n-2} + \cdots + (-1)^n p_n I_{(n)} = O_{(n)}.$

*First proof.*   We have, by Theorem 3.10.2,

$$[\text{adj}(A - \lambda I)](A - \lambda I) = |A - \lambda I| I = f(\lambda) I$$

$$= (-1)^n [I\lambda^n - p_1 I \lambda^{n-1} + \cdots + (-1)^n p_n I]$$

or

$$[\text{adj}(A - \lambda I)](\lambda I - A) = (-1)^{n+1} [I\lambda^n - p_1 I \lambda^{n-1} + \cdots + (-1)^n p_n I].$$

The right side of the last equation is a $\lambda$ matrix $F(\lambda)$ having $\lambda I - A$ as a right divisor. It follows, by Theorem 4.2.10, that $F_R(A) = O_{(n)}$. That is,

$$A^n - p_1 A^{n-1} + p_2 A^{n-2} + \cdots + (-1)^n p_n I_{(n)} = O_{(n)}.$$

*Second proof.*   Set $B = \text{adj}(A - \lambda I)$. Then each element $b_{ij}$ of $B$ is a polynomial of degree at most $n - 1$ in $\lambda$, whose coefficients are polynomial functions of the elements $a_{ij}$ of $A$. Consequently

$$B = B_0 + B_1 \lambda + B_2 \lambda^2 + \cdots + B_{n-1} \lambda^{n-1},$$

where the $B_i$ are matrices whose elements are polynomials in the $a_{ij}$. Now, by Theorem 3.10.2,

$$(A - \lambda I)B \equiv |A - \lambda I| I \equiv f(\lambda) I.$$

That is,

$$(A - \lambda I)(B_0 + B_1 \lambda + \cdots + B_{n-2} \lambda^{n-2} + B_{n-1} \lambda^{n-1})$$

$$\equiv (-1)^n [\lambda^n - p_1 \lambda^{n-1} + \cdots + (-1)^{n-1} p_{n-1} \lambda + (-1)^n p_n] I.$$

Equating corresponding coefficients of $\lambda$, we obtain

$$-B_{n-1} = (-1)^n I,$$

$$-B_{n-2} + AB_{n-1} = -(-1)^n p_1 I,$$

$$-B_{n-3} + AB_{n-2} = (-1)^n p_2 I,$$

$$\cdot \quad \cdot \quad \cdot \quad \cdot \quad \cdot \quad \cdot \quad \cdot \quad \cdot \quad \cdot$$

$$-B_1 + AB_2 = (-1)^n (-1)^{n-2} p_{n-2} I,$$

$$-B_0 + AB_1 = (-1)^n (-1)^{n-1} p_{n-1} I,$$

$$AB_0 = (-1)^n (-1)^n p_n I.$$

Premultiplying these matric relations by $A^n$, $A^{n-1}$, $A^{n-2}, \cdots, A^2, A, I$, respectively, we obtain

$$-A^n B_{n-1} = (-1)^n A^n,$$

$$-A^{n-1} B_{n-2} + A^n B_{n-1} = -(-1)^n p_1 A^{n-1},$$

$$-A^{n-2} B_{n-3} + A^{n-1} B_{n-2} = (-1)^n p_2 A^{n-2},$$

$$\cdot \quad \cdot \quad \cdot \quad \cdot \quad \cdot \quad \cdot \quad \cdot \quad \cdot \quad \cdot \quad \cdot \quad \cdot \quad \cdot \quad \cdot \quad \cdot$$

$$-A^2 B_1 + A^3 B_2 = (-1)^n (-1)^{n-2} p_{n-2} A^2,$$

$$-A B_0 + A^2 B_1 = (-1)^n (-1)^{n-1} p_{n-1} A,$$

$$A B_0 = (-1)^n (-1)^n p_n I.$$

Now adding, we find

$$O_{(n)} = (-1)^n [A^n - p_1 A^{n-1} + \cdots + (-1)^{n-1} p_{n-1} A + (-1)^n p_n I],$$

and the theorem is established.

**4.5.4** DEFINITIONS.   The equation (1) of Theorem 4.5.3 is called the *Hamilton-Cayley equation* of matrix $A$, and the polynomial on the left in (1) is called the *Hamilton-Cayley polynomial* of matrix $A$.

## PROBLEMS

**4.5–1**   Find the characteristic function and the characteristic roots of

$$A = \begin{bmatrix} 1 & -1 & 0 \\ 2 & 3 & 2 \\ 1 & 1 & 2 \end{bmatrix}.$$

**4.5–2**   Find the characteristic function and the characteristic roots of

$$A = \begin{bmatrix} 0 & i & i \\ i & 0 & i \\ i & i & 0 \end{bmatrix}.$$

**4.5–3**   Find the characteristic function and the characteristic roots of

$$A = \begin{bmatrix} 0 & 0 & -1 \\ 1 & 0 & -3 \\ 0 & 1 & -3 \end{bmatrix}.$$

**4.5–4**   Prove that $\lambda = 0$ is a characteristic root of the square matrix $A$ if and only if $A$ is singular.

**4.5–5**   If $A$ is a square complex matrix, prove that the characteristic roots of $A^*$ are the complex conjugates of those of $A$.

**4.5–6**   Find the characteristic roots of the real matrix

$$A = \begin{bmatrix} a & -b \\ b & a \end{bmatrix}.$$

**4.5–7**   When will the real matrix $\begin{bmatrix} a & b \\ c & d \end{bmatrix}$ have equal characteristic roots?

**4.5–8**   Determine the characteristic roots of the matrix $[a_{ij}]_{(n)}$ if $a_{ij} = 0$ for $i = j$ and $a_{ij} = 1$ for $i \neq j$.

**4.5–9**   Determine the characteristic roots of the matrix

$$\begin{bmatrix} a_1 & a_2 & \cdots & a_n \\ a_1 & a_2 & \cdots & a_n \\ \cdots & \cdots & \cdots & \cdots \\ a_1 & a_2 & \cdots & a_n \end{bmatrix}.$$

**4.5–10**   Let $A$ and $P$ be $n \times n$ matrices, $P$ nonsingular. Show that $A$ and $P^{-1}AP$ have the same characteristic roots.

**4.5–11**   Show that the characteristic roots of a triangular matrix are precisely the diagonal elements of the matrix.

**4.5–12**   Show that the characteristic function of

$$\begin{bmatrix} 0 & 1 & 0 \\ 0 & 0 & 1 \\ -a_3 & -a_2 & -a_1 \end{bmatrix}$$

is

$$\lambda^3 + a_1\lambda^2 + a_2\lambda + a_3.$$

**4.5–13**   Let $b(\lambda)$ be a polynomial in $\lambda$ and let $A$ be an $n \times n$ matrix. Prove that $b(A) = O$ if and only if the characteristic matrix of $A$ divides $b(\lambda)I_{(n)}$.

**4.5–14**   Let $b(\lambda)$ be a polynomial in $\lambda$ and let $A$ be an $n \times n$ matrix having characteristic roots $\lambda_1, \lambda_2, \cdots, \lambda_n$. Prove that

$$|b(A)| = b(\lambda_1)b(\lambda_2) \cdots b(\lambda_n).$$

**4.5–15**   Let $\lambda_1$ be a characteristic root of the square matrix $A$.

(a) If $k$ is a nonzero scalar, show that $k\lambda_1$ is a characteristic root of $kA$.

**(b)** If $r$ is a positive integer, show that $(\lambda_1)^r$ is a characteristic root of $A^r$.

**(c)** If $A$ is nonsingular, show that $1/\lambda_1$ is a characteristic root of $A^{-1}$.

**4.5–16**  Find the Hamilton-Cayley equation of

$$A = \begin{bmatrix} 2 & -1 & 1 \\ -1 & 2 & -1 \\ 1 & -1 & 2 \end{bmatrix}$$

and thence obtain $A^{-1}$.

## 4.6  Some results related to the characteristic function of a square matrix

In this section we establish a sequence of theorems concerning integral powers of a square matrix $A$, we extend Theorem 3.10.12 to the case where either $A$ or $B$ may be singular, and we prove an important theorem about the characteristic roots of a Hermitian matrix.

**4.6.1**  THEOREM.  *If* A *is an* n × n *matrix and* k *is a nonnegative integer, then* $A^{n+k}$ *can be expressed as a linear combination of* $I_{(n)}$, A, $\cdots$, $A^{n-1}$.

From the Hamilton-Cayley equation, we see that the theorem is true for $k = 0$. Suppose the theorem is true for $k = j$. Then we have

$$A^{n+j} = a_0 I + a_1 A + a_2 A^2 + \cdots + a_{n-1} A^{n-1}.$$

Multiplying this equation through by $A$ we find

$$A^{n+(j+1)} = a_0 A + a_1 A^2 + a_2 A^3 + \cdots + a_{n-1} A^n.$$

Now, replacing $A^n$ by its linear expression in terms of $I, A, \cdots, A^{n-1}$, we see that $A^{n+(j+1)}$ is expressible as a linear combination of $I, A, \cdots, A^{n-1}$, and the theorem follows by the principle of mathematical induction.

**4.6.2**  THEOREM.  *Any polynomial function of an* n × n *matrix* A *is expressible as a polynomial function of* A *of degree at most* n − 1.

This is an immediate consequence of Theorem 4.6.1.

**4.6.3**  THEOREM.    *If* A *is a nonsingular* n × n *matrix and if*

$$f(\lambda) \equiv (-1)^n[\lambda^n - p_1\lambda^{n-1} + p_2\lambda^{n-2} + \cdots + (-1)^n p_n]$$

*is the characteristic function of* A, *then*

$$A^{-1} = (-1)^{n-1}[A^{n-1} - p_1 A^{n-2} + p_2 A^{n-3} - \cdots + (-1)^{n-1}p_{n-1}I]/|A|.$$

For by the Hamilton-Cayley Theorem,

$$A^n - p_1 A^{n-1} + p_2 A^{n-2} - \cdots + (-1)^n p_n I = O,$$

whence, premultiplying by $A^{-1}$, we find

$$A^{n-1} - p_1 A^{n-2} + p_2 A^{n-3} - \cdots + (-1)^{n-1}p_{n-1}I = -(-1)^n p_n A^{-1}.$$

The theorem is now evident, since (by Theorem 4.5.2) $p_n = |A| \neq 0$.

*Note.*   It is to be observed that this theorem furnishes yet another way of finding the inverse of a given nonsingular matrix $A$.

**4.6.4**  THEOREM.    *If* A *is a nonsingular* n × n *matrix and if* p *is any positive integer, then* $A^{-p}$ *can be expressed as a linear combination of* $I_{(n)}, A, \cdots, A^{n-1}$.

For $A^{-p} = (A^{-1})^p$, and the theorem follows from Theorems 4.6.3 and 4.6.2.

**4.6.5**  DEFINITION.    By $r(A)/s(A)$, where $r(A)$ and $s(A)$ are polynomial functions of $A$ and where $|s(A)| \neq 0$, we mean $[r(A)][s(A)]^{-1}$.

**4.6.6**  THEOREM.    *Any rational function of an* n × n *matrix* A, *where the denominator function is nonsingular, is expressible as a polynomial function of* A *of degree at most* n − 1.

Consider the rational function $r(A)/s(A)$, where $r(A)$ and $s(A)$ are polynomials in $A$ and where $|s(A)| \neq 0$. Set $D = s(A)$. Since $|D| \neq 0$, we have (by Theorem 4.6.3) $D^{-1} = p(D)$, a polynomial in $D$. Therefore, since $D = s(A)$, $D^{-1} = q(A)$, a polynomial in $A$. It follows that $r(A)/s(A) = r(A)q(A) = t(A)$, where (by Theorem 4.6.2) $t(A)$ is a polynomial in $A$ of degree at most $n - 1$.

In Theorem 3.10.12 we proved that if $A$ and $B$ are nonsingular matrices of order $n > 1$, then $\text{adj}(AB) = (\text{adj } B)(\text{adj } A)$. We now extend this result to *any* two $\mathscr{C}$ matrices $A$ and $B$ (singular or nonsingular) of order $n > 1$.

**4.6.7**   THEOREM.    *If* A *and* B *are two* $\mathscr{C}$ *matrices of order* n > 1, *then* adj(AB) = (*adj* B)(*adj* A).

We have already, in Theorem 3.10.12, established this theorem for the case where $|A|\,|B| \neq 0$. If $|A|\,|B| = 0$, we establish the theorem by continuity considerations as follows: Subtract $\epsilon$ from each element of the principal diagonals of $A$ and $B$, where $0 < \epsilon < \delta$, $\delta$ being the smallest absolute value assumed by any nonzero characteristic root of either $A$ or $B$. Denote these matrices by $A_\epsilon$ and $B_\epsilon$, respectively. Then $|A_\epsilon|\,|B_\epsilon| \neq 0$ and adj($A_\epsilon B_\epsilon$) = (adj $B_\epsilon$)(adj $A_\epsilon$). Taking the limit as $\epsilon \to 0$, we find that adj($AB$) = (adj $B$)(adj $A$).

A very important theorem in applications is one to the effect that the characteristic roots of a real symmetric matrix are all real. We establish this as a corollary to a more general result.

**4.6.8**   THEOREM.    *The characteristic roots of a Hermitian matrix of order* n *are all real.*

Let $A$ be Hermitian and let $\lambda$ be any one of its characteristic roots. Then $|A - \lambda I| = 0$, whence the homogeneous system of linear equations

$$[A - \lambda I]X_{(n,1)} = O_{(n,1)}$$

has (by Theorem 2.8.9) a solution $X = S \neq O$. That is, $AS = \lambda S$, or $\bar{S}'AS = \lambda \bar{S}'S$. But $\bar{S}'AS$ is real (by Theorem 1.8.9). Also $\bar{S}'S$ is real and nonzero. Therefore $\lambda$ is real.

**4.6.9**   COROLLARY.    *The characteristic roots of a real symmetric matrix are all real.*

**4.6.10**   COROLLARY.    *The characteristic roots of a real skew-symmetric matrix* A *are all pure imaginary or zero.*

For $iA$ is Hermitian.

## PROBLEMS

**4.6–1**   If matrix $A$ satisfies the equation $X^2 - 5X - 2I = O$, show that

(a) $A^{-1} = (A - 5I)/2$,

**(b)** $A^4 = 145A + 54I$,

**(c)** $A^5 = 779A + 290I$.

**4.6–2**   Show that a Hermitian matrix $A$ has all positive characteristic roots if and only if the coefficients of the characteristic function $f(\lambda)$ of $A$ alternate in sign.

**4.6–3**   If $A$ is a square $\mathscr{C}$ matrix, prove that the characteristic roots of $A^*A$ are all nonnegative real numbers.

**4.6–4**   Prove that a real skew-symmetric matrix has an even or an odd number of zero characteristic roots according as its order is even or odd.

**4.6–5**   Prove that the characteristic roots of a real symmetric matrix $A$ are all equal if and only if $A$ is scalar.

**4.6–6**   Prove that if $\mu$ is an $r$-fold characteristic root of a square matrix $A$, then 0 is an $r$-fold characteristic root of $A - \mu I$.

## 4.7   Characteristic vectors of a square matrix

There are many important applications in mathematics and in the physical and social sciences wherein one must find a scalar $\lambda$ and a nonzero column vector $X$ such that for a given square matrix $A$ we have $AX = \lambda X$. This leads us to formulate the following definition.

**4.7.1**   DEFINITION.   If $A$ is a square matrix and $X$ is a nonzero column vector such that $AX = \lambda X$ for some scalar $\lambda$, then $X$ is called a *characteristic vector* (*latent vector, proper vector, eigenvector*) *of* A *corresponding to* $\lambda$.

**4.7.2**   THEOREM.   *Let* A *be a square matrix and* $\lambda$ *a scalar. Then* A *has a characteristic vector corresponding to* $\lambda$ *if and only if* $\lambda$ *is a characteristic root of* A. *There is always at least one value of* $\lambda$ *for which a characteristic vector of* A *exists.*

We are seeking a nonzero column vector $X$ such that $AX = \lambda X$, or assuming that $A$ is of order $n$, such that $(A - \lambda I_{(n)})X = O$. Now the corresponding system of $n$ homogeneous linear equations in $n$ unknowns has a nontrivial solution if and only if $|A - \lambda I| = 0$, that is, if and only if $\lambda$ is a characteristic root of $A$. Since $A$ always has at least one scalar $\lambda$ as a characteristic root, there is at least one characteristic vector of $A$.

**4.7.3** THEOREM. *Let* $\lambda_1, \lambda_2, \cdots, \lambda_k$ *be distinct characteristic roots of a matrix* A *of order* n, *and let* $X_1, X_2, \cdots, X_k$ *be characteristic vectors associated with these roots, respectively. Then* $X_1, X_2, \cdots, X_k$ *are linearly independent.*

Let $c_1, c_2, \cdots, c_k$ be scalars such that

$$c_1 X_1 + c_2 X_2 + \cdots + c_k X_k = O.$$

Premultiplying through by $A$ and recalling that $AX_j = \lambda_j X_j$, we have

$$c_1 \lambda_1 X_1 + c_2 \lambda_2 X_2 + \cdots + c_k \lambda_k X_k = O.$$

Repeating the process, we successively obtain

$$c_1 \lambda_1^2 X_1 + c_2 \lambda_2^2 X_2 + \cdots + c_k \lambda_k^2 X_k = O,$$
$$\cdots \cdots \cdots \cdots \cdots$$
$$c_1 \lambda_1^{k-1} X_1 + c_2 \lambda_2^{k-1} X_2 + \cdots + c_k \lambda_k^{k-1} X_k = O.$$

The resulting set of equations may be written in the form

$$[c_1 X_1, c_2 X_2, \cdots, c_k X_k] \begin{bmatrix} 1 & \lambda_1 & \lambda_1^2 & \cdots & \lambda_1^{k-1} \\ 1 & \lambda_2 & \lambda_2^2 & \cdots & \lambda_2^{k-1} \\ \cdots & \cdots & \cdots & \cdots & \cdots \\ 1 & \lambda_k & \lambda_k^2 & \cdots & \lambda_k^{k-1} \end{bmatrix} = O.$$

Since the $\lambda_i$ are distinct, the right factor is a nonsingular Vandermonde matrix (see Example 2 of Section 3.5). Postmultiplying by the inverse of the Vandermonde matrix, we find

$$[c_1 X_1, c_2 X_2, \cdots, c_k X_k] = O,$$

which, since no $X_i$ is zero, implies that each $c_i = 0$. It follows that the $X_i$ are linearly independent.

**4.7.4** DEFINITION. Let $A$ be a matrix of order $n$ having the $n$ characteristic roots $\lambda_1, \cdots, \lambda_n$. Let $X_1, \cdots, X_n$ be characteristic vectors of $A$ corresponding to $\lambda_1, \cdots, \lambda_n$, respectively. Then the matrix

$$[X_1, \cdots, X_n]$$

is called a *modal matrix* of $A$.

*Example.* Find a modal matrix of $A = \begin{bmatrix} 2 & -1 \\ -6 & 1 \end{bmatrix}$.

The characteristic roots of $A$ are $\lambda_1 = 4$, $\lambda_2 = -1$. We require nonzero column vectors $X_1$ and $X_2$ such that

$$AX_1 = 4X_1, \qquad AX_2 = -X_2,$$

or

$$(A - 4I_{(2)})X_1 = O, \qquad (A + I_{(2)})X_2 = O,$$

or

$$\begin{bmatrix} -2 & -1 \\ -6 & -3 \end{bmatrix} \begin{bmatrix} x_{11} \\ x_{21} \end{bmatrix} = O, \qquad \begin{bmatrix} 3 & -1 \\ -6 & 2 \end{bmatrix} \begin{bmatrix} x_{12} \\ x_{22} \end{bmatrix} = O,$$

where we have set

$$X_1 = \begin{bmatrix} x_{11} \\ x_{21} \end{bmatrix}, \qquad X_2 = \begin{bmatrix} x_{12} \\ x_{22} \end{bmatrix}.$$

We find

$$-2x_{11} - x_{21} = 0, \qquad 3x_{12} - x_{22} = 0.$$

If we take $x_{11} = x_{12} = 1$, then $x_{21} = -2$, $x_{22} = 3$, and we have the modal matrix $\begin{bmatrix} 1 & 1 \\ -2 & 3 \end{bmatrix}$.

**4.7.5** THEOREM. *A modal matrix of a matrix whose characteristic roots are distinct is nonsingular.*

This is guaranteed by Theorem 4.7.3.

We conclude the section with a theorem whose easy proof can be supplied by the reader.

**4.7.6** THEOREM. *Any linear combination of characteristic vectors of a matrix* A, *all corresponding to the same characteristic root* $\lambda_i$ *of* A, *is also a characteristic vector of* A *corresponding to the characteristic root* $\lambda_i$ *of* A, *or is a zero vector.*

## PROBLEMS

**4.7–1** Let $A$ and $P$ be $n \times n$ matrices, $P$ nonsingular. If $X$ is a characteristic vector of $A$ corresponding to $\lambda$, show that $P^{-1}X$ is a characteristic vector of $P^{-1}AP$ corresponding to $\lambda$.

**4.7–2** If $A$ and $B$ are a pair of commutative $n \times n$ matrices, if $X$ is a characteristic vector of $A$ corresponding to $\lambda$, and if $BX \neq O$, show that $BX$ is also a characteristic vector of $A$ corresponding to $\lambda$.

**4.7–3**   Find a modal matrix of $A = \begin{bmatrix} 0 & 1 & 0 \\ 0 & 0 & 1 \\ -2 & 3 & 0 \end{bmatrix}$.

**4.7–4**   If $X_1, \cdots, X_n$ are characteristic vectors of matrix $A_{(n)}$ corresponding to $\lambda_1, \cdots, \lambda_n$, respectively, and if the $X_i$ are linearly independent, show that the $\lambda_i$ are the characteristic roots of $A$.

**4.7–5**   If $X_1, \cdots, X_n$ are $n$ independent characteristic vectors of both $A_{(n)}$ and $B_{(n)}$ corresponding to the scalars $\lambda_1, \cdots, \lambda_n$, respectively, prove that $A = B$.

**4.7–6**   If $X$ is a characteristic vector of the $\mathscr{C}$ matrix $A$ corresponding to $\lambda$, if $Y$ is a characteristic vector of $A^*$ corresponding to $\mu$, and if $\bar{\mu} \neq \lambda$, show that $Y^*X = O$.

**4.7–7**   If $X$ and $Y$ are characteristic vectors of a Hermitian matrix $A$ corresponding to the distinct characteristic roots $\lambda$ and $\mu$ of $A$, show that $Y^*X = O$.

**4.7–8**   If $X$ is a characteristic vector of the $\mathscr{C}$ matrix $A$ corresponding to $\lambda$, show that $\lambda = X^*AX/X^*X$.

**4.7–9**   If $A$ is nonsingular, show that $A$ and $A^{-1}$ have the same characteristic vectors.

**4.7–10**   If the characteristic roots of $A_{(n)}$ are distinct and those of $B_{(n)}$ are distinct, prove that $AB = BA$ if and only if $A$ and $B$ have the same characteristic vectors.

**4.7–11**   Prove Theorem 4.7.6.

**4.7–12**   If $X_1$ and $X_2$ are characteristic vectors of a matrix $A$ corresponding to the characteristic roots $\lambda_1$ and $\lambda_2$, respectively, where $\lambda_1 \neq \lambda_2$, does it follow that $c_1 X_1 + c_2 X_2$, where $c_1$ and $c_2$ are scalars, is a characteristic vector of $A$?

## 4.8   The minimum function of a square matrix

We have seen in the Hamilton-Cayley Theorem that if $A$ is an $n \times n$ matrix, then $I_{(n)}, A, A^2, \cdots, A^n$ are linearly dependent. This leads to the following considerations.

**4.8.1**   DEFINITION.   Let $A$ be an $n \times n$ matrix and let $m$ be the smallest nonnegative integer such that $I_{(n)}, A, A^2, \cdots, A^m$ are linearly dependent. Then $m$ is called the *indicant*† of matrix $A$.

† Some writers use the term *index*, but since this term is also used otherwise in matrix theory (see Definitions 1.4.14 and 5.5.6), we prefer here to use the term *indicant*.

**4.8.2** THEOREM. *If* m *is the indicant of an* n × n *matrix* A, *then* m ≤ n.

This is a consequence of the Hamilton-Cayley Theorem.

*Example.* That we may have $m < n$ is illustrated by the matrix $A = I_{(2)}$. Here $m = 1, n = 2$.

**4.8.3** THEOREM. *For a given square matrix* A *of order* n *and indicant* m, *there exists a unique monic polynomial*

$$m(\lambda) \equiv \lambda^m + c_1\lambda^{m-1} + \cdots + c_{m-1}\lambda + c_m$$

*of degree* m *such that*

$$m(A) \equiv A^m + c_1A^{m-1} + \cdots + c_{m-1}A + c_mI_{(n)} = O_{(n)}.$$

Since matrix $A$ is of indicant $m$, we have that $I_{(n)}, A, A^2, \cdots, A^m$ are linearly dependent. That is, there exist scalars $a_0, a_1, \cdots, a_m$, not all zero, such that

$$a_0A^m + a_1A^{m-1} + \cdots + a_{m-1}A + a_mI_{(n)} = O_{(n)}.$$

Now $a_0 \neq 0$, since otherwise we would have $I_{(n)}, A, A^2, \cdots, A^{m-1}$ linearly dependent, which contradicts the definition of the indicant $m$ of $A$. Consequently, we may divide through by $a_0$ and obtain

$$m(A) \equiv A^m + c_1A^{m-1} + \cdots + c_{m-1}A + c_mI_{(n)} = O_{(n)}.$$

Thus there exists at least one polynomial of the described form. To show that this polynomial is unique, suppose that we also have

$$p(A) \equiv A^m + d_1A^{m-1} + \cdots + d_{m-1}A + d_mI_{(n)} = O_{(n)}.$$

Then

$$m(A) - p(A) = (c_1 - d_1)A^{m-1} + \cdots + (c_m - d_m)I_{(n)} = O_{(n)},$$

which, by the definition of the indicant $m$ of $A$, is impossible unless $c_i = d_i$ for $i = 1, \cdots, m$. It follows that $p(A) \equiv m(A)$.

**4.8.4** DEFINITION. The unique polynomial

$$m(\lambda) \equiv \lambda^m + c_1\lambda^{m-1} + \cdots + c_{m-1}\lambda + c_m$$

of Theorem 4.8.3 is called the *minimum function* of the matrix $A$.

**4.8.5**   THEOREM.   *If* A *is an* n × n *matrix, then every polynomial*
$p(\lambda)$, *such that* $p(A) = O_{(n)}$, *is exactly divisible by the minimum function*
$m(\lambda)$ *of* A.

By the division algorithm for polynomials (see Theorem 4.1.8), we
have

$$p(\lambda) \equiv m(\lambda)q(\lambda) + r(\lambda),$$

where $q(\lambda)$ and $r(\lambda)$ are polynomials and the degree of $r(\lambda)$ is less than $m$.
It follows that

$$p(A) \equiv m(A)q(A) + r(A),$$

or, since $p(A) = m(A) = O_{(n)}$, that $r(A) = O_{(n)}$. Since $r(\lambda)$ is of degree
less than $m$, this implies that $m(\lambda)$ is not the minimum polynomial of $A$
unless $r(\lambda) \equiv 0$. It follows that $p(\lambda) \equiv m(\lambda)q(\lambda)$, and the theorem is proved.

**4.8.6**   COROLLARY.   *The minimum function of a square matrix* A
*is a divisor of the characteristic function of matrix* A.

**4.8.7**   THEOREM.   *Every linear factor* $\lambda - \lambda_1$ *of the characteristic
function* $f(\lambda)$ *of an* nth-order *matrix* A *is also a factor of the minimum
function* $m(\lambda)$ *of* A.

By the division algorithm for polynomials (see Theorem 4.1.8), we
have

$$m(\lambda) \equiv (\lambda - \lambda_1)q(\lambda) + r,$$

where $r$ is a scalar. It follows that

$$m(A) = (A - \lambda_1 I_{(n)})q(A) + rI_{(n)} = O_{(n)},$$

or, rearranging and taking determinants of both sides,

$$|A - \lambda_1 I_{(n)}| \, |q(A)| = -r^n.$$

But, since $\lambda_1$ is a characteristic root of $A$, $|A - \lambda_1 I_{(n)}| = 0$. It follows that
we must have $r = 0$, whence $m(\lambda) \equiv (\lambda - \lambda_1)q(\lambda)$, and the theorem is
proved.

**4.8.8**   COROLLARY.   *If* $f(\lambda)$ *and* $m(\lambda)$ *are the characteristic function
and the minimum function, respectively, of an* n × n *matrix* A *whose* n
*characteristic roots are distinct, then* $f(\lambda) \equiv (-1)^n m(\lambda)$.

**4.8.9** DEFINITION.    A square matrix is said to be *derogatory* if and only if its indicant is less than its order.

**4.8.10** COROLLARY.    *A necessary condition for a square matrix to be derogatory is that its characteristic roots be not distinct.*

# PROBLEMS

**4.8–1** (a) Prove that any polynomial function of a matrix $A$ of indicant $m$ is expressible as a polynomial function of $A$ of degree at most $m - 1$.

(b) Prove that if $A$ is nonsingular and of indicant $m$, then $A^{-1}$ is expressible as a polynomial function of $A$ of degree at most $m - 1$.

(c) Prove that any rational function of a matrix $A$ of indicant $m$, where the denominator function is nonsingular, is expressible as a polynomial function of $A$ of degree at most $m - 1$.

**4.8–2** Prove that matrix $A$ is nilpotent if and only if its characteristic roots are all zero.

**4.8–3** Prove that the characteristic roots of an idempotent matrix $A$ are either 0 or 1, and that the rank of $A$ is the number of characteristic roots equal to 1.

**4.8–4** Show that

$$A = \begin{bmatrix} 7 & 4 & -1 \\ 4 & 7 & -1 \\ -4 & -4 & 4 \end{bmatrix}$$

is derogatory.

**4.8–5** Prove that every $2 \times 2$ nonscalar matrix is nonderogatory.

**4.8–6** Prove that a necessary and sufficient condition for a matrix to be scalar is that its minimum function be of degree one.

**4.8–7** Show that a square matrix is nonsingular if and only if the constant term in its minimum function is nonzero.

## 4.9    Finding the minimum function of a square matrix

There might be a temptation at this point to conjecture that: (1) the minimum function $m(\lambda)$ of a square matrix $A$ is simply the product of the distinct linear factors $\lambda - \lambda_i$ of the characteristic function $f(\lambda)$ of $A$,

(2) the condition in Corollary 4.8.10 is sufficient as well as necessary. Each of these conjectures is shattered by the counterexample

$$A = \begin{bmatrix} 1 & 1 \\ 0 & 1 \end{bmatrix}.$$

Here $f(\lambda) \equiv (\lambda - 1)^2$. Also, since $A - I_{(2)} \neq O_{(2)}$, $m(\lambda)$ is not $\lambda - 1$. It follows that $m(\lambda) = f(\lambda)$.

How, then, does one find the minimum function of a square matrix $A$ whose characteristic roots are not all distinct? The following theorem throws some light on the matter.

**4.9.1**   THEOREM.   *Let* $f(\lambda)$ *be the characteristic function of a square matrix* $A$ *of order* $n > 1$, *and let* $h(\lambda)$ *be the greatest common divisor of the* $(n - 1)$-*rowed minors of* $A - \lambda I$. *Then the polynomial*

$$g(\lambda) \equiv f(\lambda)/h(\lambda)$$

*is the minimum function of* $A$.

We have

$$|A - \lambda I| = f(\lambda) = g(\lambda)h(\lambda)$$

and

$$\text{adj}(A - \lambda I) = h(\lambda)B(\lambda),$$

where the greatest common divisor of the elements of $B(\lambda)$ is 1. Now (by Theorem 3.10.2)

$$(A - \lambda I)[\text{adj}(A - \lambda I)] \equiv f(\lambda)I,$$

so that

$$(A - \lambda I)h(\lambda)B(\lambda) \equiv g(\lambda)h(\lambda)I,$$

or

(1)                                     $(A - \lambda I)B(\lambda) = g(\lambda)I.$

Thus $g(\lambda)I$ has $A - \lambda I$, and hence $\lambda I - A$, as a divisor, and (by Theorem 4.2.10) $g(A) = O$. It follows (by Theorem 4.8.5) that the minimum function $m(\lambda)$ of $A$ divides $g(\lambda)$. Accordingly, set

(2)                                     $g(\lambda) = q(\lambda)m(\lambda).$

Now, since $m(A) = O$, we have (by Theorem 4.2.10) $\lambda I - A$, and hence $A - \lambda I$, is a divisor of $m(\lambda)I$, say,

(3)                                     $m(\lambda)I = (A - \lambda I)C(\lambda).$

Then, using (1), (2), and (3),

$$(A - \lambda I)B(\lambda) \equiv g(\lambda)I \equiv q(\lambda)m(\lambda)I \equiv q(\lambda)(A - \lambda I)C(\lambda),$$

and

$$B(\lambda) = q(\lambda)C(\lambda).$$

It follows that $q(\lambda)$ divides every element of $B(\lambda)$; hence $q(\lambda) = 1$, and by (2), $g(\lambda) = m(\lambda)$, as was to be proved.

**4.9.2** COROLLARY. *The minimum function of a square matrix* A *of order* n *is the invariant factor of* A $- \lambda$I *that is of highest degree.*

Since the corollary is trivial for $n = 1$, we may suppose $n > 1$. The $\lambda$ matrix $A - \lambda I$ is of rank $n$; let $g_1(\lambda), \cdots, g_n(\lambda)$ be its invariant factors in order of increasing degree. Then, employing the notation of Theorem 4.9.1, we have (by Corollary 4.3.7)

$$|A - \lambda I| = f(\lambda) = g_1(\lambda)g_2(\lambda) \cdots g_n(\lambda)$$

and

$$h(\lambda) = g_1(\lambda)g_2(\lambda) \cdots g_{n-1}(\lambda),$$

whence

$$m(\lambda) = f(\lambda)/h(\lambda) = g_n(\lambda),$$

and the corollary is established.

**4.9.3** COROLLARY. *A square matrix* A *of order* n *is nonderogatory if and only if* A $- \lambda$I *has just one nontrivial invariant factor.*

Theorems 4.8.5 and 4.9.1 were given by Frobenius in 1878. The concept of derogatory matrices was introduced by Sylvester in 1884.

The calculation of the minimum function $m(\lambda)$ of a nonzero square matrix $A$ can be laborious. One may find $m(\lambda)$ by using Theorem 4.9.1 or Corollary 4.9.2. Perhaps the most elementary way of finding $m(\lambda)$ is by the following routine:

(1) If $A = a_0 I$, then $m(\lambda) = \lambda - a_0$.

(2) If $A$ is not of the form $aI$, but $A^2 = a_1 A + a_0 I$, then $m(\lambda) = \lambda^2 - a_1\lambda - a_0$.

(3) If $A^2$ is not of the form $aA + bI$, but $A^3 = a_2 A^2 + a_1 A + a_0 I$, then $m(\lambda) = \lambda^3 - a_2\lambda^2 - a_1\lambda - a_0$.
And so on.

As an example, let us find the minimum function of

$$A = \begin{bmatrix} 1 & -1 & -1 \\ -1 & 1 & -1 \\ -1 & -1 & 1 \end{bmatrix}.$$

Clearly, $A$ is not of the form $aI$. Set

$$A^2 = \begin{bmatrix} 3 & -1 & -1 \\ -1 & 3 & -1 \\ -1 & -1 & 3 \end{bmatrix} = a_1 \begin{bmatrix} 1 & -1 & -1 \\ -1 & 1 & -1 \\ -1 & -1 & 1 \end{bmatrix} + a_0 \begin{bmatrix} 1 & 0 & 0 \\ 0 & 1 & 0 \\ 0 & 0 & 1 \end{bmatrix}.$$

Using the first two elements of the first row of each matrix, we find we must have $3 = a_1 + a_0$ and $-1 = -a_1$, or $a_1 = 1$ and $a_0 = 2$. Since these values are found to check for every element of $A^2$, we conclude that $A^2 = A + 2I$, whence $m(\lambda) = \lambda^2 - \lambda - 2$.

## PROBLEMS

**4.9–1**   Find the minimum functions of the matrices $A$ in Problems 4.5–1, 4.5–2, 4.5–3.

**4.9–2**   Prove that an $n \times n$ matrix $A$ is nonderogatory if and only if the greatest common divisor of all $(n - 1)$-rowed minors of $A - \lambda I$ is 1.

**4.9–3**   Let $A$ and $B$ be two $n \times n$ matrices and let $r(\lambda)$ and $s(\lambda)$ denote the minimum functions of $AB$ and $BA$, respectively. Prove that

(**a**) $r(\lambda) = s(\lambda)$ if at least one of $A$ and $B$ is nonsingular,

(**b**) $r(\lambda)$ and $s(\lambda)$ differ at most by a factor $\lambda$ if $A$ and $B$ are both singular.

**4.9–4**   Let $A$ be of order $(m, n)$ and $B$ of order $(n, m)$, where $m > n$. If $r(\lambda)$ and $s(\lambda)$ denote the characteristic functions of $AB$ and $BA$, respectively, prove that $r(\lambda) = \lambda^{m-n} s(\lambda)$.

**4.9–5**   (**a**) If square matrices $A_1$ and $A_2$ (not necessarily of the same order) have minimum functions $m_1(\lambda)$ and $m_2(\lambda)$, respectively, show that the minimum function $m(\lambda)$ of $A = \mathrm{diag}(A_1, A_2)$ is the least common multiple of $m_1(\lambda)$ and $m_2(\lambda)$.

(**b**) Extend part (**a**) to the direct sum of $r$ matrices.

**4.9–6**   Let

$$A = \begin{bmatrix} -5 & 4 & -6 & 3 & 8 \\ -2 & 3 & -2 & 1 & 2 \\ 4 & -3 & 4 & -1 & -6 \\ 4 & -2 & 4 & 0 & -4 \\ -1 & 0 & -2 & 1 & 2 \end{bmatrix}.$$

Show that
(a) $f(\lambda) = (\lambda - 2)^3(\lambda + 1)^2$,
(b) $m(\lambda) = (\lambda - 2)(\lambda + 1)$,
(c) the nontrivial invariant factors of $A$ are $\lambda - 2$, $\lambda^2 - \lambda - 2$, $\lambda^2 - \lambda - 2$.
(d) the nontrivial elementary divisors of $A$ are $\lambda - 2$, $\lambda - 2$, $\lambda - 2$, $\lambda + 1$, $\lambda + 1$.
(e) Find $A^{-1}$ by using the minimum function of $A$.


# ADDENDA TO CHAPTER 4


Most of the following items invite the student to establish certain important and interesting theorems of matrix theory.


### 4.1A    Elementary $\lambda$ Matrices

Let us define an $n \times n$ *elementary $\lambda$ matrix* to be any matrix obtained by performing a single elementary $\lambda$ operation on $I_{(n)}$. Establish the following theorems concerning elementary $\lambda$ matrices.


**I**  THEOREM.    *Elementary $\lambda$ matrices are nonsingular.*


**II**  THEOREM.    *The inverse of an elementary $\lambda$ matrix is an elementary $\lambda$ matrix.*


**III**  THEOREM.    *The application of an elementary $\lambda$ operation to a matrix* A *can be effected by a premultiplication or a postmultiplication of* A *by a corresponding elementary $\lambda$ matrix.*


**IV**  THEOREM.    *If $\lambda$ matrices A($\lambda$) and B($\lambda$) are $\lambda$-equivalent, there exist matrices* P *and* Q, *each a product of elementary $\lambda$ matrices, such that* A = PBQ.


**V**  THEOREM.    *A matrix* A *is a product of elementary $\lambda$ matrices if and only if* A *is square and* |A| *is a nonzero constant.*

**VI**   THEOREM.   *A $\lambda$ matrix $A(\lambda)$ has a $\lambda$ matrix for its inverse if and only if $A(\lambda)$ is a product of elementary $\lambda$ matrices.*

**VII**   THEOREM.   *A matrix $A$ can be reduced to $I$ by elementary $\lambda$ operations if and only if $|A|$ is a nonzero constant.*

### 4.2A   Systems of Linear Differential Equations with Constant Coefficients

Much of the theory of Chapter 4 finds valuable application in the solution of systems of linear differential equations with constant co-efficients. The interested student may consult A. R. Collar, W. J. Duncan, and R. A. Frazer, *Elementary Matrices and Some Applications to Dynamics and Differential Equations*, Chapter V (Cambridge University Press, 1955) and S. Perlis, *Theory of Matrices*, appendix to Chapter 7 (Addison-Wesley Publishing Company, Inc., 1952).

### 4.3A   Equivalence of Pairs of Matrices

The student is invited to establish the following theorems, which play a role in the study of bilinear forms.

**I**   THEOREM.   *If $A_1$, $A_2$, $B_1$, $B_2$ are n × n matrices with constant elements, if $B_1$ and $B_2$ are nonsingular, and if*

$$M_1(\lambda) \equiv A_1 - \lambda B_1, \qquad M_2(\lambda) \equiv A_2 - \lambda B_2$$

*are $\lambda$-equivalent, then there exist two nonsingular matrices $P$ and $Q$ with constant elements such that $M_2 = PM_1Q$.*

**II**   THEOREM.   *Using the notation of Theorem I, since $M_1(\lambda)$ and $M_2(\lambda)$ are $\lambda$-equivalent, there exist $\lambda$ matrices $P_0(\lambda)$ and $Q_0(\lambda)$ such that*

$$M_2(\lambda) = P_0(\lambda)M_1(\lambda)Q_0(\lambda).$$

*The matrices $P$ and $Q$ of Theorem I are, respectively, the left and right re-mainders when $P_0(\lambda)$ and $Q_0(\lambda)$, respectively, are divided by $M_2(\lambda)$.*

**III** Theorem. *If* $A_1$, $A_2$, $B_1$, $B_2$ *are* n × n *matrices with constant elements, if* $B_1$ *and* $B_2$ *are nonsingular, then a necessary and sufficient condition that* $A_1 \overset{E}{=} A_2$ *and* $B_1 \overset{E}{=} B_2$ *is that the two matrices*

$$M_1(\lambda) \equiv A_1 - \lambda B_1, \qquad M_2(\lambda) \equiv A_2 - \lambda B_2$$

*have the same invariant factors.*

### 4.4A  *k*th Roots of Nonsingular Matrices

Let us accept the following lemma from the theory of polynomials.

**I** Lemma. *If* f(x) *is a polynomial of degree* n > 0 *whose constant term is not zero, and if* k *is a positive integer, there exists a polynomial* g(x) *such that*

$$[g(x)]^k - x$$

*is divisible by* f(x).

We shall now, with the aid of the lemma, establish the following interesting theorem.

**II** Theorem. *If* A *is a nonsingular matrix of order* n, *and if* k *is a positive integer, there exists a matrix* B *such that* $B^k = A$ *and* B *is a polynomial in* A *of degree less than* n.

Since $A$ is nonsingular, the characteristic function $f(\lambda)$ of $A$ is a polynomial of degree $n$ whose constant term, by Theorem 4.5.2, is not zero. Hence, by the lemma, there exists a polynomial $g(\lambda)$ such that

$$[g(\lambda)]^k - \lambda \equiv f(\lambda)q(\lambda),$$

where $q(\lambda)$ is also a polynomial. From this identity it follows that

$$[g(A)]^k - A = f(A)q(A).$$

Since, by the Hamilton-Cayley Theorem (Theorem 4.5.3), $f(A) = O$, the last equation yields

$$[g(A)]^k = A.$$

Setting $B = g(A)$ we have $B^k = A$. That $B$ is a polynomial in $A$ of degree less than $n$ follows from Theorem 4.6.2.

### 4.5A    The Coefficients in the Characteristic Function

The characteristic function of an $n \times n$ matrix $A$ is defined to be

$$f(\lambda) \equiv |A - \lambda I_{(n)}| \equiv (-1)^n[\lambda^n - p_1\lambda^{n-1} + p_2\lambda^{n-2} + \cdots + (-1)^n p_n].$$

In Theorem 4.5.2 it was shown that $p_1 = \text{tr } A$ and $p_n = |A|$. One naturally wonders if the other coefficients in the characteristic function of $A$ can also be neatly and simply expressed in terms of the matrix $A$. In this connection we have the following attractive theorem.

THEOREM.    *The coefficient of* $\lambda^{n-h}$ $(1 \leq h \leq n)$ *in the characteristic function* $f(\lambda)$ *of an* $n \times n$ *matrix* $A$ *is* $(-1)^h$ *times the sum of the* h-*rowed principal minors of* $A$. *That is,*

$$p_h = \sum (\text{h-}rowed\ principal\ minors\ of\ A).$$

An indication of a proof of this theorem can be found in F. E. Hohn, *Elementary Matrix Algebra*, 2nd ed., Section 8.4 (The Macmillan Company). The student is invited to complete the details of this proof.

### 4.6A    Computation of $A^{-1}$ by the Hamilton-Cayley Equation

In Theorem 4.6.3 it was shown that if

$$f(\lambda) \equiv (-1)^n[\lambda^n - p_1\lambda^{n-1} + p_2\lambda^{n-2} + \cdots + (-1)^n p_n]$$

is the characteristic function of a nonsingular $n \times n$ matrix $A$, then

$$A^{-1} = (-1)^{n-1}[A^{n-1} - p_1 A^{n-2} + p_2 A^{n-3} - \cdots + (-1)^{n-1}p_{n-1}I_{(n)}]/|A|.$$

The student may care to justify the following procedure, based upon the preceding equation, for computing the inverse of a nonsingular $n \times n$ matrix $A$.

(1) Find $A^2, A^3, \cdots, A^{n-1}$, and the diagonal elements of $A^n$ (by punched card methods or modern high-speed electronic computer).

(2) Calculate $s_1 = \text{tr } A$, $s_2 = \text{tr } A^2, \cdots, s_n = \text{tr } A^n$.

(3) Determine

$$
\begin{aligned}
p_1 &= s_1, \\
p_2 &= (p_1 s_1 - s_2)/2, \\
p_3 &= (p_2 s_1 - p_1 s_2 + s_3)/3, \\
p_4 &= (p_3 s_1 - p_2 s_2 + p_1 s_3 - s_4)/4, \\
&\ \cdot\ \cdot\ \cdot\ \cdot\ \cdot\ \cdot\ \cdot\ \cdot\ \cdot\ \cdot \\
p_n &= (p_{n-1}s_1 - p_{n-2}s_2 + \cdots + (-1)^{n-1}s_n)/n.
\end{aligned}
$$

(4) Calculate

$$A^{-1} = (-1)^{n-1}[A^{n-1} - p_1 A^{n-2} + p_2 A^{n-3} - \cdots + (-1)^{n-1} p_{n-1} I_{(n)}]/p_n.$$

### 4.7A   Frame's Recursion Formula for Inverting a Matrix

In 1949, J. S. Frame announced a simple recursion procedure for finding the inverse of a nonsingular $n \times n$ matrix $A$ (see J. S. Frame, "A simple recursion formula for inverting a matrix," Abstract, *Bull. Amer. Math. Soc.*, 55 (1949), p. 1045). Denoting $I_{(n)}$ by $A_0$, one successively computes $c_1, A_1, c_2, A_2, \cdots, c_{n-1}, A_{n-1}, c_n$ by the two formulas

$$c_k = (1/k) \operatorname{tr}(A A_{k-1}), \qquad A_k = A A_{k-1} - c_k I.$$

Then

$$A^{-1} = A_{n-1}/c_n.$$

The student is invited to justify Frame's procedure. The student interested in the problem of matrix inversion should consult Donald Greenspan, "Methods of matrix inversion," *Amer. Math. Monthly*, 62 (1955), pp. 303–318. Here can be found a good list of references and sources for further study.

### 4.8A   Characteristic Roots of a Polynomial Function of a Matrix $A$

The student is invited to establish the following useful theorem, given by Frobenius in 1878.

THEOREM.   *If $\lambda_1, \lambda_2, \cdots \lambda_n$ are the characteristic roots, distinct or not, of an* n $\times$ n *matrix* A, *and if* g(A) *is any polynomial function of* A, *then the characteristic roots of* g(A) *are* g($\lambda_1$), g($\lambda_2$), $\cdots$, g($\lambda_n$).

# 5. SIMILARITY AND CONGRUENCE

*5.1. Similar matrices. Problems. 5.2. Similar matrices (continued). Problems. 5.3. Congruent matrices. Problems. 5.4. Canonical forms under congruency for skew-symmetric $\mathscr{C}$ matrices. Problems. 5.5. Canonical forms under congruency for symmetric $\mathscr{R}$ matrices. Problems. 5.6. Conjunctivity, or Hermitian congruence. Problems. 5.7. Orthogonal matrices and orthogonal similarity. Problems. 5.8. Unitary matrices and unitary similarity. Problems. 5.9. Normal matrices. Problems. ADDENDA. 5.1A. Companion matrices. 5.2A. Regular symmetric matrices. 5.3A. Rotations in 3-space. 5.4A. Cayley's construction of real orthogonal matrices. 5.5A. The characteristic roots of an orthogonal matrix. 5.6A. Definite, semidefinite, and indefinite real symmetric matrices. 5.7A. Gram matrices. 5.8A. Some theorems of Autonne. 5.9A. Simultaneous reduction of a pair of quadratic forms. 5.10A. Hadamard matrices.*

### 5.1 Similar matrices

In Chapter 2 we considered the important notion of *equivalence* of matrices. Two $m \times n$ matrices $A$ and $B$, it will be recalled, are *equivalent* if and only if there exist nonsingular matrices $P_{(n)}$ and $Q_{(m)}$ such that $B = QAP$. In the present chapter we shall concern ourselves with certain special types of equivalence of matrices, obtained by further restricting the two matrices $P$ and $Q$ in the relation $B = QAP$. These restricted equivalences find extensive application in various parts of science and mathematics, and just as the notion of general equivalence arises naturally in the consideration of systems of linear equations, the restricted equivalences can be shown to arise naturally in the consideration of some one or other of the applications. It is not our aim, however, to approach the restricted equivalences via applications of them.

In this first section of Chapter 5 we briefly examine the special equivalence of matrices known as *similarity*, a concept that arises naturally in the study of linear transformations.

**5.1.1** DEFINITION AND NOTATION. Square matrix $B$ is said to be *similar* to square matrix $A$, and we write $B \stackrel{S}{=} A$, if and only if there exists a nonsingular matrix $P$ such that $B = P^{-1}AP$.

**5.1.2** THEOREM. *Similarity of matrices is reflexive, symmetric, and transitive.*

Let $A$, $B$, $C$ denote square matrices. Since $A = I^{-1}AI$, we have $A \stackrel{S}{=} A$, and the relation is reflexive. Suppose $A \stackrel{S}{=} B$. Then there exists nonsingular $P$ such that $A = P^{-1}BP$, whence $B = PAP^{-1} = (P^{-1})^{-1}AP^{-1}$. That is, $B \stackrel{S}{=} A$, and the relation is symmetric. Suppose $A \stackrel{S}{=} B$ and $B \stackrel{S}{=} C$. Then there exist nonsingular $P$ and $Q$ such that $A = P^{-1}BP$, $B = Q^{-1}CQ$, whence $A = P^{-1}(Q^{-1}CQ)P = (QP)^{-1}C(QP)$. That is, $A \stackrel{S}{=} C$, and the relation is transitive.

**5.1.3** THEOREM. *Similar matrices have the same characteristic roots.*

Suppose $B \stackrel{S}{=} A$. Then there exists nonsingular $P$ such that $B = P^{-1}AP$, whence

$$|B - \lambda I| \equiv |P^{-1}AP - \lambda I| \equiv |P^{-1}AP - \lambda P^{-1}IP|$$

$$\equiv |P^{-1}(A - \lambda I)P| \equiv |P^{-1}|\,|A - \lambda I|\,|P|$$

$$\equiv |A - \lambda I|,$$

and the theorem follows.

**5.1.4** COROLLARY. *The coefficients of the characteristic function of a matrix* A *are invariant under any similarity transformation of* A. *In particular, if* B $\stackrel{S}{=}$ A, *then* tr B = tr A *and* |B| = |A|.
See Theorem 4.5.2.

**5.1.5**   THEOREM.   *A necessary and sufficient condition for an* n × n *matrix* A *to be similar to a diagonal matrix is that it possess* n *linearly independent characteristic vectors.*

Suppose $A \overset{S}{=} D = \text{diag}(d_1, d_2, \cdots, d_n)$. Then there exists a nonsingular matrix $P$ such that $P^{-1}AP = D$, or $AP = PD$. Let $C_i$ denote the $i$th column of $P$, so that

$$P = [C_1, C_2, \cdots, C_n].$$

Then we have

$$A[C_1, C_2, \cdots, C_n] = [C_1, C_2, \cdots, C_n] \text{diag}(d_1, d_2, \cdots, d_n)$$

or

$$[AC_1, AC_2, \cdots, AC_n] = [d_1 C_1, d_2 C_2, \cdots, d_n C_n].$$

It follows that

$$AC_1 = d_1 C_1, \qquad AC_2 = d_2 C_2, \qquad \cdots, \qquad AC_n = d_n C_n,$$

and $C_1, C_2, \cdots, C_n$ are characteristic vectors of $A$ (and $d_1, d_2, \cdots, d_n$ are the characteristic roots of $A$). Since $P$ is nonsingular, the columns $C_1, C_2, \cdots, C_n$ of $P$ are linearly independent.

Conversely, suppose $C_1, C_2, \cdots, C_n$ are $n$ linearly independent characteristic vectors of matrix $A$ and let $d_1, d_2, \cdots, d_n$ be the corresponding characteristic roots of $A$. Then we have

$$AC_1 = d_1 C_1, \qquad AC_2 = d_2 C_2, \qquad \cdots, \qquad AC_n = d_n C_n.$$

If we set

$$P = [C_1, C_2, \cdots, C_n],$$

the last set of equations can be written as $AP = PD$, where

$$D = \text{diag}(d_1, d_2, \cdots, d_n).$$

Since the $C_i$ are linearly independent, $P$ is nonsingular and $P^{-1}$ exists, whence, from $AP = PD$, we get $P^{-1}AP = D$ and $A \overset{S}{=} D$.

**5.1.6**   COROLLARY.   *If a matrix* A *is similar to a diagonal matrix* D, *then the elements along the diagonal of* D *are the characteristic roots of* A.

**5.1.7**   COROLLARY.   *If* P *is a nonsingular modal matrix of a matrix* A, *then* $P^{-1}AP = D$, *a diagonal matrix.*

**5.1.8** COROLLARY. *A matrix* A *whose characteristic roots are distinct is similar to a diagonal matrix* D.

For, by Theorem 4.7.5, if the characteristic roots of $A$ are distinct, then $A$ possesses a nonsingular modal matrix $P$.


## PROBLEMS


**5.1–1** Prove Corollary 5.1.6.

**5.1–2** Prove Corollary 5.1.7.

**5.1–3** Show that the following matrices are similar to diagonal matrices, and in each case find the transforming matrix and the diagonal matrix.

(a) $\begin{bmatrix} 8 & -6 & 2 \\ -6 & 7 & -4 \\ 2 & -4 & 3 \end{bmatrix}$,  (b) $\begin{bmatrix} 6 & -2 & 2 \\ -2 & 3 & -1 \\ 2 & -1 & 3 \end{bmatrix}$,

(c) $\begin{bmatrix} 8 & -8 & -2 \\ 4 & -3 & -2 \\ 3 & -4 & 1 \end{bmatrix}$,  (d) $\begin{bmatrix} 0 & 1 & 0 & 0 \\ 0 & 0 & 1 & 0 \\ 0 & 0 & 0 & 1 \\ -1 & 0 & -1 & 0 \end{bmatrix}$.

**5.1–4** If $Y$ is a characteristic vector of $B = P^{-1}AP$ corresponding to the characteristic root $\lambda_i$ of $B$, show that $X = PY$ is a characteristic vector of $A$ corresponding to the same characteristic root $\lambda_i$ of $A$.

**5.1–5** Let $A$ and $B$ be two $n \times n$ matrices possessing the same $n$ distinct characteristic roots. Show that there exist two matrices $P$ and $Q$, one of which is nonsingular, such that $A = PQ$ and $B = QP$.

**5.1–6** Let $A$ be an $n \times n$ matrix possessing $n$ distinct characteristic roots. Then (by Corollary 5.1.8) there exists a nonsingular matrix $P$ such that $P^{-1}AP$ is a diagonal matrix. Show that if $B$ is a matrix that is commutative with $A$, then $P^{-1}BP$ is a diagonal matrix.

**5.1–7** Show that any matrix $B$ similar to matrix $A$ can be obtained from $A$ by a finite sequence of pairs of elementary operations of the forms

$$o(j, k),\ o'(j, k);\qquad o(ck),\ o'(c^{-1}k);\qquad o(j + ak),\ o'(j - ak).$$

**5.1–8** Prove that two $n \times n$ diagonal matrices are similar if and only if they have the same diagonal elements, irrespective of any difference in position of these elements.

**5.1–9**  (a) Let $\lambda_j$ and $\lambda_k$ be any two characteristic roots of square matrix $A$ and let $X_k$ be a characteristic vector of $A$ corresponding to $\lambda_k$. Show that

$$(A - \lambda_j I)X_k = (\lambda_k - \lambda_j)X_k.$$

(b) Let $\lambda_1, \cdots, \lambda_n$ be the $n$ characteristic roots of an $n \times n$ matrix $A$ and let $X_1, \cdots, X_n$ be characteristic vectors of $A$ corresponding to $\lambda_1, \cdots, \lambda_n$. If $c_1, \cdots, c_n$ are scalars, show that

$$(A - \lambda_2 I)(A - \lambda_3 I) \cdots (A - \lambda_n I)(c_1 X_1 + c_2 X_2 + \cdots + c_n X_n)$$

$$= (\lambda_1 - \lambda_2)(\lambda_1 - \lambda_3) \cdots (\lambda_1 - \lambda_n)c_1 X_1.$$

(c) Fashion a proof, based on parts (a) and (b) of this problem, of Theorem 4.7.3.

**5.1–10**  (a) Let $A$ be an $n \times n$ involutoric matrix and let $X \neq O$ be an $n$th-order column vector that is not a characteristic vector of $A$. Show that $X + AX$ and $X - AX$ are independent characteristic vectors of $A$.

(b) Prove that every involutoric matrix is similar to a diagonal matrix each of whose nonzero elements is either 1 or $-1$.

**5.1–11**  Let $A$ be an $n \times n$ matrix possessing $n$ distinct characteristic roots. Show that the most general matrix $B$ that commutes with $A$ is of the form $B = f(A)$, where $f(A)$ is a polynomial function of $A$ of degree at most $n - 1$.

## 5.2   Similar matrices (continued)

One might think that the converses of Corollary 5.1.8 and Theorem 4.7.5 also hold; that is, if an $n \times n$ matrix $A$ is similar to a diagonal matrix, then the $n$ characteristic roots of $A$ are distinct, or equivalently, if an $n \times n$ matrix $A$ has $n$ linearly independent characteristic vectors, then the $n$ characteristic roots of $A$ are distinct. A simple counterexample, however, shows that these converses are not true. Thus the matrix

$$A = \begin{bmatrix} 2 & 2 & 1 \\ 1 & 3 & 1 \\ 1 & 2 & 2 \end{bmatrix}$$

has $\lambda_1 = 5$, $\lambda_2 = 1$, $\lambda_3 = 1$ for its three characteristic roots. As characteristic vectors corresponding to these respective characteristic roots, one may take

$$\{1, 1, 1\}, \qquad \{2, -1, 0\}, \qquad \{1, 0, -1\},$$

and these three characteristic vectors are linearly independent. Indeed, we have

$$\begin{bmatrix} 1 & 2 & 1 \\ 1 & -1 & 0 \\ 1 & 0 & -1 \end{bmatrix}^{-1} \begin{bmatrix} 2 & 2 & 1 \\ 1 & 3 & 1 \\ 1 & 2 & 2 \end{bmatrix} \begin{bmatrix} 1 & 2 & 1 \\ 1 & -1 & 0 \\ 1 & 0 & -1 \end{bmatrix} = \begin{bmatrix} 5 & 0 & 0 \\ 0 & 1 & 0 \\ 0 & 0 & 1 \end{bmatrix},$$

and

$$A \overset{S}{=} \operatorname{diag}(5, 1, 1).$$

The following theorem throws some light on the case where the characteristic roots of a square matrix $A$ are not distinct.

**5.2.1** THEOREM. *A necessary and sufficient condition for an* n × n *matrix* A *to be similar to a diagonal matrix is that the multiplicity of each characteristic root* $\lambda_i$ *of* A *be equal to the nullity of the matrix* A $- \lambda_i$I.

Suppose $A$ is similar to a diagonal matrix $D$. Then there exists a nonsingular matrix $P$ such that

$$P^{-1}AP = D = \operatorname{diag}(\lambda_1, \cdots, \lambda_n),$$

where $\lambda_1, \cdots, \lambda_n$ are the characteristic roots of $A$. Suppose that exactly $k$ of these characteristic roots are equal to $\lambda_i$. Then $D - \lambda_i I$ has exactly $k$ zeros in its diagonal, and is thus of rank $n - k$. But

$$P^{-1}(A - \lambda_i I)P = P^{-1}AP - \lambda_i P^{-1}IP = D - \lambda_i I,$$

whence $A - \lambda_i I$ also is of rank $n - k$. It follows that the multiplicity of $\lambda_i$ is equal to the nullity of $A - \lambda_i I$, both being equal to $k$.

Conversely, let $\lambda_1, \cdots, \lambda_s$ denote the *distinct* characteristic roots of $A$ with respective multiplicities $k_1, \cdots, k_s$, where, of course, $k_1 + \cdots + k_s = n$. Suppose, for $i = 1, \cdots, s$, the nullity of $A - \lambda_i I$ is $k_i$. Then (see Theorem 2.9.1) the homogeneous system of linear equations

$$(A - \lambda_i I)X = O$$

has $k_i$ linearly independent solutions, say, $X_{i1}, \cdots, X_{ik_i}$. Suppose there exist scalars $a_{ij}$ such that

(1) $\quad (a_{11}X_{11} + \cdots + a_{1k_1}X_{1k_1}) + (a_{21}X_{21} + \cdots + a_{2k_2}X_{2k_2})$

$$+ \cdots + (a_{s1}X_{s1} + \cdots + a_{sk_s}X_{sk_s}) = O.$$

Now

$$Y_i \equiv a_{i1}X_{i1} + \cdots + a_{ik_i}X_{ik_i}$$

is a linear combination of the characteristic vectors $X_{i1}, \cdots, X_{ik_i}$ of $A$ corresponding to the characteristic root $\lambda_i$, and is therefore (by Theorem 4.7.6) itself a characteristic vector of $A$ corresponding to the characteristic root $\lambda_i$, or is a zero vector. Since the $\lambda_i$ $(i = 1, \cdots, s)$ are distinct, it follows (by Theorem 4.7.3) that the $Y_i$ $(i = 1, \cdots, s)$, if not zero vectors, are linearly independent. But from (1) we have $Y_1 + \cdots + Y_s = O$. It follows that each $Y_i = O$, whence, since $X_{i1}, \cdots, X_{ik_i}$ are linearly independent, we must have $a_{i1} = \cdots = a_{ik_i} = 0$. That is, all the $a_{ij}$ in (1) are equal to zero, and the $n$ characteristic vectors

$$X_{11}, \cdots, X_{1k_1}; \qquad X_{21}, \cdots, X_{2k_2}; \qquad \cdots; \qquad X_{s1}, \cdots, X_{sk_s}$$

are linearly independent. We now conclude (by Theorem 5.1.5) that $A$ is similar to a diagonal matrix.

Not all square matrices are similar to diagonal matrices. For example, the matrix

$$A = \begin{bmatrix} 2 & -1 & 1 \\ 2 & 2 & -1 \\ 1 & 2 & -1 \end{bmatrix}$$

has 1, 1, 1 for its three characteristic roots. But the matrix

$$A - I = \begin{bmatrix} 1 & -1 & 1 \\ 2 & 1 & -1 \\ 1 & 2 & -2 \end{bmatrix}$$

has rank 2 and hence nullity 1. Since the nullity of $A - I$ is not equal to the multiplicity of the characteristic root 1, it follows, by Theorem 5.2.1, that $A$ is not similar to a diagonal matrix. We show, however, in the next theorem that $A$ is similar to a *triangular* matrix whose diagonal elements are the characteristic roots of $A$.

**5.2.2** Theorem.    *Any square matrix* A *is similar to an upper triangular matrix whose diagonal elements are the characteristic roots of* A.

Let $\lambda_1, \cdots, \lambda_n$ be the $n$ (not necessarily distinct) characteristic roots of $A$. Let $C_1$ be any characteristic vector of $A$ corresponding to $\lambda_1$. Then $AC_1 = \lambda_1 C_1$, $C_1 \neq O$.

Let $P_1$ be any nonsingular matrix having $C_1$ as its first column. Consider the matrix $P_1^{-1} A P_1$. Since $C_1$ is the first column of $P_1$, the first column of $P_1^{-1}AP_1$ is

$$P_1^{-1}AC_1 = P_1^{-1}\lambda_1 C_1 = \lambda_1 P_1^{-1}C_1.$$

But $P_1^{-1}C_1$ is the first column of $P_1^{-1}P_1 = I$. Therefore

$$\lambda_1 P_1^{-1}C_1 = \{\lambda_1, 0, \cdots, 0\}$$

and $P_1^{-1}AP_1$ has the form

$$\begin{bmatrix} \lambda_1 & L \\ O & B \end{bmatrix},$$

where $B$ is an $(n - 1)$-rowed nonsingular matrix with, by Theorem 5.1.3, characteristic roots $\lambda_2, \cdots, \lambda_n$.

By an argument similar to the one above, there exists a nonsingular $(n - 1)$-rowed matrix $Q$ such that

$$Q^{-1}BQ = \begin{bmatrix} \lambda_2 & M \\ O & C \end{bmatrix},$$

where $C$ is an $(n - 2)$-rowed nonsingular matrix with characteristic roots $\lambda_3, \cdots, \lambda_n$.

Define $P_2$ to be the nonsingular $n \times n$ matrix

$$\begin{bmatrix} I_{(2)} & O \\ O & Q \end{bmatrix}.$$

Now

$$P_2^{-1}P_1^{-1}AP_1P_2 = \begin{bmatrix} I_{(1)} & O \\ O & Q^{-1} \end{bmatrix} \begin{bmatrix} \lambda_1 & L \\ O & B \end{bmatrix} \begin{bmatrix} I_{(1)} & O \\ O & Q \end{bmatrix} = \begin{bmatrix} \lambda_1 & LQ \\ O & Q^{-1}BQ \end{bmatrix}.$$

Setting $LQ = [\alpha_{12} \mid L_1]$, where $L_1$ is a row vector of order $n - 2$, we then have

$$P_2^{-1}P_1^{-1}AP_1P_2 = \begin{bmatrix} \lambda_1 & \alpha_{12} & L_1 \\ 0 & \lambda_2 & M \\ O & O & C \end{bmatrix}.$$

Proceeding in this way, we obtain a sequence $P_1, \cdots, P_n$ of $n$-rowed nonsingular matrices such that

$$P_n^{-1} \cdots P_2^{-1}P_1^{-1}AP_1P_2 \cdots P_n = T,$$

where $T$ is an upper triangular matrix having $\lambda_1, \cdots, \lambda_n$ as its diagonal elements. Setting $P = P_1P_2 \cdots P_n$, we see that $P$ is nonsingular and $P^{-1}AP = T$, which proves the theorem.

## PROBLEMS

**5.2–1**  Show that the three characteristic roots of

$$A = \begin{bmatrix} 2 & 2 & 1 \\ 1 & 3 & 1 \\ 1 & 2 & 2 \end{bmatrix}$$

are 5, 1, 1. Show that

$$\{1, 1, 1\}, \qquad \{2, -1, 0\}, \qquad \{1, 0, -1\}$$

are characteristic vectors of $A$ corresponding to the characteristic roots 5, 1, 1, respectively. Show that the three characteristic vectors are linearly independent.

**5.2–2**  Verify Theorem 5.2.1 for the matrix $A$ of Problem 5.2–1.

**5.2–3**  Show that the three characteristic roots of

$$A = \begin{bmatrix} 2 & -1 & 1 \\ 2 & 2 & -1 \\ 1 & 2 & -1 \end{bmatrix}$$

are 1, 1, 1. Find, by the process described in the proof of Theorem 5.2.2, an upper triangular matrix having 1, 1, 1 as its diagonal elements and which is similar to matrix $A$.

**5.2–4**  Show that the following matrices are not similar to diagonal matrices:

(a) $\begin{bmatrix} 3 & 10 & 5 \\ -2 & -3 & -4 \\ 3 & 5 & 7 \end{bmatrix}$,  (b) $\begin{bmatrix} 2 & 3 & 4 \\ 0 & 2 & -1 \\ 0 & 0 & 1 \end{bmatrix}$,  (c) $\begin{bmatrix} 2 & 1 & 0 \\ 0 & 2 & 1 \\ 0 & 0 & 2 \end{bmatrix}$,

(d) $\begin{bmatrix} -1 & 0 & 0 & 0 \\ 1 & 2 & 1 & 0 \\ 1 & 3 & 2 & -1 \\ 0 & 5 & 3 & -1 \end{bmatrix}$.

**5.2–5**  Denoting each matrix in turn in Problem 5.2–4 by $A$, find a nonsingular matrix $P$ such that $P^{-1}AP$ is triangular.

### 5.3  Congruent matrices

Another special equivalence of matrices is that known as *congruency*, a concept that arises naturally in the study of quadratic and bilinear forms.

**5.3.1**   DEFINITION AND NOTATION.   A square matrix $B$ is said to be *congruent* to a square matrix $A$, and we write $B \overset{c}{=} A$, if and only if there exists a nonsingular matrix $P$ such that $B = P'AP$.

**5.3.2**   THEOREM.   *Congruency of matrices is reflexive, symmetric, and transitive.*

Let $A$, $B$, $C$ denote square matrices. Since $A = I'AI$, we have $A \overset{c}{=} A$, and the relation is reflexive. Suppose $A \overset{c}{=} B$. Then there exists nonsingular $P$ such that $A = P'BP$, whence $B = (P')^{-1}AP^{-1} = (P^{-1})'AP^{-1}$. That is, $B \overset{c}{=} A$, and the relation is symmetric. Suppose $A \overset{c}{=} B$ and $B \overset{c}{=} C$. Then there exist nonsingular $P$ and $Q$ such that $A = P'BP$, $B = Q'CQ$, whence $A = P'(Q'CQ)P = (QP)'C(QP)$. That is, $A \overset{c}{=} C$, and the relation is transitive.

If matrix $B$ is equivalent to matrix $A$, then one can convert $A$ into $B$ by a finite sequence of elementary row and column operations. Since congruence of two matrices is a special kind of equivalence of the two matrices, one might anticipate that, if matrix $B$ is congruent to matrix $A$, then perhaps $A$ can be converted into $B$ by a special sort of finite sequence of elementary row and column operations. The anticipation proves to be correct and leads us to formulate the following definition.

**5.3.3**   DEFINITION.   An elementary row operation applied to a square matrix, and followed by the corresponding elementary column operation, is called an *elementary cogredient operation* on the matrix.

**5.3.4**   THEOREM.   *If A and B are* n × n *matrices, then* $B \overset{c}{=} A$ *if and only if B is obtainable from A by a finite sequence of elementary cogredient operations.*

Suppose $B \overset{c}{=} A$. Then $B = P'AP$, where $P'$ is nonsingular. Since $P'$ is nonsingular, we have (by Theorem 2.2.11)

$$P' = E_1 E_2 \cdots E_k,$$

where the $E_i$ are ERT matrices. Then

$$B = E_1 E_2 \cdots E_k A E_k' \cdots E_2' E_1',$$

and it follows that $B$ is obtainable from $A$ by a sequence of $k$ elementary cogredient operations.

The converse is easily established by reversing the foregoing argument.

**5.3.5** THEOREM. *If* $B \stackrel{c}{=} A$, *then* B *is symmetric* (*skew symmetric*) *if and only if* A *is symmetric* (*skew symmetric*).

For we have $B = P'AP$, where $P$ is nonsingular, whence $B' = P'A'P$. It is now clear that $B = B'$ if $A = A'$, and that $B = -B'$ if $A = -A'$. We also have, since $P$ is nonsingular, $A = (P')^{-1}BP^{-1} = (P^{-1})'BP^{-1}$, whence $A' = (P^{-1})'B'P^{-1}$, and we see that $A = A'$ if $B = B'$, and that $A = -A'$ if $B = -B'$.

**5.3.6** THEOREM. *If*

$$A = \begin{bmatrix} A_1 & O \\ O & A_2 \end{bmatrix} \quad and \quad B = \begin{bmatrix} B_1 & O \\ O & B_2 \end{bmatrix},$$

*and if* $A_1 \stackrel{c}{=} B_1$ *and* $A_2 \stackrel{c}{=} B_2$, *then* $A \stackrel{c}{=} B$.

Since $A_1 \stackrel{c}{=} B_1$ and $A_2 \stackrel{c}{=} B_2$, we have $A_1 = P_1'B_1P_1$, $A_2 = P_2'B_2P_2$, $|P_1| \neq 0$, $|P_2| \neq 0$. Then

$$A = \begin{bmatrix} A_1 & O \\ O & A_2 \end{bmatrix} = \begin{bmatrix} P_1'B_1P_1 & O \\ O & P_2'B_2P_2 \end{bmatrix} = \begin{bmatrix} P_1' & O \\ O & P_2' \end{bmatrix}\begin{bmatrix} B_1 & O \\ O & B_2 \end{bmatrix}\begin{bmatrix} P_1 & O \\ O & P_2 \end{bmatrix}.$$

Setting

$$P = \begin{bmatrix} P_1 & O \\ O & P_2 \end{bmatrix},$$

we note that

$$P' = \begin{bmatrix} P_1' & O \\ O & P_2' \end{bmatrix}$$

and that $|P| = |P_1|\,|P_2| \neq 0$. It follows that $A \stackrel{c}{=} B$.

**5.3.7** THEOREM. *If* A *and* B *are* n × n *matrices and* $A_1$ *and* $B_1$ *are* r × r *nonsingular matrices, where* r $\leq$ n, *and if*

$$A \stackrel{c}{=} \begin{bmatrix} A_1 & O \\ O & O \end{bmatrix} \quad and \quad B \stackrel{c}{=} \begin{bmatrix} B_1 & O \\ O & O \end{bmatrix},$$

*then* A $\stackrel{c}{=}$ B *if and only if* $A_1 \stackrel{c}{=} B_1$.

Suppose $A \stackrel{c}{=} B$. Then there exists an $n \times n$ nonsingular matrix

$$P = \begin{bmatrix} P_1 & P_2 \\ P_3 & P_4 \end{bmatrix},$$

where $P_1$ is $r \times r$, such that

$$\begin{bmatrix} A_1 & O \\ O & O \end{bmatrix} = \begin{bmatrix} P_1' & P_3' \\ P_2' & P_4' \end{bmatrix}\begin{bmatrix} B_1 & O \\ O & O \end{bmatrix}\begin{bmatrix} P_1 & P_2 \\ P_3 & P_4 \end{bmatrix} = \begin{bmatrix} P_1'B_1P_1 & P_1'B_1P_2 \\ P_2'B_1P_1 & P_2'B_1P_2 \end{bmatrix}.$$

Now $A_1 = P_1'B_1P_1$, and since $A_1$ and $B_1$ are nonsingular, $P_1$ is nonsingular. It follows that $A_1 \stackrel{c}{=} B_1$.

Conversely, if $A_1 \stackrel{c}{=} B_1$, then (by Theorem 5.3.6) $A \stackrel{c}{=} B$.

**5.3.8** THEOREM. *If* A *and* B *are nth-order symmetric matrices of rank* r, *then* A $\stackrel{c}{=}$ B *if and only if a nonsingular* r-*rowed principal submatrix of* A *is congruent to a nonsingular* r-*rowed principal submatrix of* B, *in which case each nonsingular* r-*rowed principal submatrix of* A *is congruent to every nonsingular* r-*rowed principal submatrix of* B.

Since $A$ is a symmetric matrix of rank $r$, $A$ contains (by Theorem 3.9.8) a nonsingular $r$-rowed principal submatrix $A_1$. Since the rows of $A$ through $A_1$ constitute $r$ linearly independent rows of $A$, and since $A$ is symmetric, we may, by elementary cogredient operations, transform $A$ into

$$\begin{bmatrix} A_1 & O \\ O & O \end{bmatrix},$$

whence

$$A \stackrel{c}{=} \begin{bmatrix} A_1 & O \\ O & O \end{bmatrix}.$$

Similarly, $B$ contains a nonsingular $r$-rowed principal submatrix $B_1$ and

$$B \stackrel{c}{=} \begin{bmatrix} B_1 & O \\ O & O \end{bmatrix}.$$

The theorem now follows by Theorem 5.3.7.

## PROBLEMS

**5.3–1** Show that one may commute the elementary row operation and the elementary column operation of an elementary cogredient operation.

**5.3–2**   Prove that if $B \overset{C}{=} A$, then $B$ is Hermitian (skew-Hermitian) if and only if $A$ is Hermitian (skew-Hermitian).

**5.3–3**   Extend Theorem 5.3.6 to matrices $A = \text{diag}(A_1, A_2, \cdots, A_n)$ and $B = \text{diag}(B_1, B_2, \cdots, B_n)$.

**5.3–4**   If $A = [a_{ij}]_{(m,n)}$, $X = \{x_1, \cdots, x_m\}$, $Y = \{y_1, \cdots, y_n\}$, then the algebraic expression $X'AY$ is called a *bilinear form* in the two sets of variables $(x_1, \cdots, x_m)$ and $(y_1, \cdots, y_n)$. The bilinear form is said to be *symmetric, alternate, Hermitian,* or *alternate Hermitian,* according as $A$ is symmetric, skew-symmetric, Hermitian, or skew-Hermitian. The rank of the matrix $A$ is called the *rank* of the bilinear form. When $X$ and $Y$ of a bilinear form $X'AY$, in which $A$ is $n \times n$, are subjected to transformations $X = PU$ and $Y = PV$, where $P$ is nonsingular and $U$ and $V$ are $n$th-order column vectors, the bilinear form is said to be transformed *cogrediently.* When $X$ and $Y$ of a bilinear form $X'AY$, in which $A$ is $n \times n$, are subjected to transformations $X = (P^{-1})'U$ and $Y = PV$, the bilinear form is said to be transformed *contragrediently.*

(a) Show that if a bilinear form $X'AY$, in which $A$ is square, is transformed cogrediently, the bilinear form is carried into a bilinear form whose matrix is congruent to matrix $A$.

(b) Show that if a bilinear form $X'AY$, in which $A$ is square, is transformed contragrediently, the bilinear form is carried into a bilinear form whose matrix is similar to matrix $A$.

(c) Show that under either a cogredient or a contragredient transformation, a symmetric, alternate, Hermitian, or alternate-Hermitian bilinear form is carried, respectively, into a symmetric, alternate, Hermitian, or alternate-Hermitian bilinear form.

**5.3–5**   A bilinear form $X'AY$, wherein $A$ is $m \times n$, is said to be *linearly transformed* when $X$ and $Y$ are subjected to transformations $X = PU$ and $Y = QV$, where $P$ and $Q$ are nonsingular and $U$ and $V$ are column vectors of orders $m$ and $n$, respectively. Show that the rank of a bilinear form remains unchanged under any linear transformation.

**5.3–6**   Prove that the bilinear form $X'IY$ is linearly transformed into itself if and only if it is transformed contragrediently.

**5.3–7**   Prove that a nonzero bilinear form is factorable if and only if its rank is 1.

**5.3–8**   In the bilinear form $X'AY$, let $A$ be real and nonsingular. Show that

$$X'A^{-1}Y = - \begin{vmatrix} O & X' \\ Y & A \end{vmatrix} |A^{-1}|.$$

**5.3–9**   If $A$ is a symmetric and $S$ a skew-symmetric matrix of order $n$

such that $A + S$ is nonsingular, and if $T = (A + S)^{-1}(A - S)$, show that

$$T'(A + S)T = A + S \quad \text{and} \quad T'(A - S)T = A - S.$$

Deduce that a *quadratic form* $X'AX$ is left invariant by the transformation $X = TY$.

**5.3–10**    Show that $b - c$ is a congruence invariant of $\begin{bmatrix} a & b \\ c & d \end{bmatrix}$. (This was noted by Kronecker in 1883.)

## 5.4    Canonical forms under congruency for skew-symmetric $\mathscr{C}$ matrices

In this and the next section we shall obtain some canonical forms to which certain special kinds of matrices can be reduced by congruency transformations. The problem of obtaining a canonical form under congruency transformations for the general $n \times n$ matrix is difficult and not yet completely solved. The corresponding general problem for similarity transformations has been solved in various ways, but the treatments are too lengthy and complicated for consideration in our work, and we shall remain content with having obtained, in Section 5.2, a necessary and sufficient condition for a square matrix to be similar to a diagonal matrix.

In the present section we consider skew-symmetric $\mathscr{C}$ matrices under congruency transformations.

**5.4.1    THEOREM.**    *Every skew-symmetric $\mathscr{C}$ matrix* A *of rank* r *is congruent to a matrix of the form*

$$diag(\text{E}, \text{E}, \cdots, \text{E}, 0, \cdots, 0),$$

*where* $\text{E} = \begin{bmatrix} 0 & 1 \\ -1 & 0 \end{bmatrix}$ *and the number of* E's *is* r/2.

Let $A = [a_{ij}]$ be a skew-symmetric $\mathscr{C}$ matrix. The theorem is trivial if $A = O$. If $A \neq O$, there is at least one nonzero element $a_{pq}$, $p < q$. Consider the principal submatrix

$$\begin{bmatrix} 0 & a_{pq} \\ -a_{pq} & 0 \end{bmatrix}$$

lying in the $p$th and $q$th rows and the $p$th and $q$th columns of $A$. By elementary cogredient operations, wherein we interchange the first and $p$th rows and columns and the second and $q$th rows and columns, we can carry the foregoing principal submatrix into the top left corner, after which multiplication of the first row and the first column by $1/a_{pq}$ (thus performing another elementary cogredient operation) yields a matrix

$$B = [b_{ij}] = \begin{bmatrix} E & B_1 \\ -B_1' & B_2 \end{bmatrix},$$

where

$$E = \begin{bmatrix} 0 & 1 \\ -1 & 0 \end{bmatrix}.$$

We now apply a sequence of elementary cogredient operations in which we add $b_{k1}$ $(k = 3, \cdots, n)$ times the second row (column) of $B$ to its $k$th row (column) and also $-b_{k2}$ $(k = 3, \cdots, n)$ times the first row (column) of $B$ to its $k$th row (column). These further elementary cogredient operations convert $B$ into

$$C = \begin{bmatrix} E & O \\ O & C_1 \end{bmatrix}.$$

By Theorems 5.3.4 and 5.3.5, $A \overset{\text{c}}{=} C$ and $C_1$ is skew-symmetric. The process given above may now be applied to $C_1$ to yield

$$D = \begin{bmatrix} E & O \\ O & D_1 \end{bmatrix},$$

where $D_1$ is skew-symmetric. It follows, by Theorem 5.3.6, that

$$A \overset{\text{c}}{=} \begin{bmatrix} E & O & O \\ O & E & O \\ O & O & D_1 \end{bmatrix}.$$

Continuing the process until the matrix in the lower right corner becomes $E$ or a null matrix, we obtain the desired final form. That the number of $E$'s is half the rank of $A$ follows from the fact that the rank of a matrix is unchanged by a congruency transformation.

**5.4.2**  COROLLARY.  *A skew-symmetric $\mathscr{C}$ matrix always has even rank.*

**5.4.3**  COROLLARY.  *Two n-rowed skew-symmetric $\mathscr{C}$ matrices are congruent if and only if they have the same rank, that is, if and only if they are equivalent.*

**5.4.4** COROLLARY.   *A skew-symmetric $\mathscr{C}$ matrix of rank* 2s *is congruent to the matrix.*

$$B = \begin{bmatrix} O & I_{(s)} & O \\ -I_{(s)} & O & O \\ O & O & O \end{bmatrix}.$$

For $B$ is skew-symmetric and of rank $2s$, and the desired result follows by Corollary 5.4.3.

## PROBLEMS

**5.4–1**   Show that if we take

$$P = \begin{bmatrix} I_{(2)} & -E^{-1}B_1 \\ O & I_{(n-2)} \end{bmatrix}$$

in the proof of Theorem 5.4.1, then $P$ is nonsingular and $P'BP$ has the form $\text{diag}(E, C_1)$.

**5.4–2**   Prove Corollary 5.4.2.

**5.4–3**   Prove Corollary 5.4.3.

**5.4–4**   Let $P'AP = B$, where $|P| \neq 0$. Find nonsingular $Q$ such that

$$Q' \, \text{diag}(A, O) \, Q = \text{diag}(B, O).$$

**5.4–5**   Reduce the following skew-symmetric matrices to the canonical form of Theorem 5.4.1:

$$\textbf{(a)} \begin{bmatrix} 0 & -1 & -2 \\ 1 & 0 & 3 \\ 2 & -3 & 0 \end{bmatrix}, \quad \textbf{(b)} \begin{bmatrix} 0 & 1 & 4 \\ -1 & 0 & 2 \\ -4 & -2 & 0 \end{bmatrix}.$$

## 5.5   Canonical forms under congruency for symmetric $\mathscr{R}$ matrices

We now consider symmetric $\mathscr{R}$ matrices under congruency transformations. Though this is a highly special case, it is very important in applications. We start with a basic theorem concerning symmetric $\mathscr{C}$ matrices.

**5.5.1** THEOREM.   *Every symmetric $\mathscr{C}$ matrix A of rank r is congruent to a matrix of the form* $\text{diag}(d_1, d_2, \cdots, d_r, 0, \cdots, 0)$, *where all the $d_i$ are nonzero.*

The theorem is trivial if $A = O$. If $A \neq O$, we first show that $A$ is congruent to a matrix $B$ with a nonzero diagonal element. This is certainly true with $B = A$ if $A$ itself possesses a nonzero diagonal element. Suppose, then, $A$ has no nonzero diagonal element. Since $A \neq O$ and is symmetric, there exists an element $a_{hk}$ such that

$$a_{hk} = a_{hk} \neq 0, \qquad h \neq k.$$

Addition of row $k$ to row $h$, then of column $k$ to column $h$, replaces $A$ by a congruent matrix $B$ in which

$$b_{hh} = a_{hk} + a_{hk} = 2a_{hk} \neq 0.$$

Thus, in any case, $A$ is congruent to a matrix $B$ with a nonzero diagonal element.

Now, in matrix $B$, interchange row $h$ and row 1, then column $h$ and column 1, to obtain a congruent matrix $C$ of the form

$$C = \begin{bmatrix} d_1 & C_2 \\ C_2' & C_3 \end{bmatrix},$$

where $d_1 = b_{hh} \neq 0$. Next, add to row $i$, $i \neq 1$, the product of the first row by $-c_{i1}/d_1$, and then perform the corresponding elementary column operation. Since $c_{i1} = c_{1i}$, this makes the new elements in the $(i, 1)$ and $(1, i)$ positions both 0. Doing this for each $i \neq 1$ replaces $C$ by a congruent matrix

$$D = \begin{bmatrix} d_1 & O \\ O & A_1 \end{bmatrix},$$

where $A_1$ is symmetric and of order one less than that of $D$. By the transitive property of congruence, we have $A \overset{c}{=} D$.

The foregoing process may now be applied to $A_1$, yielding (if $r > 1$)

$$A_1 \overset{c}{=} D_1 = \begin{bmatrix} d_2 & O \\ O & A_2 \end{bmatrix},$$

where $d_2 \neq 0$ and $A_2$ is symmetric of order one less than that of $A_1$. By Theorem 5.3.6,

$$A \overset{c}{=} \begin{bmatrix} d_1 & 0 & 0 \\ 0 & d_2 & 0 \\ 0 & 0 & A_2 \end{bmatrix}.$$

After a finite number of applications of the process, there appears a diagonal matrix congruent to $A$, and the preservation of rank under a congruency transformation guarantees that the diagonal matrix will be of the form described in the statement of the theorem.

It is to be noted that the $d$'s in Theorem 5.5.1 are not, in general, unique, and hence the diagonal matrix is not a *canonical* form to which the symmetric $\mathscr{C}$ matrix $A$ can be reduced by a congruency transformation. If, however, the $\mathscr{C}$ matrix $A$ is actually a *real* matrix, then we can make the final form unique. We show this in Theorems 5.5.4 and 5.5.5.

**5.5.2**   DEFINITIONS.   If $A$ and $B$ are congruent $\mathscr{R}$ matrices and if in the elementary cogredient operations that allow us to pass from $A$ to $B$ we can employ only real multipliers of rows and columns, then we say $A \overset{\text{C}}{=} B$ *over* $\mathscr{R}$. If in the elementary cogredient operations that allow us to pass from $A$ to $B$ we may employ nonreal complex multipliers of rows and columns, then we say $A \overset{\text{C}}{=} B$ *over* $\mathscr{C}$.

**5.5.3**   THEOREM.   *If* A *and* B *are* $\mathscr{R}$ *matrices congruent over* $\mathscr{R}$, *then there exists a nonsingular* $\mathscr{R}$ *matrix* P *such that* B $=$ P$'$AP.
The easy proof is left to the student.

**5.5.4**   THEOREM.   *Every symmetric* $\mathscr{R}$ *matrix* A *of rank* r *is congruent over* $\mathscr{R}$ *to a matrix of the form* $diag(\mathrm{I}_{(p)}, -\mathrm{I}_{(r-p)}, \mathrm{O})$.
In the diagonal matrix of Theorem 5.5.1, the $d$'s are all real and nonzero. By elementary cogredient operations of the kind where we interchange two rows and then the corresponding two columns, we can place the positive $d$'s (say, $p$ of them) first, followed by the $r - p$ negative $d$'s. Now multiply the $i$th row and the $i$th column ($1 \leqq i \leqq p$) by $1/\sqrt{d_i}$, and the $j$th row and $j$th column ($p + 1 \leqq j \leqq r$) by $1/\sqrt{-d_j}$. This yields the desired form. The reader can check that the elementary cogredient operations used employ only real multipliers of rows and columns.

**5.5.5**   SYLVESTER'S LAW OF INERTIA.†   *The integer* p *in Theorem* 5.5.4 *is unique.*
For suppose the symmetric $\mathscr{R}$ matrix $A$ of rank $r$ and order $n$ is congruent over $\mathscr{R}$ to both

$$D_1 = \text{diag}(I_{(p)}, -I_{(r-p)}, O) \qquad \text{and} \qquad D_2 = \text{diag}(I_{(q)}, -I_{(r-q)}, O),$$

† This theorem was given by Sylvester in 1852 and rediscovered by Jacobi in 1857.

where $p \neq q$. It is a mere matter of notation to suppose $q < p$. Then $D_1 \overset{C}{=} D_2$ over $\mathscr{R}$. That is, there exists a nonsingular $\mathscr{R}$ matrix $P$ such that $D_2 = P'D_1P$. Let $f \equiv X'D_1X$, where $X = \{x_1, \cdots, x_n\}$. Then

$$(1) \qquad f = x_1{}^2 + \cdots + x_p{}^2 - x_{p+1}{}^2 - \cdots - x_r{}^2.$$

Consider the transformation $X = PY$, where $Y = \{y_1, \cdots, y_n\}$. Under this transformation we find

$$(2) \qquad f \equiv X'D_1X = Y'P'D_1PY = Y'D_2Y$$
$$= y_1{}^2 + \cdots + y_q{}^2 - y_{q+1}{}^2 - \cdots - y_r{}^2.$$

Since $P$ is nonsingular, we have $Y = P^{-1}X$, and each $y_i$ is a linear combination of $x_1, \cdots, x_n$. The equations $y_i = 0, i = 1, \cdots, q$, along with $x_j = 0, j = p + 1, \cdots, n$, may be regarded as a homogeneous system of linear equations in $x_1, \cdots, x_n$. These homogeneous equations, being $q + n - p < n$ in number, have (by Theorem 2.8.9) a nontrivial real solution:

$$(3) \qquad X_0 = \{x_{01}, \cdots, x_{0n}\} \neq \{0, \cdots, 0\}.$$

With $X = X_0$ we have, from (1),

$$f_0 = X_0'D_1X_0 = x_{01}{}^2 + \cdots + x_{0p}{}^2 \geqq 0.$$

But this value may also be computed from (2) by use of $Y_0 = P^{-1}X_0$. From (2) we see that

$$f_0 = Y_0'D_2Y_0 = -y_{0q+1}{}^2 - \cdots - y_{0r}{}^2 \leqq 0.$$

It follows that we must have $f_0 = 0$, whence $x_{01} = x_{02} = \cdots x_{0p} = 0$. Since also $x_{0p+1} = \cdots = x_{0n} = 0$, we see that $X_0 = \{0, \cdots, 0\}$, in conflict with (3). This contradiction, arising from the assumption that $p \neq q$, completes the proof.

**5.5.6** DEFINITION. The unique integer $p$ in Theorem 5.5.5 is called the *index* of the real symmetric matrix $A$.

**5.5.7** COROLLARY. *Two n-rowed symmetric $\mathscr{R}$ matrices are congruent over $\mathscr{R}$ if and only if they have the same rank and the same index.*

**5.5.8** COROLLARY. *Let A be an n-rowed symmetric $\mathscr{R}$ matrix of index p. Then $|A| > 0$ if and only if A is nonsingular and $n - p$ is even.*

Set $D = \text{diag}(I_{(p)}, -I_{(r-p)}, O)_{(n)}$, where $r$ is the rank of $A$. Then, by Theorem 5.5.4, $A \overset{C}{=} D$ over $\mathcal{R}$. Therefore there exists a nonsingular $\mathcal{R}$ matrix $P$ such that $A = P'DP$, whence $|A| = |P|^2|D| > 0$ if and only if $r = n$ and $n - p$ is even.

**5.5.9** THEOREM. *Every symmetric $\mathcal{R}$ matrix* A *of rank* $\mathfrak{r}$ *is congruent over* C *to a matrix of the form diag*$(I_{(\mathfrak{r})}, O)$.

The easy proof, paralleling that of Theorem 5.5.4, is left to the student.

## PROBLEMS

**5.5-1** Prove Theorem 5.5.3.

**5.5-2** Prove Corollary 5.5.7.

**5.5-3** Prove Theorem 5.5.9.

**5.5-4** Show that the number of $n \times n$ matrices $B$ of the form

$$\text{diag}(I_{(p)}, -I_{(r-p)}, O)$$

is $(n + 1)(n + 2)/2$.

**5.5-5** Prove that two symmetric $\mathcal{R}$ matrices are congruent over $\mathscr{C}$ if and only if they have the same rank.

**5.5-6** If $r$ is the rank and $p$ the index of a real symmetric matrix, then $s = p - (r - p)$ is called the *signature* of the matrix.

(**a**) Show that the signature of a real symmetric matrix is equal to the number of positive 1's minus the number of negative 1's found along the diagonal of the canonical form (given in Theorem 5.5.4) of the matrix.

(**b**) Prove that two $n \times n$ real symmetric matrices are congruent over $\mathcal{R}$ if and only if they have the same index and the same signature.

**5.5-7** Show that there exists a *real nonsingular linear transformation* $Y = PX$ (that is, a transformation $Y = PX$ where $X = \{x_1, \cdots, x_n\}$, $Y = \{y_1, \cdots, y_n\}$, and $P$ is a real $n \times n$ nonsingular matrix) such that the real quadratic form $X'A_{(n)}X$, in which $A = A'$, is converted into

$$y_1^2 + \cdots + y_p^2 - y_{p-1}^2 - \cdots - y_r^2,$$

where $r$ and $p$ are the rank and index, respectively, of $A$. These numbers $r$ and $p$ are also called the *rank* and the *index* of the real quadratic form.

**5.5-8** Find the rank and index of the real quadratic form

$$2x_1^2 + x_2^2 - 3x_3^2 - 8x_2x_3 - 4x_3x_1 + 12x_1x_2.$$

**5.5–9**    Find the rank and index of the quadratic form

$$4(x_1{}^2 + x_2{}^2) + 5(x_3{}^2 + x_4{}^2) + 2k(x_1x_2 + x_3x_4)$$

for all real values of $k$.

**5.5–10**    Reduce the following symmetric matrices to the canonical form of Theorem 5.5.4:

$$(a) \begin{bmatrix} 1 & 2 & 2 \\ 2 & 4 & 8 \\ 2 & 8 & 4 \end{bmatrix}, \quad (b) \begin{bmatrix} 0 & -1 & -4 \\ -1 & 0 & 2 \\ -4 & 2 & 0 \end{bmatrix}.$$

**5.5–11**    By following through the proof of Theorem 5.4.1, show that every skew-symmetric $\mathcal{R}$ matrix $A$ of rank $r$ is congruent over $\mathcal{R}$ to a matrix of the form

$$\mathrm{diag}(E, E, \cdots, E, 0, \cdots, 0),$$

where $E = \begin{bmatrix} 0 & 1 \\ -1 & 0 \end{bmatrix}$ and the number of $E$'s is $r/2$.

## 5.6    Conjunctivity, or Hermitian congruence

As might be expected, much of the theory of Hermitian and skew-Hermitian matrices runs parallel to that of real symmetric and real skew-symmetric matrices. We recall that if $A$ is a square $\mathcal{C}$ matrix, then $A^* = (\bar{A})'$ is called the *tranjugate* (transposed conjugate) of $A$, and matrix $A$ is said to be *Hermitian* if $A = A^*$ and it is said to be *skew-Hermitian* if $A = -A^*$. A real symmetric matrix is Hermitian, and a real skew-symmetric matrix is skew-Hermitian.

We shall now parallel some of the material of Sections 5.3 and 5.5, generally leaving the easy construction of analogous proofs to the student.

**5.6.1**    DEFINITION AND NOTATION.    A square $\mathcal{C}$ matrix $B$ is said to be *conjunctive*, or *Hermitely congruent*, to a square $\mathcal{C}$ matrix $A$, and we write $B \overset{\mathrm{H}}{=} A$, if and only if there exists a nonsingular $\mathcal{C}$ matrix $P$ such that $B = P^*AP$.

**5.6.2**    THEOREM.    *Conjunctivity of matrices is reflexive, symmetric, and transitive.*

**5.6.3** DEFINITION. An elementary row operation, applied to a square $\mathscr{C}$ matrix and followed by the corresponding conjugate elementary column operation (that is, by the corresponding column operation except that any multiplier $k$ of a row has been replaced by the multiplier $\bar{k}$ of the corresponding column) is called an *elementary conjunctive operation*.

*Note.* The student can easily show that the elementary transformation matrices corresponding to the two elementary operations of an elementary conjunctive operation are tranjugates of one another.

**5.6.4** THEOREM. *If* A *and* B *are* n × n $\mathscr{C}$ *matrices, then* B $\overset{H}{=}$ A *if and only if* B *is obtainable from* A *by a finite sequence of elementary conjunctive operations.*

**5.6.5** THEOREM. *If* B $\overset{H}{=}$ A, *then* B *is Hermitian (skew-Hermitian) if and only if* A *is Hermitian (skew-Hermitian).*

**5.6.6** THEOREM. *If*

$$A = \begin{bmatrix} A_1 & O \\ O & A_2 \end{bmatrix} \quad \text{and} \quad B = \begin{bmatrix} B_1 & O \\ O & B_2 \end{bmatrix},$$

*and if* $A_1 \overset{H}{=} B_1$ *and* $A_2 \overset{H}{=} B_2$, *then* A $\overset{H}{=}$ B.

**5.6.7** THEOREM. *Every Hermitian matrix* A *of rank* r *is conjunctive to a matrix of the form* $diag(d_1, d_2, \cdots, d_r, 0, \cdots, 0)$, *where all the* $d_i$ *are nonzero real numbers.*

We shall parallel the proof of Theorem 5.5.1. The theorem is trivial if $A = O$. If $A \neq O$, we first show that $A$ is conjunctive to a matrix $B$ with a nonzero element. This is certainly true with $B = A$ if $A$ itself possesses a nonzero diagonal element. Suppose, then, $A$ has no nonzero diagonal element. Since $A \neq O$ and is Hermitian, there exists an element $a_{hk}$ such that

$$a_{hk} = \bar{a}_{kh} \neq 0, \qquad h \neq k.$$

Addition of row $k$ to row $h$, then of column $k$ to column $h$, replaces $A$ by a conjunctive matrix $B$ in which

$$b_{hh} = a_{hk} + a_{kh} = 2(\text{real part of } a_{hk}).$$

If the real part of $a_{hk}$ is not zero, then $b_{hh} \neq 0$. If the real part of $a_{hk}$ is zero, then $a_{hk}$ must be pure imaginary and $a_{hk} = ic \neq 0$. In this situation we modify the construction of $B$ by adding $i$ times row $k$ to row $h$ and then $-i$ times column $k$ to column $h$. This gives

$$b_{hh} = ia_{kh} - ia_{hk} = i(a_{kh} - a_{hk})$$
$$= i(-ic - ic) = 2c \neq 0.$$

Thus, in any case, $A$ is conjunctive to a matrix $B$ with a nonzero diagonal element.

The student should now have no trouble paralleling the rest of the proof of Theorem 5.5.1. That the $d_i$ are all real follows from the fact that the final diagonal matrix is conjunctive to $A$, and hence is Hermitian; but the diagonal elements of a Hermitian matrix are real.

**5.6.8**   THEOREM.   *Every Hermitian matrix* A *of rank* r *is conjunctive to a matrix of the form* $diag(I_{(p)}, -I_{(r-p)}, O)$.

With very slight adjustment, the proof of Theorem 5.5.4 also holds here.

**5.6.9**   THEOREM.   *The integer* p *in Theorem* 5.6.8 *is unique.*

With very slight adjustment, the proof of Theorem 5.5.5 also holds here.

**5.6.10**   DEFINITION.   The unique integer $p$ of Theorem 5.6.9 is called the *index* of the Hermitian matrix $A$.

**5.6.11**   COROLLARY.   *Two* n-*rowed Hermitian matrices are conjunctive if and only if they have the same rank and index.*

We see that conjunctivity of Hermitian matrices and congruence (over $\mathcal{R}$) of real symmetric matrices are analogous theories, leading to the same canonical forms. A like analogy, however, does not hold for conjunctivity of skew-Hermitian matrices and congruence of real skew-symmetric matrices. We have seen in Section 5.4 that a real skew-symmetric matrix is congruent to a canonical form

$$diag(E, E, \cdots, E, 0, \cdots, 0),$$

where

$$E = \begin{bmatrix} 0 & 1 \\ -1 & 0 \end{bmatrix},$$

from which it follows that a real skew-symmetric matrix has even rank. But the skew-Hermitian matrices

$$[i] \quad \text{and} \quad \begin{bmatrix} i & 0 & 0 \\ 0 & 2i & 0 \\ 0 & 0 & 3i \end{bmatrix}$$

have odd rank, and therefore cannot be conjunctive to matrices of the form

$$\text{diag}(E, E, \cdots, E, 0, \cdots, 0).$$

It is easy, though, to find a canonical form to which any skew-Hermitian matrix is conjunctive. Let $A$ be a skew-Hermitian matrix. Then (see Theorem 1.8.3) $H = -iA$ is Hermitian. Therefore, by Theorem 5.6.8, there exists a nonsingular $\mathscr{C}$ matrix $P$ such that

$$P^*HP = B = \text{diag}(I_{(p)}, I_{(r-p)}, O),$$

whence

$$P^*AP = P^*(iH)P = i(P^*HP) = iB.$$

We have now established the following two theorems.

**5.6.12** THEOREM. *Every skew-Hermitian matrix* A *is conjunctive to a canonical matrix of the form* $\text{diag}(iI_{(p)}, -iI_{(r-p)}, O)$, *where* r *is the rank of* A *and* p *is the index of the Hermitian matrix* $-iA$.

**5.6.13** THEOREM. *Two* n-*rowed skew-Hermitian matrices* A *and* B *are conjunctive if and only if they have the same rank and the Hermitian matrices* $-iA$ *and* $-iB$ *have the same index.*

## PROBLEMS

**5.6–1** Prove Theorem 5.6.1.

**5.6–2** Show that the elementary transformation matrices corresponding to the two elementary operations of an elementary conjunctive operation are tranjugates of one another.

**5.6–3** Prove Theorem 5.6.4.

**5.6–4** Prove Theorem 5.6.5.

**5.6–5** Prove Theorem 5.6.6.

**5.6–6** Complete the proof of Theorem 5.6.7.

**5.6–7** Prove Theorem 5.6.8.

**5.6–8** Prove Theorem 5.6.9.

**5.6–9**   Prove Corollary 5.6.11.

**5.6–10**   Let $A$ and $B$ be two $n$-rowed skew-Hermitian matrices. Show that if the Hermitian matrices $-iA$ and $-iB$ have the same rank and the same index, then the Hermitian matrices $iA$ and $iB$ also have the same rank and the same index.

**5.6–11**   Is the following analog of Theorem 5.3.7 a true proposition? If $A$ and $B$ are $n \times n$ matrices and $A_1$ and $B_1$ are $r \times r$ nonsingular matrices, where $r \leqq n$, and if

$$A \overset{\text{H}}{=} \begin{bmatrix} A_1 & O \\ O & O \end{bmatrix} \quad \text{and} \quad B \overset{\text{H}}{=} \begin{bmatrix} B_1 & O \\ O & O \end{bmatrix},$$

then $A \overset{\text{H}}{=} B$ if and only if $A_1 \overset{\text{H}}{=} B_1$.

**5.6–12**   Is the following analog of Theorem 5.3.8 a true proposition? If $A$ and $B$ are $n$th-order Hermitian matrices of rank $r$, then $A \overset{\text{H}}{=} B$ if and only if a nonsingular $r$-rowed principal submatrix of $A$ is conjunctive to a nonsingular $r$-rowed principal submatrix of $B$, in which case each nonsingular $r$-rowed principal submatrix of $A$ is conjunctive to every nonsingular $r$-rowed principal submatrix of $B$.

**5.6–13**   Prove that the determinant and every principal minor of a Hermitian matrix are real.

**5.6–14**   Reduce the Hermitian matrix

$$\begin{bmatrix} -1 & 2 - 3i & 3 + 4i \\ 2 + 3i & 0 & 4 - 5i \\ 3 - 4i & 4 + 5i & 2 \end{bmatrix}$$

to canonical form.

**5.6–15**   Reduce the skew-Hermitian matrix

$$\begin{bmatrix} i & 2 & 3 + i \\ -2 & 2i & 0 \\ -3 + i & 0 & -i \end{bmatrix}$$

to canonical form.

## 5.7   Orthogonal matrices and orthogonal similarity

We first introduce the concept of an *orthogonal* matrix.

**5.7.1   Definition.**   A matrix $P$ is said to be *orthogonal* if and only if it is nonsingular and $P^{-1} = P'$.†

† Some writers further insist that $P$ be *real*.

**5.7.2** THEOREM. *An orthogonal matrix is square.*

**5.7.3** THEOREM. *A square matrix* P *is orthogonal if and only if* $PP' = I$.

If $P$ is orthogonal, then $P' = P^{-1}$ and $PP' = PP^{-1} = I$. Conversely, if $PP' = I$, then $P$ is nonsingular and it follows that $P' = P^{-1}$.

**5.7.4** THEOREM. *The transpose and the inverse of an orthogonal matrix are orthogonal.*

Let $P$ be orthogonal. Then $P$, $P'$, $P^{-1}$ are square and $P' = P^{-1}$. It follows that $P'(P')' = P'P = P^{-1}P = I$, whence (by Theorem 5.7.3) $P'$ is orthogonal. It also follows that $P^{-1}(P^{-1})' = P'(P')^{-1} = I$, whence (by Theorem 5.7.3) $P^{-1}$ is orthogonal.

**5.7.5** THEOREM. *The product of two or more orthogonal matrices of order* n *is an orthogonal matrix of order* n.

Let $P$ and $Q$ be orthogonal matrices of order $n$. Then $PP' = I_{(n)}$ and $QQ' = I_{(n)}$. It follows that $(PQ)(PQ)' = PQQ'P' = P(QQ')P' = PP' = I_{(n)}$, whence (by Theorem 5.7.3) $PQ$ is an orthogonal matrix of order $n$. The student can easily extend the proof to cover the product of more than two orthogonal matrices.

**5.7.6** THEOREM. *The determinant of an orthogonal matrix is equal to* $+1$ *or* $-1$.

Let $P$ be orthogonal. Then $PP' = I$, whence $|PP'| = |P|\,|P'| = |P|^2 = |I| = 1$. It follows that $|P| = \pm 1$.

**5.7.7** DEFINITIONS. An orthogonal matrix $P$ is said to be *properly orthogonal* or *improperly orthogonal*, according as $|P| = +1$ or $|P| = -1$.

**5.7.8** THEOREM. *The product of two or more properly orthogonal matrices of order* n *is a properly orthogonal matrix of order* n.

The easy proof is left to the student.

**5.7.9** DEFINITIONS. (1) An $n$th-order row or column vector $U$ is said to be *normal* if and only if $U \cdot U = 1$. (2) Two $n$th-order row or column vectors $U$ and $V$ are said to be *orthogonal* if and only if $U \cdot V = 0$. (3) A system of $k$ $n$th-order normal row (column) vectors that are orthogonal in pairs is called an *orthonormal system* of vectors. (See Definition 1.3.1 for the meaning of the inner product $U \cdot V$ of two row or column vectors of the same order $n$.)

*Examples.*    The vector $(1/3, -2/3, -2/3)$ is normal, since

$$(1/3)^2 + (-2/3)^2 + (-2/3)^2 = 1.$$

The vectors $(1/3, -2/3, -2/3)$ and $(2/3, -1/3, 2/3)$ are orthogonal, since

$$(1/3)(2/3) + (-2/3)(-1/3) + (-2/3)(2/3) = 0.$$

The student can show that the system of vectors $(1/3, -2/3, -2/3)$, $(2/3, -1/3, 2/3)$, $(-2/3, -2/3, 1/3)$ is orthonormal.

**5.7.10    THEOREM.**    *A square matrix* $P$ *is orthogonal if and only if its rows are orthonormal.*

Let $P$ be orthogonal and denote the $i$th row of $P$ by $P_i$. Since $P$ is orthogonal, we have $PP' = I$, whence $P_i \cdot P_i = 1$ and $P_i \cdot P_j = 0$ $(i \neq j)$.

Conversely, suppose $P$ is a square matrix such that for arbitrary rows $P_i$, $P_j$ $(i \neq j)$, we have $P_i \cdot P_i = 1$ and $P_i \cdot P_j = 0$. Then $PP' = I$ and $P$ is orthogonal.

**5.7.11    THEOREM.**    *A square matrix* $P$ *is orthogonal if and only if its columns are orthonormal.*

This follows from Theorem 5.7.10, inasmuch as $P$ is orthogonal if and only if $P'$ is.

**5.7.12    THEOREM.**    *If* $C_1$ *is an* $n$th-order, nonzero, real, normal *column vector, then there exists a real orthogonal matrix* $P$ *having* $C_1$ *as its first column.*

The single linear homogeneous equation

$$C_1 \cdot X = 0$$

in the $n$ unknown components of the column vector $X$ possesses (by Theorem 2.8.9) a real nonzero solution $Y_2$. Set

$$C_2 = Y_2/(Y_2 \cdot Y_2)^{1/2}.$$

Then $C_2$ is a real normal column vector orthogonal to $C_1$.

Next, the two linear homogeneous equations

$$C_1 \cdot X = 0, \qquad C_2 \cdot X = 0$$

in the $n$ unknown components of the column vector $X$ possess (if $n > 2$) a real nonzero solution $Y_3$. Set

$$C_3 = Y_3/(Y_3 \cdot Y_3)^{1/2}.$$

Then $C_3$ is a real normal column vector orthogonal to $C_1$ and $C_2$.

Proceeding in this manner, finally consider a system of $n - 1$ linear homogeneous equations

$$C_1 \cdot X = 0, \qquad \cdots, \qquad C_{n-1} \cdot X = 0$$

in the $n$ unknown components of the column vector $X$. This system possesses a real nonzero solution $Y_n$. Set

$$C_n = Y_n/(Y_n \cdot Y_n)^{1/2}.$$

Then $C_n$ is a real normal column vector orthogonal to each of $C_1, \cdots, C_{n-1}$.

It follows that $C_1, \cdots, C_n$ constitute a real orthonormal system of $n$th-order column vectors, whence (by Theorem 5.7.11) the matrix

$$P = [C_1, C_2, \cdots, C_n]$$

is a real orthogonal matrix.

We now introduce the notion of "orthogonal similarity."

**5.7.13** DEFINITION AND NOTATION. Square matrix $B$ is said to be *orthogonally similar* (or *orthogonally congruent*) to square matrix $A$, and we write $B \overset{O}{=} A$, if and only if there exists an orthogonal matrix $P$ such that $B = P'AP$ or, what is the same thing, $B = P^{-1}AP$.

**5.7.14** THEOREM. *Orthogonal similarity of matrices is reflexive, symmetric, and transitive.*

Let $A$, $B$, $C$ denote square matrices. Since $A = I'AI$, and $I$ is orthogonal, we have $A \overset{O}{=} A$, and the relation is reflexive. Suppose $A \overset{O}{=} B$. Then there exists orthogonal $P$ such that $A = P'BP$, whence $B = (P')^{-1}AP^{-1} = (P^{-1})'AP^{-1}$. Since $P^{-1}$ is orthogonal, it follows that $B \overset{O}{=} A$, and the relation is symmetric. Finally, suppose $A \overset{O}{=} B$ and $B \overset{O}{=} C$. Then there exist orthogonal $P$ and $Q$ such that $A = P'BP$ and $B = Q'CQ$, whence $A =$

$P'(Q'CQ)P = (QP)'C(QP)$. Since $QP$ is orthogonal, it follows that $A \overset{o}{=} C$, and the relation is transitive.

Since two orthogonally similar matrices are both similar and congruent, we already know, in view of earlier work, many properties possessed by orthogonally similar matrices. We shall borrow some of these properties to prove the following theorem, which is very important in analytic geometry.

**5.7.15**   THEOREM.   *Any* n × n *real symmetric matrix* A *is orthogonally similar to a diagonal matrix whose diagonal elements are the* n *characteristic roots of* A.

Let $\lambda_1, \cdots, \lambda_n$ be the $n$ (not necessarily distinct) characteristic roots of $A$. These characteristic roots are, by Corollary 4.6.9, all real.

Let $C_1$ be any characteristic vector of $A$ corresponding to the characteristic root $\lambda_1$. Then $AC_1 = \lambda_1 C_1$, $C_1 \neq O$. Since $\lambda_1$ and $A$ are real, $C_1$ can be real. Since any nonzero scalar multiple of $C_1$ is also a characteristic vector of $A$ corresponding to $\lambda_1$, we can, and do, suppose that $C_1$ is normal. Let $P_1$ be any real orthogonal matrix having $C_1$ as its first column; such a matrix exists by Theorem 5.7.12. Consider the matrix $P_1'AP_1$. Since $C_1$ is the first column of $P_1$, the first column of $P_1'AP_1$ is

$$P_1'AC_1 = P_1'\lambda_1 C_1 = \lambda_1 P_1'C_1.$$

But $P_1'C_1$ is the first column of $P_1'P_1 = I$. Therefore

$$\lambda_1 P_1'C_1 = \{\lambda_1, 0, \cdots, 0\},$$

and $P_1'AP_1$ has the form

$$\begin{bmatrix} \lambda_1 & L \\ O & B \end{bmatrix},$$

which, by Theorem 5.3.5, must be symmetric. That is, $L = O$ and $B$ is a real $(n-1)$-rowed symmetric matrix. By Theorem 5.1.3, the characteristic roots of $B$ are $\lambda_2, \cdots, \lambda_n$.

By an argument similar to the one given above, there exists a real orthogonal $(n-1)$-rowed matrix $Q$ such that

$$Q'BQ = \begin{bmatrix} \lambda_2 & O \\ O & C \end{bmatrix},$$

where $C$ is a real $(n-2)$-rowed symmetric matrix with characteristic roots $\lambda_3, \cdots, \lambda_n$.

Define $P_2$ to be the $n$-rowed matrix

$$\begin{bmatrix} I_{(1)} & O \\ O & Q \end{bmatrix}.$$

Then

$$P_2' = \begin{bmatrix} I_{(1)} & O \\ O & Q' \end{bmatrix} = \begin{bmatrix} I_{(1)} & O \\ O & Q^{-1} \end{bmatrix} = P_2^{-1},$$

and $P_2$ is a real $n$-rowed orthogonal matrix. Now

$$P_2'P_1'AP_1P_2 = P_2' \begin{bmatrix} \lambda_1 & O \\ O & B \end{bmatrix} P_2$$

$$= \begin{bmatrix} I_{(1)} & O \\ O & Q' \end{bmatrix} \begin{bmatrix} \lambda_1 & O \\ O & B \end{bmatrix} \begin{bmatrix} I_{(1)} & O \\ O & Q \end{bmatrix}$$

$$= \begin{bmatrix} \lambda_1 & O \\ O & Q'BQ \end{bmatrix}$$

$$= \begin{bmatrix} \lambda_1 & 0 & 0 \\ 0 & \lambda_2 & O \\ O & O & C \end{bmatrix}.$$

Proceeding in this manner, we obtain a sequence $P_1, \cdots, P_n$ of real $n$-rowed orthogonal matrices such that

$$P_n' \cdots P_2'P_1'AP_1P_2 \cdots P_n = \mathrm{diag}(\lambda_1, \cdots, \lambda_n).$$

Setting $P = P_1P_2 \cdots P_n$, we see that $P$ is real orthogonal and

$$P'AP = \mathrm{diag}(\lambda_1, \cdots, \lambda_n),$$

which proves the theorem.

**5.7.16  Corollary.**  *Two* n × n *real symmetric matrices are orthogonally similar if and only if they have the same characteristic roots, that is, if and only if they are similar.*

**5.7.17  Theorem.**  *If a matrix* A *is orthogonally similar to a diagonal matrix, then* A *is symmetric.*

Suppose matrix $A$ is orthogonally similar to a diagonal matrix $D$. Then there exists an orthogonal matrix $P$ such that $A = P^{-1}DP$. Since $P^{-1} = P'$, it follows that $A' = P'D'(P^{-1})' = P^{-1}DP = A$, and $A$ is symmetric.

## PROBLEMS

**5.7–1**   Give the extended part of the proof of Theorem 5.7.5.

**5.7–2**   Prove Theorem 5.7.8.

**5.7–3**   Construct a real orthogonal matrix having $\{1/9, -8/9, -4/9\}$ as its first column.

**5.7–4**   Prove that if $R_1$ is an $n$th-order, nonzero, real, normal row vector, then there exists a real orthogonal matrix $P$ having $R_1$ as its first row.

**5.7–5**   Prove Corollary 5.7.16.

**5.7–6**   If the columns or the rows of an orthogonal matrix are permuted in any way, show that the resulting matrix is orthogonal.

**5.7–7**   Show that the matrix

$$\begin{bmatrix} 0 & 2b & c \\ a & b & -c \\ a & -b & c \end{bmatrix},$$

where $a = 1/\sqrt{2}$, $b = 1/\sqrt{6}$, $c = 1/\sqrt{3}$, is orthogonal.

**5.7–8**   $P$ is a real orthogonal matrix such that all the elements of the first column, except the first element, are zero. Show that all the elements of the first row, except the first element, also are zero.

**5.7–9**   If $P_1$, $P_2$ are two square matrices, and if

$$P = \begin{bmatrix} P_1 & A \\ O & P_2 \end{bmatrix}$$

is a real orthogonal matrix, show that $P_1$ and $P_2$ are orthogonal and that $A = O$.

**5.7–10**   (a) Prove that every $2 \times 2$ real orthogonal matrix has one or the other of the two forms

$$\begin{bmatrix} \cos\theta & -\sin\theta \\ \sin\theta & \cos\theta \end{bmatrix}, \qquad \begin{bmatrix} \cos\theta & \sin\theta \\ \sin\theta & -\cos\theta \end{bmatrix}.$$

(b) Find the characteristic roots of the two matrices in part (a).

**5.7–11**   Prove that a matrix whose rows are

$$[a, b, c], \qquad [b, c, a], \qquad [c, a, b]$$

is orthogonal if and only if $a$, $b$, $c$ are the roots of an equation of one of the forms

$$x^3 + x^2 + p = 0, \qquad x^3 - x^2 + q = 0.$$

**5.7–12**  Show that the matrix $[a_{ij}]_{(4)}$, where $a_{ij} = -1/2$ if $i = j$ and $a_{ij} = 1/2$ if $i \neq j$, is orthogonal.

**5.7–13**  Prove that an orthonormal system of vectors is linearly independent.

**5.7–14**  Prove that if $X_1, \cdots, X_k$ are $k$ linearly independent real column (row) vectors, then there exists an orthonormal set $Y_1, \cdots, Y_k$ of real vectors, each of which is a linear combination of $X_1, \cdots, X_k$.

**5.7–15**  If $X$ and $Y$ are variable $n$th-order column vectors and if $P$ is a constant $n \times n$ orthogonal matrix, the transformation $Y = PX$ is called an *orthogonal transformation*. Show that such an orthogonal transformation preserves inner products of $n$th-order column vectors.

**5.7–16**  If $X'AX$, $A = A'$, is a real quadratic form, show that there exists a real orthogonal transformation that transforms the form into

$$\lambda_1 y_1^2 + \cdots + \lambda_r y_r^2,$$

where $\lambda_1, \cdots, \lambda_r$ are the nonzero characteristic roots of matrix $A$.

**5.7–17**  (a) If $A$ and $B$ are $n \times n$ nonsingular matrices such that $AA' = BB'$, prove that there exists an orthogonal matrix $P$ such that $A = BP$.

(b) If $A$ and $B$ are $n \times n$ orthogonal matrices, prove that there exists an orthogonal matrix $P$ such that $A = BP$.

**5.7–18**  If $A$ is an odd-order orthogonal matrix, show that either $A - I$ or $A + I$ is singular.

**5.7–19**  Prove that every real nonsingular matrix $A$ can be expressed as $A = QDR$, where $Q$ and $R$ are real orthogonal and $D$ is real diagonal.

**5.7–20**  Prove that any two characteristic vectors corresponding to two distinct characteristic roots of a real symmetric matrix are orthogonal.

**5.7–21**  Prove that there exists an orthonormal system of $n$ characteristic vectors for any given $n \times n$ real symmetric matrix.

**5.7–22**  Find a diagonal matrix orthogonally similar to

$$\textbf{(a)} \begin{bmatrix} 3 & -1 & 1 \\ -1 & 5 & -1 \\ 1 & -1 & 3 \end{bmatrix}, \quad \textbf{(b)} \begin{bmatrix} 6 & -2 & 2 \\ -2 & 3 & -1 \\ 2 & -1 & 3 \end{bmatrix}.$$

## 5.8  Unitary matrices and unitary similarity

The material of this section parallels that of the preceding section. We start by introducing the concept of "unitary" matrix.

**5.8.1**   DEFINITION.   A $\mathscr{C}$ matrix $P$ is said to be *unitary* if and only if it is nonsingular and $P^{-1} = P^*$.

**5.8.2**   THEOREM.   *A unitary matrix is square.*

**5.8.3**   THEOREM.   *A square $\mathscr{C}$ matrix P is unitary if and only if* $PP^* = I$.

If $P$ is unitary, then $P^* = P^{-1}$ and $PP^* = PP^{-1} = I$. Conversely, if $PP^* = I$, then also $P^*P = (PP^*)^* = I^* = I$, and it follows that $P^* = P^{-1}$.

The student should find no trouble supplying proofs for the next two theorems.

**5.8.4**   THEOREM.   *The transpose, the inverse, the conjugate, and the tranjugate of a unitary matrix are unitary.*

**5.8.5**   THEOREM.   *The product of two or more unitary matrices of order n is a unitary matrix of order n.*

**5.8.6**   THEOREM.   *The determinant of a unitary matrix has absolute value* 1.

Let $P$ be unitary. Then $PP^* = I$, whence $|PP^*| = |P||P^*| = |P||\bar{P}| = |I| = 1$.

In order to continue the parallel with the preceding section, we formulate the following definitions.

**5.8.7**   DEFINITIONS.   (1) An $n$th-order row or column $\mathscr{C}$ vector $U$ is said to be $\mathscr{C}$-normal if and only if $U \cdot \bar{U} = 1$. (2) Two $n$th-order row or column $\mathscr{C}$ vectors $U$ and $V$ are said to be $\mathscr{C}$-*orthogonal* if and only if $U \cdot \bar{V} = 0$. (3) A system of $k$ $n$th-order $\mathscr{C}$-normal row (column) $\mathscr{C}$ vectors that are $\mathscr{C}$-orthogonal in pairs is called a $\mathscr{C}$-*orthonormal system* of $\mathscr{C}$ vectors. (It is to be noted that when the $\mathscr{C}$ vectors $U$ and $V$ are *real*, these definitions coincide with those made in Definitions 5.7.9.)

*Examples.*   The vector $[1/\sqrt{2}, i/\sqrt{2}]$ is $\mathscr{C}$-normal but is not normal, and it is orthogonal to itself but is not $\mathscr{C}$-orthogonal to itself.

**5.8.8** THEOREM. *A square $\mathscr{C}$ matrix* P *is unitary if and only if its rows are $\mathscr{C}$-orthonormal.*

Let $P$ be unitary and denote the $i$th row of $P$ by $P_i$. Since $P$ is unitary, we have $P\bar{P}' = PP^* = I$, whence $P_i \cdot \bar{P}_i = 1$ and $P_i \cdot \bar{P}_j = 0$ $(i \neq j)$.

Conversely, suppose $P$ is a square $\mathscr{C}$ matrix such that for arbitrary rows $P_i, P_j (i \neq j)$ we have $P_i \cdot \bar{P}_i = 1$ and $P_i \cdot \bar{P}_j = 0$. Then $PP^* = P\bar{P}' = I$ and $P$ is unitary.

**5.8.9** THEOREM. *A square $\mathscr{C}$ matrix* P *is unitary if and only if its columns are $\mathscr{C}$-orthonormal.*

This follows from Theorem 5.8.8, inasmuch as $P$ is unitary if and only if $P'$ is unitary.

**5.8.10** THEOREM. *If* $C_1$ *is an nth-order, nonzero, $\mathscr{C}$-normal column $\mathscr{C}$ vector, then there exists a unitary matrix* P *having* $C_1$ *as its first column.*

The single linear homogeneous equation

$$\bar{C}_1 \cdot X = O$$

in the $n$ unknown components of the column $\mathscr{C}$ vector $X$ possesses (by Theorem 2.8.9) a nonzero solution $Y_2$. Set

$$C_2 = Y_2/(Y_2 \cdot \bar{Y}_2)^{1/2}.$$

Then $C_2$ is a $\mathscr{C}$-normal column $\mathscr{C}$ vector $\mathscr{C}$-orthogonal to $C_1$.

Next, the two linear homogeneous equations

$$\bar{C}_1 \cdot X = O, \qquad \bar{C}_2 \cdot X = O$$

in the $n$ unknown components of the column $\mathscr{C}$ vector $X$ possess (if $n > 2$) a nonzero solution $Y_3$. Set

$$C_3 = Y_3/(Y_3 \cdot \bar{Y}_3)^{1/2}.$$

Then $C_3$ is a $\mathscr{C}$-normal column $\mathscr{C}$ vector $\mathscr{C}$-orthogonal to $C_1$ and $C_2$.

Proceeding in this manner, finally consider a system of $n - 1$ linear homogeneous equations

$$\bar{C}_1 \cdot X = O, \qquad \cdots, \qquad \bar{C}_{n-1} \cdot X = O$$

in the $n$ unknown components of the column $\mathscr{C}$ vector $X$. This system possesses a nonzero solution $Y_n$. Set

$$C_n = Y_n/(Y_n \cdot \bar{Y}_n)^{1/2}.$$

Then $C_n$ is a $\mathscr{C}$-normal column $\mathscr{C}$ vector $\mathscr{C}$-orthogonal to each of $C_1, \cdots,$ $C_{n-1}$.

It follows that $C_1, \cdots, C_n$ constitute a $\mathscr{C}$-orthonormal system of $n$th-order column $\mathscr{C}$ vectors, whence (by Theorem 5.8.9) the matrix

$$P = [C_1, C_2, \cdots, C_n]$$

is a unitary matrix.

We now introduce the notion of "unitary similarity."

**5.8.11** DEFINITION AND NOTATION. A square $\mathscr{C}$ matrix $B$ is said to be *unitarily similar* (or *unitarily conjunctive*) to a square $\mathscr{C}$ matrix $A$, and we write $B \overset{U}{=} A$, if and only if there exists a unitary matrix $P$ such that $B = P^*AP$ or, what is the same thing, $B = P^{-1}AP$.

**5.8.12** THEOREM. *Unitary similarity of matrices is reflexive, symmetric, and transitive.*

Let $A$, $B$, $C$ denote square $\mathscr{C}$ matrices. Since $A = I^*AI$ and $I$ is unitary, we have $A \overset{U}{=} A$, and the relation is reflexive. Suppose $A \overset{U}{=} B$. Then there exists unitary $P$ such that $A = P^*BP$, whence $B = (P^*)^{-1}AP^{-1}$ $= (P^{-1})^*AP^{-1}$. Since $P^{-1}$ is unitary, it follows that $B \overset{U}{=} A$, and the relation is symmetric. Finally, suppose $A \overset{U}{=} B$ and $B \overset{U}{=} C$. Then there exist unitary $P$ and $Q$ such that $A = P^*BP$ and $B = Q^*CQ$, whence $A = P^*(Q^*CQ)P$ $= (QP)^*C(QP)$. Since $QP$ is unitary, it follows that $A \overset{U}{=} C$, and the relation is transitive.

**5.8.13** THEOREM. *Any* n × n *Hermitian matrix* A *is unitarily similar to a real diagonal matrix whose diagonal elements are the* n *characteristic roots of* A.

Let $\lambda_1, \cdots, \lambda_n$ be the $n$ (not necessarily distinct) characteristic roots of $A$. These characteristic roots are, by Theorem 4.6.8, all real.

Let $C_1$ be any characteristic vector of $A$ corresponding to the characteristic root $\lambda_1$. Then $AC_1 = \lambda_1 C_1$, $C_1 \neq O$. Since any nonzero scalar multiple of $C_1$ is also a characteristic vector of $A$ corresponding to $\lambda_1$, we can, and do, suppose that $C_1$ is $\mathscr{C}$-normal. Let $P_1$ be any unitary matrix having $C_1$ as its first column; such a matrix exists by Theorem 5.8.10. Consider the matrix $P_1^*AP_1$. Since $C_1$ is the first column of $P_1$, the first column of $P_1^*AP_1$ is

$$P_1^*AC_1 = P_1^*\lambda_1 C_1 = \lambda_1 P_1^*C_1.$$

But $P_1{}^*C_1$ is the first column of $P_1{}^*P_1 = I$. Therefore

$$\lambda_1 P_1{}^*C_1 = \{\lambda_1, 0, \cdots, 0\},$$

and $P_1{}^*AP_1$ has the form

$$\begin{bmatrix} \lambda_1 & L \\ O & B \end{bmatrix},$$

which, by Theorem 5.6.5, must be Hermitian. That is, $L = O$ and $B$ is an $(n - 1)$-rowed Hermitian matrix. By Theorem 5.1.3, the characteristic roots of $B$ are $\lambda_2, \cdots, \lambda_n$.

By an argument similar to that given above, there exists an $(n - 1)$-rowed unitary matrix $Q$ such that

$$Q^*BQ = \begin{bmatrix} \lambda_2 & O \\ O & C \end{bmatrix},$$

where $C$ is an $(n - 2)$-rowed Hermitian matrix with characteristic roots $\lambda_3, \cdots, \lambda_n$.

Define $P_2$ to be the $n$-rowed matrix

$$\begin{bmatrix} I_{(1)} & O \\ O & Q \end{bmatrix}.$$

Then

$$P_2{}^* = \begin{bmatrix} I_{(1)} & O \\ O & Q^* \end{bmatrix} = \begin{bmatrix} I_{(1)} & O \\ O & Q^{-1} \end{bmatrix} = P_2{}^{-1},$$

and $P_2$ is an $n$-rowed unitary matrix. Now

$$
\begin{aligned}
P_2{}^*P_1{}^*AP_1P_2 &= P_2{}^* \begin{bmatrix} \lambda_1 & O \\ O & B \end{bmatrix} P_2 \\
&= \begin{bmatrix} I_{(1)} & O \\ O & Q^* \end{bmatrix} \begin{bmatrix} \lambda_1 & O \\ O & B \end{bmatrix} \begin{bmatrix} I_{(1)} & O \\ O & Q \end{bmatrix} \\
&= \begin{bmatrix} \lambda_1 & O \\ O & Q^*BQ \end{bmatrix} \\
&= \begin{bmatrix} \lambda_1 & 0 & 0 \\ 0 & \lambda_2 & 0 \\ 0 & 0 & C \end{bmatrix}.
\end{aligned}
$$

Proceeding in this manner, we obtain a sequence $P_1, \cdots, P_n$ of $n$-rowed unitary matrices such that

$$P_n{}^* \cdots P_2{}^*P_1{}^*AP_1P_2 \cdots P_n = \operatorname{diag}(\lambda_1, \cdots, \lambda_n).$$

Setting $P = P_1 P_2 \cdots P_n$, we see that $P$ is unitary and

$$P^*AP = \text{diag}(\lambda_1, \cdots, \lambda_n),$$

which proves the theorem.

**5.8.14** COROLLARY. *Two* n × n *Hermitian matrices are unitarily similar if and only if they have the same characteristic roots, that is, if and only if they are similar.*

**5.8.15** THEOREM. *If a $\mathscr{C}$ matrix* A *is unitarily similar to a real diagonal matrix, then* A *is Hermitian.*

Suppose $\mathscr{C}$ matrix $A$ is unitarily similar to a real diagonal matrix $D$. Then there exists a unitary matrix $P$ such that $A = P^{-1}DP$. Since $P^{-1} = P^*$, it follows that $A^* = P^*D^*(P^{-1})^* = P^{-1}DP = A$, and $A$ is Hermitian.

## PROBLEMS

**5.8–1** Prove Theorem 5.8.4.

**5.8–2** Prove Theorem 5.8.5.

**5.8–3** Prove Corollary 5.8.14.

**5.8–4** If the columns or the rows of a unitary matrix are permuted in any way, show that the resulting matrix is unitary.

**5.8–5** If $R_1$ is an $n$th-order nonzero $\mathscr{C}$-normal row $\mathscr{C}$ vector, prove that there exists a unitary matrix $P$ having $R_1$ as its first row.

**5.8–6** If $P$ is a unitary matrix such that all elements of the first column (except the first element) are zero, show that all elements of the first row, except the first element, also are zero.

**5.8–7** Prove that a $\mathscr{C}$-orthonormal system of $\mathscr{C}$ vectors is linearly independent.

**5.8–8** Show that

$$\begin{bmatrix} a + ib & -c + id \\ c + id & a - ib \end{bmatrix},$$

where $a$, $b$, $c$, $d$ are real, is unitary if $a^2 + b^2 + c^2 + d^2 = 1$.

**5.8–9** Prove that any two characteristic vectors corresponding to two distinct characteristic roots of a Hermitian matrix are $\mathscr{C}$-orthogonal.

**5.8–10** Prove that there exists a $\mathscr{C}$-orthonormal system of $n$ characteristic vectors for any given $n \times n$ Hermitian matrix.

**5.8–11**   If $P$ is a unitary matrix and $D$ is a real diagonal matrix such that $P^{-1}AP = D$, show that $A$ is Hermitian.

**5.8–12**   Show that the three row vectors

$$[1 + i, i, 1]/4, \qquad [i, 1 - i, 0]/3, \qquad [1 - i, 1, 3i]/12$$

are $\mathscr{C}$-orthonormal.

**5.8–13**   Prove that if matrix $P$ is both unitary and Hermitian, then it is also involutoric.

**5.8–14**   If $A$ is unitary and if $B = AP$, where $P$ is nonsingular, show that $PB^{-1}$ is unitary.

**5.8–15**   Prove that if $U$ is unitary and $U^*AU$ and $U^*BU$ are both diagonal matrices, then $AB = BA$.

## 5.9   Normal matrices

In this concluding section of Chapter 5 we shall characterize those $\mathscr{C}$ matrices $A$ that are unitarily similar to a diagonal matrix $D$. We have seen, in Theorems 5.8.13 and 5.8.15, that if $D$ is real, then the class of matrices $A$ coincides with the class of Hermitian matrices. If $D$ is not restricted to be real, we shall show that the class of matrices $A$ coincides with the class of so-called *normal* matrices.

**5.9.1**   DEFINITION.   A $\mathscr{C}$ matrix A is said to be *normal* if and only if it commutes with its tranjugate, that is, if and only if $AA^* = A^*A$.

**5.9.2**   THEOREM.   *Normal matrices include Hermitian, skew-Hermitian, unitary, real orthogonal, real symmetric, real skew-symmetric, and complex diagonal matrices as special instances.*

The routine verification is left to the student.

*Note.*   It should be noted that there are normal matrices, such as

$$A = \begin{bmatrix} 0 & 1 + i \\ 1 + i & 0 \end{bmatrix},$$

which do not belong to any of the special types mentioned in Theorem 5.9.2.

**5.9.3**   SCHUR'S THEOREM.   *Every square $\mathscr{C}$ matrix A is unitarily similar to an upper triangular matrix.*

The proof will follow the lines of the proofs of Theorems 5.2.2, 5.7.15, and 5.8.13.

Let $\lambda_1, \cdots, \lambda_n$ be the $n$ (not necessarily distinct) characteristic roots of $A$. Let $C_1$ be any $\mathscr{C}$-normal characteristic vector of $A$ corresponding to $\lambda_1$. Let $P_1$ be any unitary matrix having $C_1$ as its first column; such a matrix exists by Theorem 5.8.10. As in the proof of Theorem 5.8.13, $P_1{}^*AP_1$ has the form

$$\begin{bmatrix} \lambda_1 & L \\ O & B \end{bmatrix},$$

where $B$ is a $\mathscr{C}$ matrix of order $n-1$, with characteristic roots $\lambda_2, \cdots, \lambda_n$. By a similar argument, there exists an $(n-2)$-rowed unitary matrix $Q$ such that

$$Q^*BQ = \begin{bmatrix} \lambda_2 & M \\ O & C \end{bmatrix},$$

where $C$ is a $\mathscr{C}$ matrix of order $n-2$, with characteristic roots $\lambda_3, \cdots, \lambda_n$. Define $P_2$ to be the $n \times n$ $\mathscr{C}$ matrix

$$\begin{bmatrix} I_{(1)} & O \\ O & Q \end{bmatrix},$$

which, as in the proof of Theorem 5.8.13, is unitary. Now

$$P_2{}^*P_1{}^*AP_1P_2 = \begin{bmatrix} I_{(1)} & O \\ O & Q^* \end{bmatrix}\begin{bmatrix} \lambda_1 & L \\ O & B \end{bmatrix}\begin{bmatrix} I_{(1)} & O \\ O & Q \end{bmatrix} = \begin{bmatrix} \lambda_1 & LQ \\ O & Q^*BQ \end{bmatrix}.$$

Setting $LQ = [\alpha_{12} \mid L_1]$, where $L_1$ is a row vector of order $n-2$, we then have

$$P_2{}^*P_1{}^*AP_1P_2 = \begin{bmatrix} \lambda_1 & \alpha_{12} & L_1 \\ 0 & \lambda_2 & M \\ O & O & C \end{bmatrix}.$$

Proceeding in this manner, we obtain a sequence $P_1, \cdots, P_n$ of $n$-rowed unitary matrices such that

$$P_n{}^* \cdots P_2{}^*P_1{}^*AP_1P_2 \cdots P_n = T,$$

where $T$ is an upper triangular matrix having $\lambda_1, \cdots, \lambda_n$ as its diagonal elements. Setting $P = P_1P_2 \cdots P_n$, we see that $P$ is unitary and $P^*AP = T$, which proves the theorem.

**5.9.4** TheoreM. *Every matrix unitarily similar to a normal matrix is normal.*

Suppose $B = P^*AP$, where $A$ is normal and $P$ is unitary. Then $B^* = P^*A^*P$, and

$$BB^* = P^*APP^*A^*P = P^*AA^*P = P^*A^*AP = P^*A^*PP^*AP = B^*B,$$

and $B$ is normal.

**5.9.5   THEOREM.   *Every normal upper triangular matrix is diagonal.***
Let $A = [a_{ij}]_{(n)}$, where $a_{ij} = 0$ when $i > j$, be a normal upper triangular matrix. The (1, 1)th element of $AA^*$ is

$$\sum_{j=1}^{n} a_{1j}\bar{a}_{1j} = \sum_{j=1}^{n} |a_{1j}|^2$$

and the (1, 1)th element of $A^*A$ is

$$\sum_{i=1}^{n} a_{i1}\bar{a}_{i1} = \sum_{i=1}^{n} |a_{i1}|^2 = |a_{11}|^2,$$

since $a_{ij} = 0$ when $i > j$. Since $AA^* = A^*A$, it follows that $a_{12} = \cdots = a_{1n} = 0$. The (2,2)th element of $AA^*$ is

$$\sum_{j=1}^{n} a_{2j}\bar{a}_{2j} = \sum_{j=1}^{n} |a_{2j}|^2$$

and the (2,2)th element of $A^*A$ is

$$\sum_{i=1}^{n} a_{i2}\bar{a}_{i2} = \sum_{i=1}^{n} |a_{i2}|^2 = |a_{22}|^2,$$

since $a_{12} = 0$ and $a_{ij} = 0$ when $i > j$. Since $AA^* = A^*A$, it follows that $a_{23} = \cdots = a_{2n} = 0$. Proceeding in this manner, we arrive at the desired result.

**5.9.6   TOEPLITZ'S THEOREM.   *A necessary and sufficient condition for a $\mathscr{C}$ matrix to be similar to a diagonal matrix is that the $\mathscr{C}$ matrix be normal.***
Let $\mathscr{C}$ matrix $A$ be unitarily similar to a diagonal matrix $D$. Then there exists a unitary matrix $P$ such that $P^*AP = D$. Since $P^* = P^{-1}$, it follows that $A = PDP^*$ and $A^* = PD^*P^*$. Therefore

$$AA^* = PDP^*PD^*P^* = PDD^*P^* = PD^*DP^* = PD^*P^*PDP^* = A^*A,$$

and $A$ is normal.

Conversely, let $A$ be normal. By Schur's Theorem 5.9.3, there exists an upper triangular matrix $T$ such that $A \overset{U}{=} T$. By Theorem 5.9.4, $T$ is normal. By Theorem 5.9.5, $T$ is diagonal.

## PROBLEMS

**5.9–1**  Prove Theorem 5.9.2.

**5.9–2**  Show that $\begin{bmatrix} 0 & 1+i \\ 1+i & 0 \end{bmatrix}$ is normal.

**5.9–3**  If $A$ is normal and nonsingular, prove that $A^{-1}$ is normal.

**5.9–4**  Show that the $\mathscr{C}$ matrix $\begin{bmatrix} 0 & a \\ b & 0 \end{bmatrix}$ is normal if and only if $a\bar{a} = b\bar{b}$.

**5.9–5**  Show that any $\mathscr{C}$ matrix of the form $\begin{bmatrix} a & c \\ c & a \end{bmatrix}$ is normal.

**5.9–6**  Prove that every normal matrix with real characteristic roots is Hermitian.

**5.9–7**  Show that a normal matrix is similar to its transpose.

**5.9–8**  Prove that two normal matrices are unitarily similar if and only if they are similar.

**5.9–9**  If $C$ is a characteristic vector of a normal matrix $A$ corresponding to a characteristic root $\lambda$, prove that $C$ is also a characteristic vector of $A^*$ corresponding to the characteristic root $\bar{\lambda}$.

**5.9–10**  Prove that characteristic vectors corresponding to two distinct characteristic roots of a normal matrix are $\mathscr{C}$-orthogonal.

**5.9–11**  Prove that for any positive integer $k$ and any normal matrix $A$ there is a normal matrix $B$ such that $B^k = A$ and $\rho(B) = \rho(A)$.

**5.9–12**  Prove that if $U$ is unitary, then $A$ is normal if and only if $U^*AU$ is normal.

**5.9–13**  If $A$ and $B$ are normal and $AB = BA$, show that $AB$ is normal.

**5.9–14**  If $A$ is normal, show that $A^p$, where $p$ is any positive integer, is also normal.

**5.9–15**  If $A$ is normal and $g(A)$ is any polynomial in $A$ with complex coefficients, prove that $g(A)$ is normal.

## ADDENDA TO CHAPTER 5

Each of the following items can serve as material for a "junior" research project.

### 5.1A   Companion matrices

Every $n \times n$ $\mathscr{C}$ matrix $A$ possesses a characteristic equation; it is a certain monic polynomial equation $f(\lambda) = 0$ of degree $n$ with complex coefficients. One naturally wonders if, given a monic polynomial equation $f(\lambda) = 0$ of degree $n$ with complex coefficients, one can easily construct an $n \times n$ $\mathscr{C}$ matrix $A$ having $f(\lambda) = 0$ as its characteristic equation. An affirmative answer can be given, for if the monic polynomial equation is

$$f(\lambda) \equiv \lambda^n + a_1\lambda^{n-1} + a_2\lambda^{n-2} + \cdots + a_n = 0,$$

the matrix

$$A = \begin{bmatrix} 0 & 1 & 0 & \cdots & 0 \\ 0 & 0 & 1 & \cdots & 0 \\ \cdots & \cdots & \cdots & \cdots & \cdots \\ 0 & 0 & 0 & \cdots & 1 \\ -a_n & -a_{n-1} & -a_{n-2} & \cdots & -a_1 \end{bmatrix}$$

has $f(\lambda) = 0$ as its characteristic equation. To prove this, in the matrix $A - \lambda I_{(n)}$ add $\lambda$ times the last column to the $(n-1)$th column, in the resulting matrix add $\lambda$ times the $(n-1)$th column to the $(n-2)$th column, and so on, to obtain the matrix

$$B = \begin{bmatrix} 0 & 1 & 0 & \cdots & 0 \\ 0 & 0 & 1 & \cdots & 0 \\ \cdots & \cdots & \cdots & \cdots & \cdots \\ 0 & 0 & 0 & \cdots & 1 \\ -f(\lambda) & * & * & \cdots & * \end{bmatrix}.$$

Now

$$|A - \lambda I_{(n)}| = |B| = (-1)^n f(\lambda),$$

whence $f(\lambda) = 0$ is the characteristic equation of $A$.

The matrix $A$ given above is called the *companion matrix* of the monic polynomial $f(\lambda)$. (If $n = 1$, then $f(\lambda) \equiv \lambda + a_1$ and $A$ is taken as the $1 \times 1$ matrix $[-a_1]$.) It can be shown that *the companion matrix of a monic*

*polynomial has the monic polynomial as its minimum function as well as its
characteristic function.*

Companion matrices play a prominent role in the study of canonical
matrices for the similarity relation. Herein lies a rich and rather extensive
"junior" research project.

### 5.2A   Regular Symmetric Matrices

If $A = [a_{ij}]$ is an $n \times n$ matrix, we define the *leading principal minors*
of $A$ to be

$$p_0 = 1, \qquad p_1 = a_{11}, \qquad p_2 = \begin{vmatrix} a_{11} & a_{12} \\ a_{21} & a_{22} \end{vmatrix},$$

$$p_3 = \begin{vmatrix} a_{11} & a_{12} & a_{13} \\ a_{21} & a_{22} & a_{23} \\ a_{31} & a_{32} & a_{33} \end{vmatrix}, \qquad \cdots, \qquad p_n = |A|.$$

A symmetric matrix of rank $r$ is said to be *regular* if no two consecutive $p$'s
in the sequence $p_0, p_1, \cdots, p_r$ are zero. The interested student may care
to try to establish the following theorems.

**I   THEOREM.**   *If* A *is an* n $\times$ n *nonsingular symmetric matrix, then
by cogredient interchanges of rows and columns,* A *can be converted into a
nonsingular symmetric matrix in which not both* $p_{n-1}$ *and* $p_{n-2}$ *are zero.*

**II   THEOREM.**   *If* A *is an* n $\times$ n *symmetric matrix and if* $p_{n-2}p_n \neq 0$
*but* $p_{n-1} = 0$, *then* $p_{n-2}$ *and* $p_n$ *have opposite signs.*

**III   THEOREM.**   *If* A *is an* n $\times$ n *symmetric matrix of rank* r, *then by
cogredient interchanges of rows and columns,* A *can be converted into a
regular symmetric matrix of rank* r.

### 5.3A   Rotations in 3-Space

Consider three-dimensional Euclidean space with two sets, $Ox, Oy,
Oz$ and $Ox', Oy', Oz'$, of rectangular Cartesian axes having the same origin
$O$. The student will recall from his work in solid analytic geometry that if a
point $P$ has coordinates $(a, b, c)$ relative to the first set of axes, and coordinates $(a', b', c')$ relative to the second set, then

$$a' = a \cos(x'x) + b \cos(x'y) + c \cos(x'z),$$
$$b' = a \cos(y'x) + b \cos(y'y) + c \cos(y'z),$$
$$c' = a \cos(z'x) + b \cos(z'y) + c \cos(z'z),$$

where $\cos(x'x)$ is the cosine of the angle between the $x$-axis and the $x'$-axis, etc. Consider the real matrix

$$A = \begin{bmatrix} \cos(x'x) & \cos(x'y) & \cos(x'z) \\ \cos(y'x) & \cos(y'y) & \cos(y'z) \\ \cos(z'x) & \cos(z'y) & \cos(z'z) \end{bmatrix}.$$

Since the elements of each row (column) of $A$ are the direction cosines of one of the new (old) coordinate axes relative to the old (new) axes, each row (column) is a normal vector. Also, since the axes of each set are mutually perpendicular, any two rows (columns) of $A$ are orthogonal. It follows that $A$ is a real orthogonal matrix. It can be shown that $A$ is properly or improperly orthogonal, according as the two systems of axes have like or unlike dispositions.

Generalizing to $n$ dimensions, it follows that a real orthogonal matrix of order $n$ may be associated with a transformation of $n$-dimensional Euclidean space from one set of mutually perpendicular axes to another such set having the same origin. Herein lies one of the important applications of real orthogonal matrices, and also the reason for the name "orthogonal."

### 5.4A   Cayley's Construction of Real Orthogonal Matrices

A determinant whose matrix is orthogonal is known as an *orthogonant*, and orthogonal matrices were studied from this point of view before there existed an independent theory of matrices. Over 120 papers devoted to orthogonants were published during the nineteenth century. In one of these papers, published in 1846, Cayley offered the following very simple way of constructing real orthogonants or, what is the same thing, real orthogonal matrices.

**I   THEOREM.**   *If* S *is a real skew-symmetric matrix, then* I − S *is nonsingular and the matrix*

$$A = (I + S)(I - S)^{-1}$$

*is a real orthogonal matrix.*

The matrix $I - S$ is nonsingular, for the equation $|I - S| = 0$ implies that 1 is a characteristic root of the matrix $S$, which is impossible because the characteristic roots of a real skew-symmetric matrix are zero or pure imaginary. Now we have

$$A' = [(I - S)^{-1}]'(I + S)' = [(I - S)']^{-1}(I + S)'.$$

But

$$(I - S)' = I' - S' = I + S, \qquad (I + S)' = I' + S' = I - S.$$

Therefore

$$A' = (I + S)^{-1}(I - S),$$

whence

$$A'A = (I + S)^{-1}(I - S)(I + S)(I - S)^{-1}$$

$$= (I + S)^{-1}(I + S)(I - S)(I - S)^{-1} = I,$$

and $A$ is real orthogonal.

We can also prove the converse of the preceding theorem.

**II  THEOREM.**  *Every real orthogonal matrix* A *that does not have* $-1$ *as a characteristic root can be expressed as*

$$A = (I + S)(I - S)^{-1}$$

*by a suitable choice of a real skew-symmetric matrix* S.

Let $A$ be a real orthogonal matrix that does not have $-1$ as a characteristic root. Then $|A + I| \neq 0$, whence $A + I$ is nonsingular. Consider the real matrix $S$ defined by

$$(1) \qquad\qquad S = (A + I)^{-1}(A - I).$$

Then

$$S' = (A - I)'[(A + I)^{-1}]' = (A - I)'[(A + I)']^{-1}$$

$$= (A' - I)(A' + I)^{-1} = (A' + I)^{-1}(A' - I) \quad \text{[see Prob. 2.2–17 (b)]}$$

$$= (A' + A'A)^{-1}(A' - A'A) = [A'(I + A)]^{-1}A'(I - A)$$

$$= (I + A)^{-1}(A')^{-1}A'(I - A) = (I + A)^{-1}(I - A) = -S,$$

and $S$ is skew-symmetric. But, from (1),

$$A - I = (A + I)S,$$

whence

$$A(I - S) = I + S.$$

Now $I - S$ is nonsingular, for if $|I - S| = 0$, we must have 1 as a characteristic root of $S$, which is impossible because the characteristic roots of a real skew-symmetric matrix are zero or pure imaginary. From the last displayed equation we then obtain

$$A = (I + S)(I - S)^{-1},$$

and the theorem is established.

Corresponding to the two preceding theorems for real orthogonal matrices, we have the following for unitary matrices.

**III** THEOREM.    *If* S *is a skew-Hermitian matrix, then* I − S *is nonsingular and the matrix*

$$A = (I + S)(I - S)^{-1}$$

*is a unitary matrix.*

**IV** THEOREM.    *Every unitary matrix* A *that does not have* −1 *as a characteristic root can be expressed as*

$$A = (I + S)(I - S)^{-1}$$

*by a suitable choice of a skew-Hermitian matrix* S.

The student can construct proofs of Theorems III and IV, paralleling the proofs of Theorems I and II. As a corollary the student can then obtain

**V** THEOREM.    *If* H *is a Hermitian matrix, then*

$$A = (H + iI)^{-1}(H - iI) = (H - iI)(H + iI)^{-1}$$

*is unitary, and every unitary matrix* A *that does not have* −1 *as a characteristic root can be expressed in this form.*

### 5.5A    The Characteristic Roots of an Orthogonal Matrix

A polynomial $f(x)$ of degree $n$ is said to be *reciprocal* if and only if

$$f(x) \equiv \pm x^n f(1/x).$$

It follows that if $r$ is a zero of a reciprocal polynomial, then $1/r$ is also a zero of the polynomial, and this accounts for the name "reciprocal."

It is interesting that *the characteristic function of an orthogonal matrix is reciprocal.* To show this, let $P$ be an orthogonal matrix. Then we have

$$P - \lambda I \equiv \lambda P(I/\lambda - P') \equiv -\lambda P(P' - I/\lambda).$$

Therefore, taking determinants and recalling that $|P| = \pm 1$,

$$f(\lambda) \equiv |P - \lambda I| \equiv (-\lambda)^n |P| \, |P' - I/\lambda|$$

$$\equiv \pm \lambda^n |P - I/\lambda| \equiv \pm \lambda^n f(1/\lambda).$$

As a consequence of the preceding theorem, it follows that *if an orthogonal $\mathscr{C}$ matrix is of odd order, then it has $+1$ or $-1$ as one of its characteristic roots.*

Essentially the two results given above are found in an 1854 paper of Francesco Brioschi (1824–1897) and, with an improved proof, in an 1854 paper of F. Faa di Bruno (1825–1888). If an orthogonal matrix is real, we have the following attractive allied property: *The characteristic roots of a real orthogonal matrix are of unit moduli.* To prove this, let $\lambda$ be a characteristic root of the real orthogonal matrix $P$, and let $C$ be a corresponding characteristic vector of $P$. Then

$$PC = \lambda C,$$

where $C$ is real and nonzero. It follows that

$$C^* P^* = \bar{\lambda} C^*,$$

or, since $P^* = \bar{P}' = P'$,

$$C^* P' = \bar{\lambda} C^*.$$

Therefore

$$C^* P' P C = \lambda \bar{\lambda} C^* C,$$

or, since $P'P = I$,

$$C^* C = \lambda \bar{\lambda} C^* C.$$

Since $C^* C$ is a nonzero real number, we find that $\lambda \bar{\lambda} = 1$, and the desired result follows.

The student may care to generalize the last proof so as to apply to an arbitrary unitary matrix.

### 5.6A    Definite, Semidefinite, and Indefinite Real Symmetric Matrices

Let $A$ be a real symmetric $n \times n$ matrix of rank $r$ and index $p$. Then $A$ is said to be

*positive definite* if and only if $r = n$, $p = n$,
*negative definite* if and only if $r = n$, $p = 0$,
*positive semidefinite* if and only if $r < n$, $p = r$,
*negative semidefinite* if and only if $r < n$, $p = 0$,
*indefinite* in every other case.

If $A$ is a real symmetric $n \times n$ matrix and if $X$ is a real variable $n$th-order column vector, then the real quadratic form $X'AX$ is said to be *positive definite*, *negative definite*, *positive semidefinite*, *negative semidefinite*, or *indefinite*, according as $A$ is positive definite, negative definite, positive semidefinite, negative semidefinite, or indefinite.

**I**  THEOREM.  *If $A_{(n)}$ is positive definite, then $A_{(n)} \overset{c}{=} I_{(n)}$; if $A_{(n)}$ is negative definite, then $A_{(n)} \overset{c}{=} -I_{(n)}$; if $A_{(n)}$ is positive semidefinite of rank $r$, then $A_{(n)} \overset{c}{=} diag(I_{(r)}, O_{(n-r)})$; if $A_{(n)}$ is negative semidefinite of rank $r$, then $A_{(n)} \overset{c}{=} diag(-I_{(r)}, O_{(n-r)})$.*

**II**  THEOREM.  *If $A$ is a real symmetric matrix and if $B \overset{c}{=} A$, then $B$ is positive definite, negative definite, positive semidefinite, negative semidefinite, or indefinite, according as $A$ is positive definite, negative definite, positive semidefinite, negative semidefinite, or indefinite.*

**III**  THEOREM.  *A real definite quadratic form $X'AX$ has the value zero if $X = O$; for every nonzero value of $X$ it has a positive or negative value, according as the form is positive or negative definite.*

**IV**  THEOREM.  *A real positive semidefinite form is nonnegative; a real negative semidefinite form is nonpositive.*

**V**  THEOREM.  *If $X'AX$ is a real semidefinite form, there exists a nonzero $X$ such that $X'AX = O$.*

**VI**  THEOREM.  *Every real indefinite form assumes positive as well as negative values.*

**VII**   THEOREM.   *If A is a real symmetric semidefinite matrix, then* $|A| = 0$; *if A is a real symmetric positive definite matrix, then* $|A| > 0$.

**VIII**   *A necessary and sufficient condition for a real symmetric matrix A to be positive definite is that all the leading principal minors of A be positive.*

**IX.**   THEOREM.   *A necessary and sufficient condition for a real symmetric matrix A to be negative definite is that the leading principal minors of A, starting with that of the first order, be alternately negative and positive.*

Theorems VIII and IX are more difficult to prove than the other seven theorems; these last two criteria were given by Frobenius in 1876.

We leave to the student the extension of the foregoing definitions and theorems to the case where $A$ is an $n \times n$ Hermitian matrix of rank $r$ and index $p$.

## 5.7A   Gram Matrices

If $A$ is any $\mathcal{R}$ matrix, then the real symmetric matrix $A'A$ is called the *Gram matrix* of $A$, named after a J. P. Gram who foreshadowed the concept in an 1881 paper published in *Crelle's Journal*.

**I**   THEOREM.   *The Gram matrix of a real matrix A is positive definite or positive semidefinite, according as the rank of A is equal to or less than the number of its columns.*

**II**   THEOREM.   *The Gram matrix of a real matrix A is positive definite or positive semidefinite, according as the columns of A constitute a linearly independent or a linearly dependent system of vectors.*

**III**   THEOREM.   *The columns of a real matrix A are or are not linearly independent according as* $|A'A| > 0$ *or* $|A'A| = 0$.

**IV**   THEOREM.   *If any principal minor of the Gram matrix of a real matrix is zero, then the Gram matrix is singular.*

**V**  THEOREM.  *A real matrix and its Gram matrix have the same rank.*

**VI**  THEOREM.  *Every positive definite or positive semidefinite real symmetric matrix is the Gram matrix of some real matrix.*

**VII**  THEOREM.  *The columns of a real matrix* A *are mutually orthogonal if and only if the Gram matrix of* A *is diagonal.*

If $A$ is any $\mathscr{C}$ matrix, one may define the *Gram matrix* of $A$ to be the Hermitian matrix $A^*A$. The construction of analogs of the foregoing theorems for this generalization is left to the student.

### 5.8A  Some Theorems of Autonne

The following three theorems were established by L. Autonne in 1902 and 1903.

**I**  THEOREM.  *If* A *is a positive semidefinite Hermitian matrix and* m *is any positive integer, then there exists a unique positive semidefinite Hermitian matrix* B *such that* $B^m = A$, *and rank* B = *rank* A.

**II**  THEOREM.  *If* A *is a positive semidefinite Hermitian matrix, there exists a unique positive semidefinite Hermitian matrix* P *of the same rank as* A *such that* A = P\*P.

**III**  THEOREM.  *Every nonsingular* $\mathscr{C}$ *matrix is uniquely expressible as a product of a unitary matrix and a positive definite Hermitian matrix.*

If one defines an *inversion* to be an improper real orthogonal matrix $S$ such that $|\lambda I - S| = (\lambda - 1)^{n-1}(\lambda + 1)$, where $n$ is the order of $S$, Autonne proved (also in 1903) the following theorem:

**IV**  THEOREM.  *Every real orthogonal matrix is a product of inversions.*

In 1915, Autonne proved the following attractive theorem, the real case of which had been treated earlier by E. Beltrami, C. Jordan, and J. J. Sylvester.

**V  THEOREM.**  *If* A *is a nonsingular $\mathscr{C}$ matrix, there exist unitary matrices* U *and* V *such that*

$$UAV = diag(u_1, u_2, \cdots, u_n),$$

*where* $u_1, u_2, \cdots, u_n$ *are the positive square roots of the characteristic roots of* AA*.*

### 5.9A  Simultaneous Reduction of a Pair of Quadratic Forms

If $X'AX$ and $X'BX$ are two $n$th-order quadratic forms, it is sometimes possible to choose a linear transformation $X = PY$ that transforms both forms into diagonal forms. In this connection we state the following two theorems, whose proofs will be left to the student.

**I  THEOREM.**  *If* A *and* B *are two* n × n *symmetric $\mathscr{C}$ matrices such that the roots of the equation*

$$|A - \lambda B| = 0$$

*are distinct, there exists a $\mathscr{C}$ matrix* P *such that both* P'AP *and* P'BP *are diagonal.*

**II  THEOREM.**  *If* A *and* B *are two* n × n *symmetric $\mathscr{R}$ matrices such that* B *is positive definite, there exists a nonsingular $\mathscr{R}$ matrix* P *such that*

$$P'AP = diag(\lambda_1, \cdots, \lambda_n), \qquad P'BP = I_{(n)},$$

*where* $\lambda_1, \cdots, \lambda_n$ *are the roots of the equation*

$$|A - \lambda B| = 0.$$

The student may state and prove two analogous theorems for Hermitian matrices.

### 5.10A Hadamard Matrices

An *Hadamard matrix* [named after the French mathematician Jacques Hadamard (1865–1963)] is a square matrix $H$ such that: (1) the elements of $H$ are 1's and $-1$'s, (2) the elements in any two different rows or in any two different columns of $H$ coincide in exactly half the number of positions.

The student is invited to establish the following theorems.

**I** THEOREM. *Any two different rows (columns) of an Hadamard matrix are orthogonal.*

**II** THEOREM. *An Hadamard matrix is of even order.*

**III** THEOREM. *If H is an Hadamard matrix of order* n, *then* $HH' = nI_{(n)}$.

**IV** LEMMA. *The following is an algebraic identity*:

$$(a_1 + b_1)(a_1 + c_1) + (a_2 + b_2)(a_2 + c_2) + \cdots + (a_n + b_n)(a_n + c_n)$$
$$= (a_1^2 + a_2^2 + \cdots + a_n^2) + (a_1b_1 + a_2b_2 + \cdots + a_nb_n)$$
$$+ (a_1c_1 + a_2c_2 + \cdots + a_nc_n) + (b_1c_1 + b_2c_2 + \cdots + b_nc_n).$$

**V** THEOREM. *Let*

$$H = \begin{bmatrix} a_1 & a_2 & a_3 & \cdots & a_n \\ b_1 & b_2 & b_3 & \cdots & b_n \\ c_1 & c_2 & c_3 & \cdots & c_n \\ \cdots & \cdots & \cdots & \cdots & \cdots \end{bmatrix}$$

*represent an Hadamard matrix of order* n > 3. *Then*

(1) $(a_1 + b_1)(a_1 + c_1) + \cdots + (a_n + b_n)(a_n + c_n) = n$,

(2) $(a_i + b_i)(a_i + c_i) = 0$ *or* 4, i = 1, $\cdots$, n.

**VI** THEOREM. *Except for the* $2 \times 2$ *Hadamard matrices, the order of an Hadamard matrix is a multiple of* 4.

**VII** THEOREM.   *If* H *is an Hadamard matrix of order* n, *then*

$$\begin{bmatrix} H & H \\ H & -H \end{bmatrix}$$

*is an Hadamard matrix of order* 2n.

It is believed that the reverse of Theorem VI is also true: *There exists an Hadamard matrix of order* n *whenever* n *is a multiple of* 4. This reverse conjecture has not yet (July 1978) been established, though Hadamard matrices have been constructed for many multiples of 4, in particular for all less than 268. The interested student may care to consult S. W. Golomb and L. D. Baumert, "The search for Hadamard matrices," *The American Mathematical Monthly,* vol. 70 (1963), pp. 12–17.

### 5.11A   Equitable Matrices

An *equitable matrix* is a real square matrix $[a_{ij}]$ with positive elements such that $a_{ij}a_{jk} = a_{ik}$ for all $i$, $j$, $k$. F. D. Parker has applied equitable matrices to a problem in money exchange and to an important problem in group theory.† The student may care to establish independently the following theorems about equitable matrices.

**I** THEOREM.   *If* M *is an equitable matrix of order* n, *then* $M^2 = nM$.

**II** THEOREM.   *If* S *is a square matrix of order* n *consisting entirely of unity elements, then* S *is an equitable matrix, and any equitable matrix (of order* n) *is similar to* S.

**III** THEOREM.   *All equitable matrices of order* n *are similar, and their characteristic roots are* n, 0, $\cdots$, 0.

**IV** DEFINITION.   *If* $A = [a_{ij}]$, $B = [b_{ij}]$ are $n \times n$ matrices, then the *Hadamard product* of $A$ and $B$ is the $n \times n$ matrix $C = [c_{ij}]$ such that $c_{ij} = a_{ij}b_{ij}$.

---

† F. D. Parker, "Matrices in the market place," *Mathematics Magazine,* vol. 38 (1965), pp. 125–128; F. D. Parker, "When is a loop a group?", *American Mathematical Monthly,* vol. 72 (1965), pp. 765–766.

**V** THEOREM. *Equitable matrices form a commutative group under Hadamard multiplication.*

**VI** THEOREM. *A matrix that diagonalizes an equitable matrix* $A = [a_{ij}]$ *of order* n *is*

$$R = \begin{bmatrix} 1 & 0 & 0 & \cdots & 0 & -1 \\ a_{21} & a_{21} & 0 & \cdots & 0 & 0 \\ a_{31} & -a_{31} & a_{31} & \cdots & 0 & 0 \\ a_{41} & 0 & -a_{41} & \cdots & 0 & 0 \\ \cdots & \cdots & \cdots & \cdots & \cdots & \cdots \\ a_{n1} & 0 & 0 & \cdots & -a_{n1} & a_{n1} \end{bmatrix}.$$

# 6. TOWARD ABSTRACTION

*6.1. Number rings and number fields. Problems. 6.2. General rings and general fields. Problems. 6.3. Matrix representation. Problems. 6.4. k-vector spaces over a field F. Problems. 6.5. General vector spaces. Problems. 6.6. Linear transformations of vector spaces. Problems. 6.7. Jordan and Lie algebras. Problems.*

## 6.1 Number rings and number fields

In building up an algebra of rectangular arrays of numbers in Chapter 1, we introduced operations on the arrays called *addition*, *Cayley multiplication*, and *scalar multiplication*. In performing these operations, we had to add and multiply the numbers themselves, and in later parts of the book we had to perform subtractions and divisions with these numbers. These facts, along with a desire on some occasions to limit the kinds of numbers appearing in the arrays, leads to the following definitions.

**6.1.1** DEFINITIONS. A nonempty set $R$ of complex numbers is called a *number ring* if and only if each of the familiar operations of addition, subtraction, and multiplication performed on any two numbers in $R$ yields a number in $R$. A set $F$ of complex numbers, consisting of more than one number, is called a *number field* if and only if each of the familiar operations of addition, subtraction, multiplication, and division (except by 0) performed on any two numbers in $F$ yields a number in $F$.

More briefly put, we say that a nonempty set $R$ of complex numbers

276

is a number ring if and only if $R$ is *closed* under the familar operations of addition, subtraction, and multiplication, and that a set $F$ of at least two complex numbers is a number field if and only if $F$ is *closed* under the familiar operations of addition, subtraction, multiplication, and permissible division. Clearly, a number field is also a number ring, but a number ring need not be a number field.

A number ring that contains the number 1 is called a *number ring with unity*.

*Examples.* Some examples of number rings that are not also number fields are:

(1) The set consisting of the single number 0. This is called the *trivial number ring*.

(2) The set of all integers. This is called the *integral number ring*.

(3) The set of all even integers.

(4) The set of all numbers of the form $a + b\sqrt{2}$, where $a$ and $b$ are integers.

(5) The set of all numbers of the form $a + bi$, where $a$ and $b$ are integers and $i$ is the imaginary unit. This is called the *ring of Gaussian integers*.

(6) The set of all numbers of the form $a + b\sqrt{2}i$, where $a$ and $b$ are integers and $i$ is the imaginary unit.

Of these number rings, (2), (4), (5), (6) are number rings with unity.

Some examples of number fields are:

(1') The set of all rational numbers. This is called the *rational number field*.

(2') The set of all real numbers. This is called the *real number field*.

(3') The set of all complex numbers. This is called the *complex number field*.

(4') The set of all numbers of the form $a + b\sqrt{2}$, where $a$ and $b$ are rational numbers.

(5') The set of all numbers of the form $a + bi$, where $a$ and $b$ are rational numbers and $i$ is the imaginary unit.

(6') The set of all numbers of the form $a + b\sqrt{2}i$, where $a$ and $b$ are rational numbers and $i$ is the imaginary unit.

**6.1.2** DEFINITIONS. If $R_1$ and $R_2$ are two number rings and if $R_1$ is a subset of $R_2$, then $R_1$ is called a *subring* of $R_2$. If, further, $R_1$ is distinct from $R_2$, then $R_1$ is called a *proper subring* of $R_2$.

If $F_1$ and $F_2$ are two number fields and if $F_1$ is a subset of $F_2$, then $F_1$ is called a *subfield* of $F_2$. If, further, $F_1$ is distinct from $F_2$, then $F_1$ is called a *proper subfield* of $F_2$.

**6.1.3**  THEOREM.  *Every number ring* R *contains the trivial number ring as a subring; every number field* F *contains the rational number field as a subfield.*

Since $R$ is nonempty, it contains at least one number $a$. It then contains $a - a = 0$.

Since $F$ contains at least two numbers, it contains a number $b \neq 0$. It then contains $b/b = 1$. By repeated additions and subtractions, it now follows that $F$ contains all the integers, and therefore (by permissible divisions) all the rational numbers.

**6.1.4**  DEFINITION.  A number ring or a number field that contains the complex conjugate of each of its elements is said to be *closed under conjugation*.

*Examples.*  All examples of number rings and number fields following Definitions 6.1.1 are closed under conjugation. We leave it to the student to wonder whether a number ring or a number field is necessarily closed under conjugation. (See Problem 6.1–16.)

**6.1.5**  DEFINITION.  A number field $F$ is said to be *algebraically closed* if the roots of every polynomial equation (in one variable) with coefficients in $F$ are numbers in $F$.

**6.1.6**  THEOREM.  *The complex number field is algebraically closed.*

This is essentially the fundamental theorem of algebra.

*Note.*  It can be shown that the complex number field contains infinitely many algebraically closed subfields. See, for example, Louis Weisner, *Introduction to the Theory of Equations* (New York: The Macmillan Company, 1938).

**6.1.7**  DEFINITION.  If the elements of a matrix are chosen from some set $S$, then we say the matrix is *over S*.

In the previous chapters of this book we assumed, unless explicitly stated otherwise, that the elements of our matrices were arbitrary complex numbers or arbitrary polynomials (in one variable) with complex coefficients. Actually, all our general definitions and all our general theorems, with their proofs, hold for matrices over more restricted sets of numbers. Thus, essentially, most of Chapter 1, with Sections 1.7 and 1.8 excepted, holds for matrices over an arbitrary but fixed number ring $R$, and certainly the whole chapter, with Sections 1.7 and 1.8 excepted, holds for matrices over an arbitrary but fixed number ring $R$ with unity. Sections 1.7 and 1.8 hold for matrices over any arbitrary but fixed number ring $R$ closed under conjugation. Though much of Chapters 2 and 3 holds for matrices over an arbitrary fixed number ring with unity, all material of these chapters holds for matrices over an arbitrary fixed number field $F$. Again, the general parts of Chapters 4 and 5 hold if the matrices are taken over an arbitrary fixed number field $F$ and if the $\lambda$ matrices are such that their polynomial elements have coefficients in the number field $F$. Sections 5.6, 5.8, and 5.9 hold for matrices over an arbitrary but fixed number field $F$ which is closed under conjugation. In certain parts of Chapters 4 and 5 involving the characteristic roots of a matrix, it may be desirable to restrict the number field $F$ to one that is algebraically closed.

## PROBLEMS

**6.1–1**  Verify that the examples (1), (2), (3), (4), (5), (6) following Definitions 6.1.1 are number rings that are not also number fields.

**6.1–2**  Verify that the examples (1′), (2′), (3′), (4′), (5′), (6′) following Definitions 6.1.1 are number fields.

**6.1–3**  For each of the examples in Problem 6.1–1, list the other examples that are proper subrings.

**6.1–4**  For each of the examples in Problem 6.1–2, list the other examples that are proper subfields.

**6.1–5**  Show that the following sets of numbers are not number fields, and determine those that are also not number rings:

(a) All nonnegative integers.

(b) All odd integers.

(c) All rational numbers $\leq 1,000,000$.

(d) All numbers of the form $b\sqrt{2}$, where $b$ is rational.

(e) The four numbers $1$, $-1$, $i$, $-i$.

(f) The number $0$.

**(g)** All numbers of the form $a + b\sqrt[3]{3}$, where $a$ and $b$ are rational numbers.

**(h)** All numbers of the form $a + b\sqrt{2} + c\sqrt{3}$, where $a$, $b$, $c$ are rational numbers.

**6.1–6** **(a)** Show that every nontrivial number ring contains an infinite number of elements.

**(b)** Show that the real number field and the complex number field are the only number fields that contain all the real numbers.

**6.1–7** Prove that any subset $S$ of a number ring $R$ is a subring of $R$ if and only if for arbitrary numbers $a$ and $b$ of $S$, $a - b$ and $ab$ are in $S$.

**6.1–8** Show that a set $S$ of at least two complex numbers is a number field if $a - b$, $ab$, $1/a$ are in $S$ whenever $a$ and $b$ are in $S$, where $a \neq 0$ in the case of $1/a$.

**6.1–9** **(a)** Show that the numbers common to two intersecting number rings comprise a number ring.

**(b)** Show that the numbers common to two intersecting number fields comprise a number field.

**6.1–10** Show that the set of all numbers of the form $a + b\sqrt[3]{2} + c\sqrt[3]{4}$, where $a$, $b$, $c$ are rational numbers, is a number field.

**6.1–11** Let $a_1, \cdots, a_n$ be a given set of complex numbers and let $F(a_1, \cdots, a_n)$ denote the set of all numbers each of which can be obtained from $a_1, \cdots, a_n$ by a finite number of additions, subtractions, multiplications, and permissible divisions. Clearly, $F(a_1, \cdots, a_n)$ is the smallest number field containing $a_1, \cdots, a_n$; it is called the *number field generated by* $a_1, \cdots, a_n$.

**(a)** Show that $F(1)$ is the rational number field.

**(b)** Show that $F(\sqrt{2})$ consists of all numbers of the form $a + b\sqrt{2}$, where $a$ and $b$ are rational.

**(c)** Show that $F(\sqrt{2}, \sqrt{3}) = F(\sqrt{2}, \sqrt{6}) = F(\sqrt{3}, \sqrt{6})$.

**6.1–12** Using the notation of Problem 6.1–11, show that

**(a)** if $a$ is in $F(b)$, then every number of $F(a)$ is in $F(b)$;

**(b)** if $a$ is in $F(b)$ and $b$ is in $F(a)$, then $F(a) = F(b)$.

**6.1–13** Using the notation of Problem 6.1–11, show that

**(a)** $\sqrt{2}$ is not in $F(\sqrt{3})$;

**(b)** $i$ is not in $F(\sqrt{-3})$.

**6.1–14** What is the smallest number field over which each of the following matrices can be taken?

**(a)** $\begin{bmatrix} 2/3 & \sqrt{2} \\ -3 & -1/2 \end{bmatrix}$, **(b)** $\begin{bmatrix} 2 & 2-i \\ i & -3 \end{bmatrix}$.

**6.1–15**  If $\lambda$ is a root of a given quadratic equation with rational coefficients, show that the set of all numbers of the form $a + \lambda b$, where $a$ and $b$ are rational numbers, is a number field.

**6.1–16**  Prove that the number field $F(\sqrt[3]{2}\omega)$, where $\omega$ is one of the imaginary cube roots of unity, is not closed under conjugation.

## 6.2  General rings and general fields

A good deal of matrix theory can be extended to matrices with elements chosen from more general systems, called simply *rings* and *fields*, of which number rings and number fields are special but very important examples.

**6.2.1**  DEFINITIONS.   A *ring* $R$ is a nonempty set of elements $a$, $b$, $c, \cdots$, along with two binary operations, which we shall call *addition* (denoted by $a + b$) and *multiplication* (denoted by $a \times b$, $a \cdot b$, or $ab$), such that if $a$, $b$, $c$ are any three elements of $R$ then

**P1:** a + b *and* ab *are uniquely defined elements of* R (the closure property for addition and multiplication).
**P2:** a + b = b + a (the commutative law of addition).
**P3:** a + (b + c) = (a + b) + c (the associative law of addition).
**P4:** *The equation* a + x = b *always has a solution in* R.
**P5:** a(bc) = (ab)c (the associative law of multiplication).
**P6:** a(b + c) = ab + ac *and* (b + c)a = ba + ca (the left and right distributive laws of multiplication over addition).

If, in addition, we have

**P7:** ab = ba (the commutative law of multiplication),

then ring $R$ is said to be *commutative*, or *Abelian*.
It can be shown that every ring $R$ has a unique element $z$, called the *zero element* of $R$, which has the properties that for every element $a$ of $R$

$$a + z = z + a = a, \qquad az = za = z.$$

It can also be shown that the solution $x$ guaranteed by P4 is unique. It follows that for each element $a$ of $R$, there is a unique element $\bar{a}$ of $R$ such that $a + \bar{a} = z$. The element $\bar{a}$ is called the *additive inverse* of the element $a$. If $R$ contains an element $e$ such that $ae = a$ for every $a$ in $R$, then $e$ is called a *right unity element* of $R$. Similarly, an element $f$ of $R$ such that $fa = a$ for every $a$ in $R$ is called a *left unity element* of $R$. There are rings that have neither a right unity element nor a left unity element, some that have a right unity element but no left unity element, some that have a left unity element but no right unity element, and some that have both a right unity element and a left unity element. If, however, $R$ has both a right unity element $e$ and a left unity element $f$, it can be shown that the two are identical and unique. Such a ring is called a *ring with unity*, and the unique right and left unity element is called the *unity element* of the ring.

**6.2.2** DEFINITIONS.    A commutative ring $F$ of at least two distinct elements in which

**P8:** *The equation* $ax = b$ *always has a solution in* F *when* $a \neq z$,

is called a *field*.

It can be shown that every field is a ring with unity $u \neq z$, and it can be shown that the solution $x$ guaranteed by P8 is unique. It follows that for each element $a$ of $F$ there is a unique element $a^{-1}$ of $F$ such that $aa^{-1} = u$. The element $a^{-1}$ is called the *multiplicative inverse* of the element $a$. It can further be shown for any field that if $ab = z$, then either $a = z$ or $b = z$. A particularly simple field is a field containing only two distinct elements $z$ and $u$, where

$$
\begin{array}{ll}
z + z = z, & z \times z = z, \\
z + u = u + z = u, & z \times u = u \times z = z, \\
u + u = z, & u \times u = u.
\end{array}
$$

Such a field is called a *field of characteristic 2*. In a field not of characteristic 2, it can be shown that the equation $a + a = z$ implies $a = z$; in a field of characteristic 2, however, $u + u = z$ and $u \neq z$, and the implication does not hold. Because of this difference between fields of characteristic 2 and fields not of characteristic 2, there exist some valuable theorems which can be shown to hold in all fields except those of characteristic 2, these exceptional fields there appearing in the role of black sheep.†

It is interesting that, if we bar fields (and rings) of characteristic 2, then every definition and theorem (with its proof) stated to hold for matrices over a fixed but arbitrary number ring $R$, or a fixed but arbitrary

---

† Webster defines a *black sheep* as "one in a family or company whose worthlessness or infamous conduct causes him to stand out from the rest as a cause of shame and disgrace to them," and in the American College Dictionary we find the definition, "a person worthless despite good background."

number ring $R$ with unity, or a fixed but arbitrary number field $F$, also hold for matrices over, respectively, a fixed but arbitrary general ring $R$, or a fixed but arbitrary general ring $R$ with unity, or a fixed but arbitrary general field $F$. Thus, essentially, most of Chapter 1, with Sections 1.7 and 1.8 excepted, holds for matrices over a fixed but arbitrary general ring $R$, and certainly the whole chapter, with Sections 1.7 and 1.8 excepted, holds for matrices over a fixed but arbitrary general ring $R$ with unity. Sections 1.7 and 1.8 must be excepted because they involve the notion of conjugation, which is a concept specifically associated with complex matrices. All the material of Chapters 2 and 3 holds for matrices over a fixed but arbitrary general field $F$. In this connection, note that the given proof of Theorem 3.3.6 requires that $F$ be not of characteristic 2; this is why the theorem was later reestablished (in Theorem 3.4.8) with a proof that circumvents the objectionable case. The general parts of Chapters 4 and 5 hold if the matrices are taken over an arbitrary fixed general field $F$ and if the $\lambda$ matrices are such that their polynomial elements have coefficients in the general field $F$. Of course much of Chapter 5 must be excepted, since much of Chapter 5 concerns itself with special number fields. In this connection, note that though Section 5.4 was specifically limited to skew-symmetric $\mathscr{C}$ matrices, all material of this section actually holds for matrices over any arbitrary but fixed general field $F$ that is not of characteristic 2. In the proof of Theorem 5.4.1 we used the fact that in a skew-symmetric $\mathscr{C}$ matrix $A = [a_{ij}]$, the diagonal elements $a_{pp}$ are all equal to zero. Now, if $A$ is a skew-symmetric matrix over an arbitrary but fixed general field $F$, then (by definition) $a_{pq} = -a_{qp}$. In particular $a_{pp} = -a_{pp}$, or $a_{pp}(u + u) = z$, where $u$ and $z$ are the unity and zero elements, respectively, of $F$. If $F$ is not of characteristic 2, then $u + u \neq z$, and it follows that $a_{pp} = z$. On the other hand, if $F$ is of characteristic 2, then $u + u = z$, and $a_{pp}$ may be different from $z$. A similar discussion applies to Theorem 5.5.1; this theorem (with its proof) actually holds for any symmetric matrix $A$ over an arbitrary but fixed general field $F$ that is not of characteristic 2. The reason for excepting fields of characteristic 2 is apparent in the early part of the proof of the theorem, where we conclude that

$$b_{hh} = a_{hk} + a_{kh} = a_{hk} + a_{hk} \neq z;$$

this inequality does not follow if $a_{hk} = u$, and we are in a field of characteristic 2.

When studying matrices over general rings and general fields, the game is to codify or classify all definitions and theorems according to whether they hold for matrices over an arbitrary but fixed general ring $R$, an arbitrary but fixed general ring $R$ with unity, an arbitrary but fixed general commutative ring $R$, an arbitrary but fixed general field $F$, or an arbitrary but fixed general field $F$ not of characteristic 2.

## PROBLEMS

**6.2–1**   Show that a number ring is an example of a general ring, and that a number field is an example of a general field.

**6.2–2**   Show that the three elements 0, 1, 2 constitute a ring if addition and multiplication are defined by the following tables:

| + | 0 | 1 | 2 |
|---|---|---|---|
| 0 | 0 | 1 | 2 |
| 1 | 1 | 2 | 0 |
| 2 | 2 | 0 | 1 |

| × | 0 | 1 | 2 |
|---|---|---|---|
| 0 | 0 | 0 | 0 |
| 1 | 0 | 1 | 2 |
| 2 | 0 | 2 | 1 |

**6.2–3**   Show that the set of all polynomials in the single variable $\lambda$ and having coefficients in a given ring $R$, along with ordinary addition and multiplication of polynomials, forms a ring.

**6.2–4**   (a) Show that the set of elements $a$, $b$, $c$, $d$ constitutes a commutative ring if addition and multiplication are defined by the following tables:

| + | a | b | c | d |
|---|---|---|---|---|
| a | a | b | c | d |
| b | b | a | d | c |
| c | c | d | a | b |
| d | d | c | b | a |

| × | a | b | c | d |
|---|---|---|---|---|
| a | a | a | a | a |
| b | a | b | c | d |
| c | a | c | d | b |
| d | a | d | b | c |

**(b)** Does this ring have a unity?

**6.2–5**   (a) Show that the set of elements $a$, $b$, $c$, $d$ constitutes a noncommutative ring if addition and multiplication are defined by the following tables:

| + | a | b | c | d |
|---|---|---|---|---|
| a | a | b | c | d |
| b | b | a | d | c |
| c | c | d | a | b |
| d | d | c | b | a |

| × | a | b | c | d |
|---|---|---|---|---|
| a | a | a | a | a |
| b | a | b | c | d |
| c | a | a | a | a |
| d | a | b | c | d |

**(b)** Does this ring have a unity?

**6.2–6**  Show that the set of all ordered triples $(a, b, c)$ of integers, with addition and multiplication of the triples defined by

$$(a, b, c) + (d, e, f) = (a + d, b + e, c + f)$$

and

$$(a, b, c)(d, e, f) = (ad, bd + ce, cf),$$

constitutes a noncommutative ring with unity.

**6.2–7**  Show that the set of all $n \times n$ matrices over a ring $R$ constitutes a noncommutative ring $R_n$ with respect to matrix addition and multiplication. Show that if $R$ has a unity element, then so does $R_n$.

**6.2–8**  Let $U$ be a given set and let $R$ be the set of all subsets of $U$, including the empty set and the entire set $U$. Show that $R$, along with addition and multiplication of sets defined as set union and set intersection, respectively, constitutes a commutative ring with unity.

**6.2–9**  If $a$ and $b$ are any integers, define

$$a \oplus b = a + b - 1, \qquad a \otimes b = a + b - ab.$$

Show that with respect to these *new* definitions of "addition" and "multiplication," the set of all integers is a commutative ring with unity.

**6.2–10**  (a) Show that if, in a ring $R$, $a + z = a$ for some element $a$ of $R$, then $b + z = b$ for every element $b$ of $R$.

(b) Show that there is a unique element $z$ of $R$ such that $a + z = a$ for every element $a$ of $R$.

(c) Show that the equation $a + x = b$ has a unique solution $x$ in $R$.

(d) Show that for each element $a$ of $R$ there is a unique element $\bar{a}$ of $R$ such that $a + \bar{a} = z$.

(e) Show that $az = za = z$ for every element $a$ of $R$.

(f) Show that if $R$ has a right unity element $e$ and a left unity element $f$, then $e = f = u$, say, where $u$ is unique.

**6.2–11**  Which of the examples in Problems 6.2–2 through 6.2–9 are also fields?

**6.2–12**  (a) Show that every field $F$ is a ring with a unity element $u$.

(b) Show that $u \neq z$.

(c) Show that the equation $ax = b$ has a unique solution in $F$ whenever $a \neq z$.

(d) Show that for each element $a \neq z$ in $F$ there exists a unique element $a^{-1}$ of $F$ such that $aa^{-1} = a^{-1}a = u$.

(e) Show that if $ab = z$ in $F$, then either $a = z$ or $b = z$.

**6.2–13**  (a) Show that a field of characteristic 2 is indeed a field.

(b) Show that in a field of characteristic 2, the equation $a = \bar{a}$ does not imply that $a = z$.

## 6.3    Matrix realization

Rings, commutative rings, rings with unity, and fields are examples of abstract algebraic systems, each system being defined by a basic set of postulates. It is the province of the subject called *abstract algebra* to study various algebraic systems and their interrelations. Once an algebraic system has been postulationally defined, it is useful and edifying to have one or more concrete examples of the system. It is very interesting that algebraists have so frequently found, for their algebraic systems, examples in which the elements of the systems are interpreted as matrices of one sort or another. Such an example is called a *matrix realization* of the concerned system.

We preface our illustrations of matrix realization by first introducing the important abstract algebraic system known as a *group*.

**6.3.1** DEFINITIONS.    A *group* is a set $G$ of elements in which a binary operation $*$ is defined, satisfying the following four postulates:

**G1:** *For each* a *and* b *in* G, a $*$ b *is a uniquely defined element of* G.
**G2:** *For all* a, b, c *in* G, a $*$ (b $*$ c) = (a $*$ b) $*$ c.
**G3:** *There exists an element* i *of* G *such that, for all* a *in* G, a $*$ i = a. (The element *i* is called a *right identity element* of *G*.)
**G4:** *For each element* a *of* G *there exists an element* $a^{-1}$ *of* G *such that* a $*$ $a^{-1}$ = i. (The element $a^{-1}$ is called a *right inverse element* of *a*.)

If, in addition to the foregoing four postulates, the following postulate is satisfied, the group is called a *commutative*, or an *Abelian*, *group*.

**G5:** *For all* a *and* b *in* G, a $*$ b = b $*$ a.

A group for which Postulate G5 does not hold is called a *noncommutative*, or *non-Abelian*, *group*.

It is an easy matter to find examples of commutative groups, but it is somewhat more difficult to find examples of noncommutative groups. We leave to the student the easy establishment of the following two theorems, furnishing matrix realizations of both types of groups.

**6.3.2**   THEOREM.   *The set* G *of all* $2 \times 2$ $\mathcal{R}$ *matrices, along with addition of matrices, is a commutative group.*

**6.3.3**   THEOREM.   *The set* G *of all* $2 \times 2$ *nonsingular* $\mathcal{R}$ *matrices, along with Cayley multiplication of matrices, is a noncommutative group.*

Since, in the definition of a group $G$ we postulated the existence of a *right* identity element for $G$ and the existence of a *right* inverse element for each element of $G$, we shall alternatively call a group an (R, R) *system*.

If, in the definition of a group $G$, we should replace G3 by

**G3′**: *There exists an element* i *of* G *such that, for all* a *in* G, i * a = a,

we shall call the resulting system an (L, R) *system*, since in this system we postulate the existence of a *left* identity element for $G$ and a *right* inverse element for each element of $G$.

If, in the definition of a group $G$, we should replace G4 by

**G4′**: *For each element* a *of* G *there exists an element* $a^{-1}$ *of* G *such that* $a^{-1}$ * a = i,

we shall call the resulting system an (R, L) *system*, since in this system we postulate the existence of a *right* identity element for $G$ and a *left* inverse element for each element of $G$.

If, in the definition of a group $G$, we should replace G3 by G3′ and G4 by G4′, we shall call the resulting system an (L, L) *system*, since in this system we postulate the existence of a *left* identity element for $G$ and a *left* inverse element for each element of $G$.

Now it can be shown that an (R, R) system is also an (L, L) system, that the right and left identity elements are unique and equal, and that the right and left inverse elements of each element are unique and equal. It is then natural to wonder if there exist (L, R) systems that are not (R, L) systems, and (R, L) systems that are not (L, R) systems. The existence of such systems is guaranteed by the following matrix realizations.

**6.3.4**   THEOREM.   *The set* G *of all* $2 \times 2$ $\mathcal{R}$ *matrices of the form*

$$\begin{bmatrix} 0 & 0 \\ a & b \end{bmatrix}, \qquad b \neq 0,$$

*along with Cayley multiplication of matrices, is an* (L, R) *system that is not an* (R, L) *system.*

The student can show that here

$$\begin{bmatrix} 0 & 0 \\ 0 & 1 \end{bmatrix}$$

is a left identity element and that

$$\begin{bmatrix} 0 & 0 \\ 0 & 1/b \end{bmatrix}$$

is a right inverse of

$$\begin{bmatrix} 0 & 0 \\ a & b \end{bmatrix}.$$

**6.3.5** **THEOREM.** *The set* G *of all* $2 \times 2$ $\mathscr{R}$ *matrices of the form*

$$\begin{bmatrix} 0 & a \\ 0 & b \end{bmatrix}, \qquad b \neq 0,$$

*along with Cayley multiplication of matrices, is an* (R, L) *system that is not an* (L, R) *system.*

The student can show that here

$$\begin{bmatrix} 0 & 0 \\ 0 & 1 \end{bmatrix}$$

is a right identity element and that

$$\begin{bmatrix} 0 & 0 \\ 0 & 1/b \end{bmatrix}$$

is a left inverse of

$$\begin{bmatrix} 0 & a \\ 0 & b \end{bmatrix}.$$

A simple abstract algebraic system, which seems to be of increasing importance in mathematics and which is closely allied to a group, is the so-called *loop*. A *loop* is a set $L$ of elements, along with a binary operation $*$, satisfying the two postulates:

**L1:** *For each* a *and* b *in* L, *there are unique* x *and* y *in* L *such that* $a * x = b$ *and* $y * a = b$.

**L2:** *There exists an element* e *in* L *such that* $a * e = e * a = a$ *for every* a *in* L.

It is easily shown that a group is a loop whose binary operation is associative; that is, a group is an associative loop. One naturally wonders if there are nonassociative loops. The following matrix realization shows that there are; the verification is left to the student.

**6.3.6** Theorem. *The set* L *of all* 2 × 2 $\mathscr{R}$ *matrices of the form*

$$\begin{bmatrix} a & 0 \\ b & c \end{bmatrix}, \quad a > 0, \quad c > 0,$$

*along with Jordan multiplication of matrices,*

$$A * B = (AB + BA)/2,$$

*is a nonassociative loop.*

As a final example we consider the abstract algebraic system known as a *division ring* (*pseudo-field*, *sfield*), which is postulationally defined by all the postulates for a field of Definition 6.2.2, except Postulate P7. That is, a field is a commutative division ring. One naturally wonders if there are noncommutative division rings. The following matrix realization shows that there are; the verification is left to the student.

**6.3.7** Theorem. *The set* D *of all* 2 × 2 $\mathscr{C}$ *matrices of the form*

$$\begin{bmatrix} u & v \\ -v & u \end{bmatrix},$$

*along with addition and Cayley multiplication of matrices, is a noncommutative division ring.*

## PROBLEMS

**6.3–1** Establish the following fundamental theorems about a group *G*.

(a) If *a*, *b*, *c* are in *G* and $a * c = b * c$, then $a = b$.

(b) For all *a* in *G*, $i * a = a * i$.

(c) A group has a unique identity element.

(d) For each element *a* of *G*, $a^{-1} * a = a * a^{-1}$.

(e) If *a*, *b*, *c* are in *G* and $c * a = c * b$, then $a = b$.

(f) Each element of a group has a unique inverse element.

(g) If *a* is in *G*, then $(a^{-1})^{-1} = a$.

**(h)** If $a$ and $b$ are in $G$, then there exist unique elements $x$ and $y$ of $G$ such that $a * x = b$ and $y * a = b$.

**6.3–2**  Show that the set of two numbers 1, $-1$, under ordinary multiplication, constitutes a group.

**6.3–3**  **(a)** Do the even integers form a group with respect to addition?

**(b)** Do the odd integers form a group with respect to addition?

**(c)** Do all the rational numbers form a group with respect to multiplication?

**(d)** Let $a * b = a - b$, where $a$ and $b$ are integers. Do the integers form a group with respect to this operation?

**(e)** Do all the integral multiples of 3 form a group with respect to addition?

**6.3–4**  Let $G$ be the set of all rotations

$$R: \begin{cases} x' = x \cos \theta - y \sin \theta, \\ y' = x \sin \theta + y \cos \theta, \end{cases}$$

of the plane about the origin, and let $R_2 * R_1$ denote the result of performing first rotation $R_1$ and then rotation $R_2$. Show that $G$, along with the operation $*$, constitutes an infinite commutative group.

**6.3–5**  Let $G$ be the set of five integers 0, 1, 2, 3, 4, and let $a * b$ denote the remainder obtained by dividing the ordinary product of $a$ and $b$ by 5. Does $G$, with the operation $*$, constitute a group?

**6.3–6**  Show that Postulates G3 and G4 for a group may be replaced by

**G3″**: *If* a *and* b *are any elements of* G, *then there exist elements* x *and* y *of* G *such that* a $*$ x $=$ b *and* y $*$ a $=$ b.

**6.3–7**  Prove Theorem 6.3.2.

**6.3–8**  Prove Theorem 6.3.3.

**6.3–9**  Prove Theorem 6.3.4.

**6.3–10**  Prove Theorem 6.3.5.

**6.3–11**  **(a)** Show that the identity element $e$ of a loop is unique.

**(b)** Show that in a loop $a * x = a * y$ or $x * a = y * a$ implies $x = y$.

**(c)** Show that an associative loop is a group.

**6.3–12**  Prove Theorem 6.3.6.

**6.3–13**  Prove Theorem 6.3.7.

**6.3–14**  Study the matrix representation of complex numbers described in Addendum 1.3A.

**6.3–15**  Give a matrix realization of a ring with left unity elements but no right unity elements.

**6.3–16**   A ring $R$ is said to be *regular* if for each element $a$ of $R$ there is an element $a'$ of $R$ such that $aa'a = a$. Regular rings were introduced by John von Neumann (1903–1957) in 1936, who showed that if $R$ is a regular ring with unity, then the set of all $n \times n$ matrices over $R$, along with addition and Cayley multiplication of matrices, is also a regular ring with unity.

(a) Show that a division ring is a regular ring.

(b) Obtain a matrix realization of a regular ring.

**6.3–17**   Show that the set of all $4 \times 4$ $\mathscr{R}$ matrices of the form

$$\begin{bmatrix} a & b & c & d \\ -b & a & -d & c \\ -c & d & a & -b \\ -d & -c & b & a \end{bmatrix},$$

along with addition and Cayley multiplication of matrices, is a noncommutative division ring.

## 6.4   *k*-vector spaces over a field *F*

Up to this point, a matrix has been regarded as an ordered rectangular array of its elements, but it can also be thought of as an ordered array of its row vectors, or as an ordered array of its column vectors. This latter viewpoint is often very helpful, as it furnishes a valuable tie between matrices and vector spaces.

**6.4.1   Definitions and notation.**   The set of all $k$th-order row vectors over a field $F$ will be denoted by $V_k(F)$, the set of all $k$th-order column vectors over $F$ will be denoted by $V'_k(F)$, and we shall refer to $V_k(F)$ and $V'_k(F)$ as *k-vector spaces over F*. Any nonempty subset $S$ of vectors of $V_k(F)$ or of $V'_k(F)$ will be called a *subspace* of $V_k(F)$ or of $V'_k(F)$, respectively, if and only if $S$ is closed under vector addition and multiplication by scalars of $F$. $S$ will be called a *row space* or a *column space*, according as it is a subset of $V_k(F)$ or of $V'_k(F)$.

The student should find no trouble in proving the following theorem.

**6.4.2   Theorem.**   *The set* S *of all linear combinations over* F *of any given system of* s *fixed vectors of a* k-*vector space over* F *is a subspace of the* k-*vector space.*

We now formulate some convenient definitions.

**6.4.3** DEFINITION. The subspace $S$ of a $k$-vector space over $F$ which is the set of all linear combinations over $F$ of a given system of $s$ fixed vectors of the $k$-vector space is said to be *spanned* by the given system of $s$ vectors.

**6.4.4** DEFINITION. A system of vectors is called a *basis* of a subspace $S$ of a $k$-vector space over $F$ if and only if: (1) $S$ is spanned by the system of vectors, (2) the system of vectors is linearly independent.

It can be shown that a basis of a subspace $S$ of a $k$-vector space over $F$ can be selected from a set of vectors that span $S$, except for the *trivial subspace* consisting of the zero vector alone. It follows that every subspace $S$, except the trivial subspace, has a basis. It can be shown, moreover, that the number of vectors in any basis of $S$ is the same as in any other basis of $S$. These facts motivate the following definition.

**6.4.5** DEFINITION. The number of vectors in any basis of a subspace of a $k$-vector space over $F$ is called the *dimension* of $S$. The trivial subspace is said to have dimension zero.

We formulate one more definition.

**6.4.6** DEFINITION. An $m$-plane is any set of vectors of the form $v + x$, where $v$ is any fixed vector of a $k$-vector space over $F$ and $x$ is an arbitrary vector of a subspace of dimension $m$ of the $k$-vector space.

If $A$ is an $m \times n$ matrix over a field $F$, the subspace of $V_n(F)$ spanned by the rows of $A$ is called the *row space* of $A$, and the subspace of $V'_m(F)$ spanned by the columns of $A$ is called the *column space* of $A$. We now restate some results, already established earlier in our book, in the picturesque and suggestive language of vector space theory.

**6.4.7** THEOREM. *If* A *is a matrix over a field* F, *then the row space and the column space of* A *have the same dimension.*

**6.4.8** THEOREM. *If* A *and* B *are* m × n *and* n × p *matrices over a field* F, *then the dimension of the row (column) space of* AB *does not exceed the dimension of the row (column) space of either* A *or* B.

**6.4.9** THEOREM. *If* A *and* B *are* m × n *matrices over a field* F, *then the dimension of the row (column) space of* A + B *does not exceed the sum of the dimensions of the row (column) spaces of* A *and* B.

**6.4.10** THEOREM. *If* A *is an* m × n *matrix over a field* F *and if* X *and* C *are* n*th-order column vectors, the latter over* F, *then the system of equations* AX = C *has a solution* X *in* F *if and only if* C *belongs to the column space of* A.

**6.4.11** THEOREM. *If, in the system* AX = C *of Theorem 6.4.10, the column space of* A *is of dimension* r, *then the solutions of the system belong to a subspace of* $V_n(F)$ *of dimension* n − r *if* C = O *and to an* (n − r)*-plane of* $V_n(F)$ *if* C ≠ O.

## PROBLEMS

**6.4–1** (a) Let $e_i$ denote the $k$th-order row vector, all of whose elements except the $i$th are the zero element of a field $F$ and whose $i$th element is the unity element of $F$. Show that $e_1, e_2, \cdots, e_k$ constitute a basis for $V_k(F)$.

(b) Show that $V_k(F)$ and $V'_k(F)$ are of dimension $k$.

**6.4–2** Determine which of the following sets of vectors $x = [x_1, \cdots, x_k]$ are subspaces of $V_k(\mathscr{R})$, where $\mathscr{R}$ is the real number field:

(a) All $x$ in $\mathscr{R}$ where $x_1$ is an integer.

(b) All $x$ in $\mathscr{R}$ where either $x_1 = 0$ or $x_2 = 0$.

(c) All $x$ in $\mathscr{R}$ where $x_1 = 0$.

(d) All $x$ in $\mathscr{R}$ where $x_1 = x_2 = 0$.

(e) All $x$ in $\mathscr{R}$ such that $3x_1 + 4x_2 = 0$.

(f) All $x$ in $\mathscr{R}$ such that $7x_1 - x_2 = 1$.

**6.4–3** (a) Show that the set containing only the zero row vector of order $k$ is a subspace of $V_k(F)$.

(b) Show that the zero row vector of order $k$ belongs to every subspace of $V_k(F)$.

**6.4–4.** Prove Theorem 6.4.2.

**6.4–5**   If $S$ and $T$ are subspaces of $V_k(F)$, show that $S \cup T$ and $S \cap T$ are also subspaces of $V_k(F)$.

**6.4–6**   Theorems 6.4.7 through 6.4.11 are alternative statements of theorems previously established in the text. Find the original statements of these theorems.

**6.4–7**   Represent the vectors of $V_3(\mathscr{R})$, where $\mathscr{R}$ is the real number field, by points in Cartesian three-dimensional space. Describe the geometrical representation of a 0-plane, a 1-plane, and a 2-plane of $V_3(\mathscr{R})$.

**6.4–8**   Let $S$ be a subspace of $V_k(\mathscr{R})$, where $\mathscr{R}$ is the real number field. Denote by $C(S)$ the set of all row vectors orthogonal to every row vector of $S$.

(**a**) Show that $C(S)$ is a subspace of $V_k(\mathscr{R})$.

(**b**) Show that the dimension of $S$ plus the dimension of $C(S)$ is $k$.

(**c**) Show that $C(C(S)) = S$.

(**d**) Show that any row vector $x$ in $V_k(\mathscr{R})$ has a unique representation of the form $x = x' + x''$, where $x'$ is in $S$ and $x''$ is in $C(S)$.

**6.4–9**   Show that the solutions of the homogeneous linear system of equations $AX = O$, where $A$ is an $m \times n$ real matrix and $X$ and $O$ are column vectors, constitute the subspace $C(S)$ of $V_n(\mathscr{R})$, where $S$ is the row space of $A$.

**6.4–10**   If $S$ and $T$ are subspaces of $V_k(F)$, prove that

$$d(S \cup T) = d(S) + d(T) - d(S \cap T),$$

where $d(S \cup T)$ denotes the dimension of $S \cup T$, etc.


## 6.5   General vector spaces

The $k$-vector spaces over a field $F$ of the preceding section are important instances of more general systems called simply *vector spaces*.


**6.5.1**   DEFINITIONS AND NOTATION.   A *vector space* (or *linear space*) $V(F)$ is a system consisting of

(1) a field $F$ of elements called *scalars*, with unity element $u$;

(2) a set $V$ of objects called *vectors*;

(3) a binary operation over $V$, called *vector addition*, such that with respect to this operation $V$ is a commutative group;

(4) a binary operation, called *scalar multiplication*, which associates

with each scalar $c$ of $F$ and each vector $\alpha$ of $V$ a vector $c\alpha$ of $V$ such that, for arbitrary scalars $a$ and $b$ and arbitrary vectors $\alpha$ and $\beta$,

(a) $(a + b)\alpha = a\alpha + b\alpha$,
(b) $a(\alpha + \beta) = a\alpha + a\beta$,
(c) $(ab)\alpha = a(b\alpha)$,
(d) $u\alpha = \alpha$.

*Note.*   We are denoting both addition of the scalars of $F$ and addition of the vectors of $V$ by the familiar plus sign; it is always clear from the context which addition the sign indicates. Again, we are denoting by juxtaposition both the product of two scalars of $F$ and the product of a scalar of $F$ with a vector of $V$; it is always clear from the context which type of product is meant.

*Examples.*   As some examples of vector spaces we have:

(1) The field $F$ of all real numbers and the additive commutative group $V$ of all complex numbers, along with familiar multiplication of a complex number by a real number.

(2) The field $F$ of all rational numbers and the additive commutative group $V$ of all polynomials (in one variable $x$) with real coefficients, along with familiar multiplication of a real polynomial by a rational number.

(3) The field $F$ of all rational numbers and the additive commutative group $V$ of all real polynomials of degree at most 3, along with familiar multiplication of a real polynomial by a rational number.

(4) The field $F$ of all rational numbers and the additive commutative group $V$ of all infinite sequences of real numbers, along with familiar multiplication of a real sequence by a rational number.

(5) The field $F$ of all real numbers and the additive commutative group $V$ of all geometric vectors lying in a plane and radiating from a common origin (addition of geometric vectors being defined by the familiar "parallelogram law"), along with the familiar multiplication of a geometric vector by a real number. (This vector space is the model that motivated the abstract study of general vector spaces.)

(6) A field $F$ of scalars and the additive commutative group $V$ of ordered $k$-tuples of elements of $F$, along with multiplication of an ordered $k$-tuple of elements of $F$ by an element of $F$. (This vector space is the $k$-vector space $V_k(F)$.)

(7) The field $F$ of all real numbers and the additive commutative group $V$ of all $n \times n$ complex matrices, along with scalar multiplication of a complex matrix by a real number. (This is a matrix representation of a vector space in that the vectors of the space are interpreted as matrices.)

**6.5.2** DEFINITION. A set $S$ of vectors of a vector space $V(F)$ is said to *span* $V(F)$ if and only if every vector of $V(F)$ can be expressed as a linear combination over $F$ of the vectors of $S$.

**6.5.3** DEFINITION. A set $S$ of vectors of a vector space $V(F)$ is called a *basis* of $V(F)$ if and only if the set $S$ spans $V(F)$ and is linearly independent over $F$.

It can be shown that every vector space $V(F)$, except a *trivial vector space* consisting of a zero vector alone, possesses a basis, and it can be shown that any vector of $V(F)$ is *uniquely* expressible as a linear combination over $F$ of the basis vectors of $V(F)$. It can also be shown that if there are $k$ vectors in a basis of a vector space $V(F)$, then there are $k$ vectors in every basis of $V(F)$. This motivates the following definition.

**6.5.4** DEFINITION. If a vector space $V(F)$ contains a finite basis, then the number of vectors in the basis is called the *dimension* of $V(F)$. A trivial vector space is said to have dimension zero.

We can now establish the following theorem, which brings out the great importance of $k$-vector spaces over a field $F$.

**6.5.5** THEOREM. *Any* k-*dimensional vector space* V(F) *is isomorphic to the vector space* $V_k$(F).

Let $\{\alpha_1, \alpha_2, \cdots, \alpha_k\}$ be a basis for $V(F)$ and let $\alpha$ be an arbitrary vector of $V(F)$. Then

$$\alpha = c_1\alpha_1 + c_2\alpha_2 + \cdots + c_k\alpha_k,$$

where the $c$'s are unique elements of $F$. The student can now show that the one-to-one correspondence

$$\alpha \leftrightarrow (c_1, c_2, \cdots, c_k)$$

establishes an isomorphism between $V(F)$ and $V_k(F)$. That is, if

$$\alpha \leftrightarrow (c_1, c_2, \cdots, c_k) \qquad \text{and} \qquad \beta \leftrightarrow (d_1, d_2, \cdots, d_k),$$

then

$$\alpha + \beta \leftrightarrow (c_1, c_2, \cdots, c_k) + (d_1, d_2, \cdots, d_k)$$

and, for any element $c$ of $F$,

$$c\alpha \leftrightarrow (cc_1, cc_2, \cdots, cc_k).$$

# PROBLEMS

**6.5–1** Verify that Examples (1) through (7) following Definition 6.5.1 are vector spaces.

**6.5–2** Establish the following elementary properties of a vector space $V(F)$, wherein $z$ denotes the zero element of $F$ and $\theta$ denotes the zero element of $V$.

**(a)** $a\theta = \theta$ for every scalar $a$.

**(b)** $z\alpha = \theta$ for every vector $\alpha$.

**(c)** $a(-\alpha) = -a(\alpha) = -(a\alpha)$ for every scalar $a$ and every vector $\alpha$.

**(d)** If $a$ is a scalar, $\alpha$ a vector, and $a\alpha = \theta$, then either $a = z$ or $\alpha = \theta$.

**6.5–3** A subsystem of a vector space $V(F)$, which is itself a vector space with respect to the operations of $V(F)$, is called a *subspace* of $V(F)$.

**(a)** Prove that a nonempty subset $S$ of vectors of a vector space $V(F)$ constitutes a subspace of $V(F)$ if and only if $S$ is closed under the operations of vector addition and scalar multiplication of $V(F)$.

**(b)** Prove that the set $S$ of all linear combinations over $F$ of any given system of fixed vectors of a vector space $V(F)$ is a subspace of $V(F)$.

**6.5–4** Complete the proof of Theorem 6.5.

**6.5–5** Show that a field can be considered as a vector space of dimension 1 over itself.

## 6.6 Linear transformations of vector spaces

In this section we point out the intimate connection between $n \times n$ matrices over a field $F$ and linear transformations of an $n$-dimensional vector space over $F$ into itself. Since a linear transformation is a special kind of "mapping" of one set of elements into another set of elements, we commence with a formal definition of this latter basic concept.

**6.6.1** DEFINITIONS AND NOTATION. A *mapping $M$* of a set $A$ into a set $B$ is a correspondence that associates with each element $\alpha$ of $A$ a unique element $\beta$ of $B$. This mapping may be indicated by writing

$$M: \alpha \to \beta,$$

and we call $\beta$ the *image* of $\alpha$ under this mapping. To indicate that $\beta$ is the image of $\alpha$ under mapping $M$, we write $\beta = \alpha M$ (read, "$\beta$ is the image of $\alpha$ under $M$").

*Examples.*  (1) Consider the sets $A = \{1, 2, 3, 4\}$ and $B = \{x, y, z\}$. Then the associations

$$1 \rightarrow x, \qquad 2 \rightarrow y, \qquad 3 \rightarrow x, \qquad 4 \rightarrow y$$

define a mapping $M$ of $A$ into $B$. Here we have $x = 1M, y = 2M, x = 3M, y = 4M$.

(2) The mapping $M$ of the set of natural numbers into itself that associates with each natural number $n$ its square $n^2$ (that is, the mapping $M$ such that $nM = n^2$ for every natural number $n$) may be indicated by

$$M: n \rightarrow n^2.$$

**6.6.2** DEFINITION.  A *linear transformation* of a vector space $V(F)$ into a vector space $W(F)$ over the same field $F$ is a mapping $T$ of $V$ into $W$ such that for arbitrary $\alpha_1$ and $\alpha_2$ of $V$ and arbitrary $c$ of $F$,

$$(\alpha_1 + \alpha_2)T = \alpha_1 T + \alpha_2 T, \qquad (c\alpha_1)T = c(\alpha_1 T).$$

*Examples.*  (1) Let $V(F)$ and $W(F)$ each be the vector space whose vectors are real polynomials $f(x)$ and whose field is the rational number field. Then the mapping

$$T: f(x) \rightarrow f'(x),$$

where $f'(x)$ denotes the derivative with respect to $x$ of $f(x)$, is a linear transformation of $V(F)$ into $W(F)$, since for all real polynomials $f(x)$ and $g(x)$,

$$[f(x) + g(x)]' = f'(x) + g'(x),$$

and for every real polynomial $f(x)$ and every rational number $c$,

$$[cf(x)]' = cf'(x).$$

(2) The mapping

$$T: (c_1, c_2, c_3) \rightarrow c_1\lambda + c_2\mu + c_3\nu,$$

where $c_1$, $c_2$, $c_3$ are arbitrary elements of some field $F$ and $\lambda$, $\mu$, $\nu$ are fixed elements of $F$, is a linear transformation of $V_3(F)$ into $V_1(F)$, since for any two triples $(c_1, c_2, c_3)$, $(d_1, d_2, d_3)$ of elements of $F$,

$$[(c_1, c_2, c_3) + (d_1, d_2, d_3)]T = (c_1 + d_1)\lambda + (c_2 + d_2)\mu + (c_3 + d_3)\nu$$

$$= (c_1\lambda + c_2\mu + c_3\nu) + (d_1\lambda + d_2\mu + d_3\nu)$$

$$= (c_1, c_2, c_3)T + (d_1, d_2, d_3)T,$$

and for any triple $(c_1, c_2, c_3)$ of elements of $F$ and an arbitrary element $c$ of $F$,

$$[c(c_1, c_2, c_3)]T = (cc_1, cc_2, cc_3)T$$

$$= (cc_1)\lambda + (cc_2)\mu + (cc_3)\nu$$

$$= c(c_1\lambda + c_2\mu + c_3\nu)$$

$$= c[(c_1, c_2, c_3)T].$$

**6.6.3** THEOREM. *If* $V(F)$ *and* $W(F)$ *are vector spaces over the same field* F, *and if* $\{\alpha_1, \cdots, \alpha_n\}$ *is a basis for* $V(F)$ *and* $\{\beta_1, \cdots, \beta_n\}$ *is an arbitrary set of* n *vectors of* W, *then there is one and only one linear transformation* T *of* $V(F)$ *into* $W(F)$ *such that* $\alpha_1 T = \beta_1, \cdots, \alpha_n T = \beta_n$. *Moreover, if*

$$\alpha = c_1\alpha_1 + \cdots + c_n\alpha_n,$$

*then*

$$\alpha T = c_1\beta_1 + \cdots + c_n\beta_n.$$

Consider a mapping $T$ of $V(F)$ into $W(F)$ wherein

$$\alpha = c_1\alpha_1 + \cdots + c_n\alpha_n \rightarrow c_1\beta_1 + \cdots + c_n\beta_n.$$

We note that for each $i = 1, \cdots, n$, $\alpha_i T = \beta_i$. We now prove that $T$ is a linear transformation of $V(F)$ into $W(F)$. Toward this end, let $\alpha' = d_1\alpha_1 + \cdots + d_n\alpha_n$ be a second vector of $V(F)$. Then

$$(\alpha + \alpha')T = \left(\sum_{i=1}^{n} c_i\alpha_i + \sum_{i=1}^{n} d_i\alpha_i\right)T = \left[\sum_{i=1}^{n} (c_i + d_i)\alpha_i\right]T$$

$$= \sum_{i=1}^{n} (c_i + d_i)\beta_i = \sum_{i=1}^{n} c_i\beta_i + \sum_{i=1}^{n} d_i\beta_i = \alpha T + \alpha' T.$$

Also, for arbitrary $c$ of $F$,

$$(c\alpha)T = \left[c\sum_{i=1}^{n} c_i\alpha_i\right]T = \left[\sum_{i=1}^{n} (cc_i)\alpha_i\right]T$$

$$= \sum_{i=1}^{n} (cc_i)\beta_i = c\sum_{i=1}^{n} c_i\beta_i = c(\alpha T).$$

Finally, $T$ is the only linear transformation of $V(F)$ into $W(F)$ such that $\alpha_i T = \beta_i$, $i = 1, \cdots, n$. For suppose $T'$ is another linear transformation of $V(F)$ into $W(F)$ such that $\alpha_i T' = \beta_i$, $i = 1, \cdots, n$. Then

$$\alpha T' = \left( \sum_{i=1}^{n} c_i \alpha_i \right) T' = \sum_{i=1}^{n} \left[ (c_i \alpha_i) T' \right]$$

$$= \sum_{i=1}^{n} [c_i (\alpha_i T')] = \sum_{i=1}^{n} c_i \beta_i = \alpha T.$$

Since $\alpha$ represents an arbitrary vector of $V(F)$, it follows that $T' = T$.

**6.6.4** DEFINITION. Let $T$ be a linear transformation of an $n$-dimensional vector space $V(F)$ into itself and let $\{\alpha_1, \cdots, \alpha_n\}$ be a basis in $V(F)$. Then there exist unique $t_{ij}$ in $F$, $i, j, = 1, \cdots, n$, such that

$$\alpha_1 T = t_{11} \alpha_1 + t_{12} \alpha_2 + \cdots + t_{1n} \alpha_n,$$

$$\alpha_2 T = t_{21} \alpha_1 + t_{22} \alpha_2 + \cdots + t_{2n} \alpha_n,$$

$$\cdot \quad \cdot \quad \cdot \quad \cdot \quad \cdot \quad \cdot \quad \cdot \quad \cdot$$

$$\alpha_n T = t_{n1} \alpha_1 + t_{n2} \alpha_2 + \cdots + t_{nn} \alpha_n.$$

The matrix $[t_{ij}]_{(n)}$ is called the *matrix associated with* T *for the basis* $\{\alpha_1, \cdots, \alpha_n\}$.

Now, before proving the famous isomorphism theorem that gives the intimate connection between $n \times n$ matrices over a field $F$ and linear transformations of an $n$-dimensional vector space over $F$ into itself, we define "addition" and "multiplication" of such linear transformations and state a pertinent theorem whose straightforward proof will be left to the student.

**6.6.5** DEFINITIONS AND NOTATION. Let $S$ and $T$ be two linear transformations of a vector space $V(F)$ into itself. By the *sum* $S + T$ of these two linear transformations, we mean the mapping of $V$ into $V$ defined by

$$S + T: \alpha(S + T) = \alpha S + \alpha T$$

for every $\alpha$ of $V$. By the *product* $ST$ of the two linear transformations, we mean the mapping of $V$ into $V$ defined by

$$ST: \alpha(ST) = (\alpha S)T$$

for every $\alpha$ of $V$.

**6.6.6** THEOREM. *If* S *and* T *are two linear transformations of a vector space* V(F) *into itself, then* S + T *and* ST *are also linear transformations of* V(F) *into itself.*

**6.6.7** THE ISOMORPHISM THEOREM. *The set of linear transformations of an* n-*dimensional vector space* V(F) *into itself is a ring with unity under the operations of addition and multiplication of such linear transformations, and this ring is isomorphic to the ring of all* n × n *matrices with elements in* F.

Let $S$ and $T$ be any two linear transformations of $V(F)$ into itself and let $\{\alpha_1, \cdots, \alpha_n\}$ be a fixed basis of $V(F)$. Suppose $[s_{ij}]$ and $[t_{ij}]$ are the matrices associated with $S$ and $T$ for the fixed basis. Consider the one-to-one correspondence $A \leftrightarrow [a_{ij}]$ between linear transformations $A$ of $V(F)$ into itself and the associated matrices $[a_{ij}]$ for the fixed basis. Now

$$\alpha_i S = \sum_{j=1}^{n} s_{ij}\alpha_j, \qquad \alpha_i T = \sum_{j=1}^{n} t_{ij}\alpha_j,$$

whence

$$\alpha_i(S + T) = \alpha_i S + \alpha_i T = \sum_{j=1}^{n} s_{ij}\alpha_j + \sum_{j=1}^{n} t_{ij}\alpha_j = \sum_{j=1}^{n} (s_{ij} + t_{ij})\alpha_j,$$

or

$$S + T \leftrightarrow [s_{ij} + t_{ij}] = [s_{ij}] + [t_{ij}].$$

Also

$$\alpha_i(ST) = (\alpha_i S)T = \left(\sum_{j=1}^{n} s_{ij}\alpha_j\right)T = \sum_{j=1}^{n} (s_{ij}\alpha_j T)$$

$$= \sum_{j=1}^{n} s_{ij}(\alpha_j T) = \sum_{j=1}^{n} s_{ij}\left(\sum_{k=1}^{n} t_{jk}\alpha_k\right)$$

$$= \sum_{j=1}^{n} \left(\sum_{k=1}^{n} s_{ij}t_{jk}\alpha_k\right) = \sum_{k=1}^{n} \left(\sum_{j=1}^{n} s_{ij}t_{jk}\alpha_k\right)$$

$$= \sum_{k=1}^{n} \left(\sum_{j=1}^{n} s_{ij}t_{jk}\right)\alpha_k = \sum_{k=1}^{n} u_{ik}\alpha_k,$$

where

$$u_{ik} = \sum_{j=1}^{n} s_{ij}t_{jk}.$$

That is,

$$ST \leftrightarrow [u_{ij}] = [s_{ij}][t_{ij}].$$

Thus the one-to-one correspondence is an isomorphism relative to the addition and multiplication operations. Since the set of matrices, under addition and multiplication, is a ring with unity, it follows that the set of linear transformations is also a ring with unity, and the theorem is established.

As a consequence of Theorem 6.6.7, we see that the study of linear transformations of an $n$-dimensional vector space $V(F)$ into itself is equivalent to the study of $n \times n$ matrices over $F$. That is, any theorem about such matrices can be phrased as a theorem about linear transformations, and conversely. For this reason there is an advantage in simultaneously studying linear transformations of an $n$-dimensional vector space $V(F)$ into itself and $n \times n$ matrices over $F$; each viewpoint assists and advances the other.

There is another matter in connection with Theorem 6.6.7 that we should like to point out. While Theorem 6.6.7 has established an isomorphism between a ring of linear transformations and a ring of matrices, the actual correspondences depend on the basis chosen for the vector space. If the basis is changed, the matrix that corresponds to a given linear transformation $T$ will change. An important problem in the theory of vector spaces is the determination of a basis that will simplify the associated matrix of a given linear transformation. In matrix theory this is the problem of reducing a matrix to its various canonical forms.

The one-to-one correspondence between a linear transformation $T$ of an $n$-dimensional vector space $V(F)$ into itself and the associated matrix of $T$ for a fixed basis of $V(F)$ is very much like the one-to-one correspondence between an ordered pair of real numbers and the associated point of a plane for a fixed Cartesian coordinate system of the plane. Just as the latter one-to-one correspondence yields a method (known as *Cartesian analysis*) for the study of the geometry of the points of a plane, so does the former one-to-one correspondence yield a method for studying $n \times n$ matrices over a field $F$. The method by which matrices have been studied in our book might be called the *direct method*, and the alternative method furnished by the preceding one-to-one correspondence, the *linear transformation method*. These two methods compare with the two ways of studying geometry, known as the *synthetic method* and the *analytic method*. In each case, the first of the two methods is the more elementary one.

## PROBLEMS

**6.6–1**   A mapping $T$ of $V_2(\mathscr{R})$ into $V_3(\mathscr{R})$, where $\mathscr{R}$ is the real number field, is defined by $(x, y) \rightarrow (x', y', z')$, where

$$x' = x - y, \qquad y' = x + y, \qquad z' = 2x + y.$$

Show that this mapping is a linear transformation of $V_2(\mathscr{R})$ into $V_3(\mathscr{R})$.

**6.6–2**   Which of the following mappings of $V_2(\mathscr{R})$ into itself are linear transformations?

(a) $(x, y) \rightarrow (0, 0)$.

(b) $(x, y) \rightarrow (x + 1, y - x)$.

(c) $(x, y) \rightarrow (x, y)$.

(d) $(x, y) \rightarrow (2x + y, x - y)$.

**6.6–3**   Prove that a mapping $T$ of $V_2(\mathscr{R})$ into itself defined by $(x, y) \rightarrow (x', y')$ is a linear transformation of $V_2(\mathscr{R})$ into itself if and only if there exist real numbers $a_1, a_2, b_1, b_2$ such that

$$x' = a_1 x + a_2 y, \qquad y' = b_1 x + b_2 y.$$

**6.6–4**   Prove that a linear transformation of $V_2(\mathscr{R})$ into itself is completely determined by the images $(a_1, b_1)$, $(a_2, b_2)$ of the unit vectors $(1, 0)$, $(0, 1)$.

**6.6–5**   Show that rotation of the Cartesian plane about the origin through an angle $\theta$ is a linear transformation of $V_2(\mathscr{R})$ into itself.

**6.6–6**   Show that an expansion of the Cartesian plane, in which each point is moved radially out to a point $k$ times as far from the origin as originally, is a linear transformation of $V_2(\mathscr{R})$ into itself.

**6.6–7**   Let $V(\mathscr{R})$ be the vector space of complex numbers over the real number field $\mathscr{R}$. Show that the mapping $a + bi \rightarrow a - bi$ of each complex number into its conjugate complex number is a linear transformation of $V(\mathscr{R})$ into itself, and determine the associated matrix for the basis $\{1, i\}$.

**6.6–8**   Find the associated matrix, for the basis $\{(1, 1), (-2, 3)\}$, of the linear transformation of $V_2(\mathscr{R})$ into itself, defined by $(1, 0) \rightarrow (3, 0)$, $(0, 1) \rightarrow (-2, 1)$.

**6.6–9**   Prove Theorem 6.6.6.

**6.6–10**   Prove the associative and distributive laws for multiplication of $n \times n$ matrices over a field $F$, by using the corresponding properties for linear transformations of an $n$-dimensional vector space $V(F)$ into itself.

**6.6–11** Prove that the set of linear transformations of an $n$-dimensional vector space $V(F)$ into itself is a commutative group under addition of linear transformations, by using the corresponding property for $n \times n$ matrices over $F$.

**6.6–12** Let $A$ and $B$ be the associated matrices of a linear transformation $T$ of $V_n(F)$ into itself for bases $\{\alpha_1, \cdots, \alpha_n\}$, $\{\beta_1, \cdots, \beta_n\}$, respectively. Show that $A$ and $B$ are similar matrices.

**6.6–13** Devise a method, similar to that of the text, for setting up a one-to-one correspondence between the set of linear transformations of an $n$-dimensional vector space $V(F)$ into an $m$-dimensional vector space $W(F)$ and the set of $m \times n$ matrices over the field $F$.

## 6.7 Jordan and Lie algebras

Though a set $S$, along with any operations and relations defined on $S$, is often broadly called an *algebraic system*, the term *an algebra* has come to have a much more reserved technical meaning. The definition of *an algebra* has not yet become uniform in the literature, but the following version will suit our purposes.

**6.7.1** DEFINITIONS. An *algebra* (sometimes called a *linear algebra*) is a vector space $V(F)$ in which a vector multiplication $\alpha\beta$ is defined for every pair of vectors $\alpha$ and $\beta$ of $V$ such that for arbitrary $\alpha$, $\beta$, $\gamma$ in $V$ and arbitrary $c$ in $F$,

(1) $(c\alpha)\beta = \alpha(c\beta) = c(\alpha\beta)$,
(2) $\alpha(\beta + \gamma) = (\alpha\beta) + (\alpha\gamma)$,
(3) $(\beta + \gamma)\alpha = (\beta\alpha) + (\gamma\alpha)$.

An algebra for which $\alpha\beta = \beta\alpha$ for all vectors $\alpha$ and $\beta$ is called a *commutative algebra*; otherwise it is called a *noncommutative algebra*. An algebra for which $\alpha(\beta\gamma) = (\alpha\beta)\gamma$ for all vectors $\alpha$, $\beta$, $\gamma$ is called an *associative algebra*; otherwise it is called a *nonassociative algebra*.

The first noncommutative algebra was the quaternion algebra created by Sir William Rowan Hamilton in 1843. Shortly after, other noncommutative algebras were devised by Hermann Günther Grassmann, and in 1857 Arthur Cayley laid the foundation for the noncommutative algebra of $n \times n$ matrices over a field. But it is only in recent years that nonassociative algebras have been considered. In this concluding section of Chapter 6 we point out two of these nonassociative algebras. Each is a nonassociative algebra with some additional required structure, and each possesses a simple matrix realization.

**6.7.2** DEFINITION. A *Jordan algebra* is a nonassociative algebra such that for arbitrary vectors $\alpha$ and $\beta$, the vector multiplication further satisfies
(1) $\alpha\beta = \beta\alpha$,
(2) $\alpha(\beta\alpha^2) = (\alpha\beta)\alpha^2$.

**6.7.3** DEFINITION. A *Lie algebra* is a nonassociative algebra such that for arbitrary vectors $\alpha$ and $\beta$ the vector multiplication further satisfies
(1) $\alpha\beta = -\beta\alpha$,
(2) $(\alpha\beta)\gamma + (\beta\gamma)\alpha + (\gamma\alpha)\beta = \theta$, where $\theta$ is the zero vector.

We now state two theorems, whose proofs will be left to the student, which give a simple matrix representation of a Jordan algebra and of a Lie algebra. These matrix realizations constitute part of the subject matter of Addendum 1.7A.

**6.7.4** THEOREM. *The vector space of all* n × n F *matrices over a field* F *along with* Jordan multiplication *of matrices,*

$$A * B = (AB + BA)/2,$$

*is a Jordan algebra.*

**6.7.5** THEOREM. *The vector space of all* n × n F *matrices over a field* F *along with* Lie multiplication *of matrices,*

$$A \times B = AB - BA,$$

*is a Lie algebra.*

## PROBLEMS

**6.7-1** The *order* of an algebra is the dimension of the vector space of the algebra. Show that a ring and a field can each be considered as an algebra of order 1.

**6.7–2**    Show that the set of all $n \times n$ matrices over a field $F$, along with addition of matrices, and Cayley and scalar multiplication of matrices, is an algebra of order $n$.

**6.7–3**    Prove Theorem 6.7.4.

**6.7–4**    Prove Theorem 6.7.5.

**6.7–5**    Show that the set of all geometrical vectors in 3-space, along with addition of vectors, vector multiplication of vectors, and multiplication of vectors by real scalars, constitutes a Lie algebra of order 3.

# 7. EPILEGOMENON

As intimated in the Prolegomenon, matrix study is today very active and very extensive. There is ample material to engage an interested student for a long time, and there are many outstanding unsolved problems to consider.

There are essentially three avenues for further study in matrix work, and which avenue one might prefer depends upon one's particular interests. We can refer to these three avenues as *computational work*, *applications*, and *advanced theory*. Of course the three avenues are necessarily somewhat entwined with one another.

With the advent of the large electronic computing machines, it is now quite feasible to consider computational problems involving large matrices, and every computing center has its standard procedures for inverting large matrices, for finding characteristic roots of matrices, for obtaining bounds for the characteristic roots of a matrix, and so on. A number of matrix texts have been developed which are purposely slanted toward computer practices. Here one encounters methods for solving exactly and approximately large systems of simultaneous linear equations, numerical matrix methods for solving systems of linear ordinary differential equations with either constant or variable coefficients, numerical techniques for obtaining high powers of a matrix and for finding any dominant characteristic root of a matrix, procedures for bounding and evaluating large-order determinants, computational practices connected with various iterated methods, and other important matters. Many of the practical theorems and procedures of computational work bear the names of their discoverers. Thus one meets the Gauss-Doolittle process, the Crout process, the Gauss-Seidel method, Aitken's triple product, the theorems of Schmidt-Mises-Geiringer, Frobenius, Reich, Stein-Rosenberg, the methods and procedures of Cesari, Jacobi, Kacmarz, Cimmino, Magnier, Jahn, Givens, Leverier, Krylov-Duncan, Hessenberg, and Samuelson, the Van der Corput device, the Smith algorithm, and Hotelling's and Wielandt's deflation.

Applications of matrices can be found in all fields of science and engineering. Here one finds elegant and compact matrix treatments of such things as the kinematics and dynamics of physical systems, friction theory, flutter problems, electromagnetic theory, quantum mechanics, relativity theory, the analytical treatment of various geometries, transformation theory, the simplex method of linear programming, linear inequalities, communication theory, small oscillations of mechanical systems, the theory of elasticity, and the analysis of structures. As in the case of computational work, matrix texts are appearing which are written specifically for the physicist, the mechanical engineer, the electrical engineer, and the social scientist. There seems little doubt that the present rapid mathematizing of so much of the social sciences will lead to the development of many new applications of matrix theory. Growing applications are being found in such seemingly diverse fields as business and graph theory. Linear algebra and matrix theory is coming to vie with the calculus as an essential early college course in mathematics.

There is much advanced matrix theory for the interested and prepared student. In our work we have considered only matrix algebra, and have said nothing about matrix calculus. Very interesting is the theory of infinite sequences and series of matrices. There is an extensive theory of functions of matrices. The solution of matrix equations is difficult and only spottily developed; the solution of even so simple-appearing an equation as $AX = XB$, where $X$, $A$, $B$ all represent $n \times n$ matrices, presents problems. There is an extensive literature connected with canonical forms of matrices, with the theory of invariant factors and elementary divisors, with commutative matrices, with the finding of $m$th roots of both singular and nonsingular matrices, and so on. The abstract study of matrices via linear transformations of one finite-dimensional vector space into another finite-dimensional vector space over the same field has been extensively developed. The matrix study of bilinear, quadratic, and Hermitian forms is vast, and there is a considerable theory of special matrices, such as matrices with nonnegative elements and matrices whose elements are all 0's and 1's. A very interesting field is the study of infinite matrices (matrices containing an infinite number of rows and of columns) and their application to sequence spaces. And there is some extension of the theory of plane matrices to solid, and higher-dimensional, matrices.

The Bibliography in this book can furnish a student a start in further matrix work. In the Bibliographies of many of these texts, the interested student will find references to periodical literature; this latter literature is very extensive and is rapidly growing.

# BIBLIOGRAPHY

AITKEN, A. C., *Determinants and Matrices*, 8th ed. New York: Interscience Publishers, Inc., 1954.

ALBERT, A. A., *Introduction to Algebraic Theories*. Chicago: University of Chicago Press, 1941.

AMUNDSON, N. R., *Mathematical Methods in Chemical Engineering: Matrices and Their Applications*. Englewood Cliffs, N. J.: Prentice-Hall, Inc,, 1966.

AYRES, JR., F., *Theory and Problems of Matrices* (Schaum's Outline Series). New York: Schaum Publishing Co., 1962.

BEAUMONT, R. A., and R. W. BALL, *Introduction to Modern Algebra and Matrix Theory*. New York: Holt, Rinehart & Winston, Inc., 1954.

BELLMAN, R., *Introduction to Matrix Analysis*. New York: McGraw-Hill Book Co., Inc., 1960.

BERBERIAN, S. K., *Introduction to Hilbert Space*. New York: Oxford University Press, Inc., 1961.

BODEWIG, E., *Matrix Calculus*, 2nd ed. New York: Interscience Publishers, Inc., 1950.

BORG, S. F., *Matrix-Tensor Methods in Continuum Mechanics*. Princeton, N. J.: D. Van Nostrand Co., Inc., 1962.

BOWMAN, F., *An Introduction to Determinants and Matrices*. Princeton, N. J.: D. Van Nostrand Co., Inc., 1962.

BROWNE, E. T., *Introduction to the Theory of Determinants and Matrices*. Chapel Hill, N. C.: University of North Carolina Press, 1958.

CAMPBELL, H. G., *An Introduction to Matrices, Vectors, and Linear Programming*. New York: Appleton-Century-Crofts, Inc., 1964.

COOKE, R. G., *Infinite Matrices and Sequence Spaces*. New York: The Macmillan Co., 1950.

CURTIS, C. W., *Linear Algebra: An Introductory Approach*. Boston: Allyn and Bacon, Inc., 1963.

————, and I. REINER, *Representation Theory of Finite Groups*. New York: Interscience Publishers, Inc., 1963.

DAVIS, P. J., *Mathematics of Matrices*. New York: Blaisdell Publishing Co., 1965.

DE VEUBEKE, F. (ed.), *Matrix Methods of Structural Analysis*. New York: The Macmillan Company, 1964.

DICKSON, L. E., *Linear Algebras* (Cambridge Tracts in Mathematics and Mathematical Physics, No. 16). New York: Cambridge University Press, Inc., 1930.

DRESDEN, A., *Solid Analytical Geometry and Determinants*. New York: Dover Publications, Inc., 1964.

DWYER, P. S., *Linear Computations*. New York: John Wiley & Sons, Inc., 1951.

EISENMAN, R. L., *Matrix Vector Analysis*. New York: McGraw-Hill Book Co., Inc., 1963.

FADDEEV, D. K., and V. N. FADDEEVA, *Numerical Methods in Linear Algebra*. San Francisco, W. H. Freeman and Co., Publishers, 1963.

FADDEEVA, V. N., *Computational Methods of Linear Algebra*. New York: Dover Publications, Inc., 1959.

FERRAR, W. L., *Algebra, a Text-Book of Determinants, Matrices, and Algebraic Forms*. Oxford: Clarendon Press, 1941.

————, *Finite Matrices*. Oxford: Clarendon Press, 1951.

FINKBEINER, II, D. T., *Matrices and Linear Transformations*. San Francisco: W. H. Freeman and Co., Publishers, 1960.

FLEGG, H. G., *Boolean Algebra and Its Applications, Including Boolean Matrix Algebra*. New York: John Wiley & Sons, Inc., 1964.

FRAZER, R. A., W. J. DUNCAN, and A. R. COLLAR, *Elementary Matrices and Some Applications to Dynamics and Differential Equations*. New York: Cambridge University Press, Inc., 1950.

FULLER, L. E., *Basic Matrix Theory*. Englewood Cliffs, N. J.: Prentice-Hall, Inc., 1962.

GANTMACHER, F. R., *The Theory of Matrices*, 2 vols. New York: Chelsea Publishing Co., 1959.

GEL'FAND, I. M., *Lectures on Linear Algebra* (tr. by A. Shenitzer). New York: Interscience Publishers, Inc., 1961.

GREEN, H. S., *Matrix Mechanics*. Groningen, Holland: P. Noordhoff, Ltd., 1965.

GREUB, W. H., *Linear Algebra*, 2nd ed. New York: Academic Press, Inc., 1963.

HADLEY, G., *Linear Algebra*. Reading, Mass.: Addison-Wesley Publishing Co., Inc., 1959.

HALL, G. G., *Matrices and Tensors*. New York: Pergamon Press, Inc., 1963.

HALMOS, P. R., *Finite-Dimensional Vector Spaces*, 2nd ed. Princeton, N. J.: D. Van Nostrand Co., Inc., 1958.

————, *Introduction to Hilbert Space*. New York: Chelsea Publishing Co., 1951.

HAMBURGER, H. L., and M. E. GRIMSHAW, *Linear Transformations in n-Dimensional Vector Space, an Introduction to the Theory of Hilbert Spaces*. New York: Cambridge University Press, Inc., 1951.

HAUSNER, M., *A Vector Space Approach to Geometry*. Englewood Cliffs, N. J.: Prentice-Hall, Inc., 1965.

HEADING, J., *Matrix Theory for Physicists*. New York: Longmans, Green and Co., 1958.

HIGMAN, B., *Applied Group-Theoretic and Matrix Methods*. Oxford: Clarendon Press, 1955.

HOFFMAN, K., and R. KUNZE, *Linear Algebra*. Englewood Cliffs, N. J.: Prentice-Hall, Inc., 1961.

HOHN, F. E., *Elementary Matrix Algebra*, 2nd ed. New York: The Macmillan Co., 1964.

HORST, P., *Matrix Algebra for Social Scientists*. New York: Holt, Rinehart & Winston, Inc., 1963.

HOUSEHOLDER, A. S., *Theory of Matrices in Numerical Analysis*. Boston: Ginn and Company, 1964.

HUELSMAN, L. P., *Circuits, Matrices, and Linear Vector Spaces*. New York: McGraw-Hill Book Co., Inc., 1963.

JAEGER, A., *Introduction to Analytic Geometry and Linear Algebra*. New York: Holt, Rinehart & Winston, Inc., 1960.

JONES, B. W., *The Arithmetic Theory of Quadratic Forms* (Carus Mathematical Monographs, No. 10). Buffalo, N. Y.: The Mathematical Association of America, 1950.

KUIPER, N. H., *Linear Algebra and Geometry*. Amsterdam, Holland: North-Holland Publishing Co., 1962.

LITTLEWOOD, D. E., *The Theory of Group Characters and Matrix Representations of Groups*. Oxford: Clarendon Press, 1950.

MCMINN, S. J., *Matrices for Structural Analysis*. New York: John Wiley & Sons, Inc., 1962.

MacDuffee, C. C., *The Theory of Matrices*, 2nd ed. New York: Chelsea Publishing Co., 1946.

————, *Vectors and Matrices* (Carus Mathematical Monographs, No. 7). Buffalo, N. Y.: Mathematical Association of America, 1943.

Marcus, M., and H. Minc, *A Survey of Matrix Theory and Matrix Inequalities*. Boston: Allyn and Bacon, Inc., 1964.

————, *Introduction to Linear Algebra*. New York: The Macmillan Company, 1965.

Mirsky, L., *An Introduction to Linear Algebra*. New York: Oxford University Press, Inc., 1955.

Mitrinovic, D. S., *Elementary Matrices* (Tutorial Text No. 3). Groningen, The Netherlands: P. Noordhoff Ltd., 1965.

Muir, T., *The Theory of Determinants in the Historical Order of Development* (four volumes bound as two). New York: Dover Publications, Inc., 1960.

————, and W. H. Metzler, *A Treatise on the Theory of Determinants*. New York: Dover Publications, Inc., 1960.

Murdoch, D. C., *Linear Algebra for Undergraduates*. New York: John Wiley & Sons, Inc., 1957.

Murnaghan, F. D., *The Theory of Group Representations*. Baltimore, Md.: Johns Hopkins University Press, 1938.

Narayan, S., *A Text Book of Matrices*. Delhi, India: S. Chand and Co., 1957.

Nering, E. D., *Linear Algebra and Matrix Theory*. New York: John Wiley & Sons, Inc., 1963.

Nye, J. F., *Physical Properties of Crystals: Their Representation by Tensors and Matrices*. Oxford: Clarendon Press, 1957.

O'Meare, O. T., *Introduction to Quadratic Forms*. New York: Academic Press, Inc., 1963.

Paige, L. J., and J. D. Swift, *Elements of Linear Algebra*. Boston: Ginn and Company, 1961.

Parker, W. V., and J. C. Eaves, *Matrices*. New York: The Ronald Press Co., 1960.

Pease, M. C. iii, *Methods of Matrix Algebra*. New York: Academic Press, Inc., 1965.

Pedoe, D., *A Geometrical Introduction to Linear Algebra*. New York: John Wiley & Sons, Inc., 1963.

Perlis, S., *Theory of Matrices*. Reading, Mass.: Addison-Wesley Publishing Co., Inc., 1952.

PESTEL, E. C., and F. A. LECKIE, *Matrix Methods in Elastomechanics.* New York: McGraw-Hill Book Co., Inc., 1963.

PIPES, L., *Matrix Methods for Engineering.* Englewood Cliffs, N. J.: Prentice-Hall, Inc., 1962.

SCHOOL MATHEMATICS STUDY GROUP, *Introduction to Matrix Algebra,* preliminary edition. Ann Arbor, Mich.: Cushing-Malloy, Inc., 1960.

SCHREIER, O., and E. SPERNER, *Introduction to Modern Algebra and Matrix Theory,* tr. by M. Davis and M. Hausner. New York: Chelsea Publishing Co., 1951.

SCHWARTZ, J. T., *Introduction to Matrices and Vectors.* New York: McGraw-Hill Book Co., Inc., 1961.

SCHWERDTFEGER, H., *Introduction to Linear Algebra and the Theory of Matrices.* Groningen, Holland: P. Noordhoff, 1950.

SCOTT, R. F., and G. B. MATHEWS, *The Theory of Determinants and Their Applications.* New York: Cambridge University Press, Inc., 1904.

SHILOV, G. E., *An Introduction to the Theory of Linear Spaces,* tr. by R. A. Silverman. Englewood Cliffs, N. J.: Prentice-Hall, Inc., 1961.

SMILEY, M. F., *Algebra of Matrices.* Boston: Allyn and Bacon, Inc., 1965.

SMIRNOV, V. I., *Linear Algebra and Group Theory,* tr. by R. A. Silverman. New York: McGraw-Hill Book Co., Inc., 1961.

STIGANT, S. A., *The Elements of Determinants, Matrices and Tensors for Engineers.* New York: Dover Publications, Inc., 1964.

STOLL, R. R., *Linear Algebra and Matrix Theory.* New York: McGraw-Hill Book Co., Inc., 1952.

THRALL, R. M., and L. TORNHEIM, *Vector Spaces and Matrices.* New York: John Wiley & Sons, Inc., 1957.

TURNBULL, H. W., *Theory of Determinants, Matrices, and Invariants.* New York: Dover Publications, Inc., 1960.

————, and A. C. AITKEN, *An Introduction to the Theory of Canonical Matrices.* Dover Publications, Inc., 1961.

VARGA, R., *Matrix Iterative Analysis.* Englewood Cliffs, N. J.: Prentice-Hall, Inc., 1962.

WADE, T. L., *The Algebra of Vectors and Matrices.* Reading, Mass.: Addison-Wesley Publishing Co., Inc., 1951.

WEDDERBURN, J. H. M., *Lectures on Matrices.* Providence, R. I.: American Mathematical Society, 1934.

WELD, L. G., *Determinants.* New York: John Wiley & Sons, Inc., 1896.

YEFIMOV, N. V., *Quadratic Forms and Matrices, an Introductory Approach,* tr. by A. Shenitzer. New York: Academic Press, Inc., 1964.

ZELINGER, G., *Basic Matrix Algebra and Transistor Circuits.* New York: The Macmillan Company, 1963.

# INDEX

Addition of matrices, 17
  associative law, 18
  cancellation laws, 18
  commutative law, 18
Adjacent transposition, 110
Adjoint, 153
  $k$th, 156
Adjugate, 153
Affine transformation, 105
Albert, A. A., 54
Algebra, 304
  associative, 304
  commutative, 304
  Jordan, 54, 305
    matrix realization of, 305
  Lie, 305
    matrix realization of, 305
  linear, 304
  nonassociative, 304
  noncommutative, 304
  order of, 305
Algebraic complement, 136
Algebraic system, 304
$\alpha$ matrices, 11
Alternant, 126, 127
Alternating function, 127
Annihilation matrices, 11
Anticommutative matrices, 23
Argand matrices, 11
Associated matrices:
  of a conic, 104
  of a conicoid, 104

Associated matrices (*Contd.*):
  of a linear transformation, 300
Associative algebra, 304
Associative law:
  for addition of matrices, 18
  for Cayley multiplication of matrices, 22
  for scalar multiplication, 18
Augmented matrix, 86
Autonne, L., 271
  theorems of, 271

Basis of $k$-vector space over $F$, 292
Basis of vector space, 296
Baumert, L. D., 274
Beltrami, E., 272
Bessel function matrices, 11
$\beta$ matrices, 11
Bilinear form(s), 44, 234
  alternate, 234
  alternate Hermitian, 234
  cogredient transformation of, 234
  contragredient transformation of, 234
  equivalent, 45
  Hermitian, 234
  linear transformation of, 234
  symmetric, 234
Binet, J. P. M., 140$n$
Borel matrices, 11
Box matrix, 103

Box matrix (*Contd.*):
  depth layers, 103
  depth rank, 103
  length layers, 103
  length rank, 103
  width layers, 103
  width rank, 103
Brioschi, F., 268
Bruno, F. Faa di, 268

Cancellation law for addition of
  matrices, 18
Canonical matrix, 74
Catalan, E. C., 125$n$
Cauchy, A. L., 123$n$, 127, 140$n$,
  199
Cauchy's theorem, 156
Cayley, A., 10, 200, 265, 304
Cayley product, 20
Cellitti, C., 58
Characteristic determinant, 199
Characteristic equation, 199
Characteristic function, 199
Characteristic matrix, 199
Characteristic root, 199
  of a polynomial function of a
    matrix, 221
Characteristic vector, 207
Chio, F., 129
Chio's pivotal condensation process,
  130
Circulant, 125
Classification of conicoids, 106–
  107
Classification of conics, 106
Classification of pairs of lines,
  104
Classification of triples of planes,
  104
$\mathscr{C}$ matrix, 16
Cofactor, 119, 136
Cogredient transformation of bi-
  linear form, 234
Cojoint product of matrices, 159
Collar, A. R., 102$n$, 218
Column-equivalent matrices, 70
Column rank of a matrix, 77

Column space, 291
  of a matrix, 292
Column vector, 15
Communication matrices, 11
Commutative algebra, 304
Commutative law for addition of
  matrices, 18
Companion matrix, 263
Complement, 136
  algebraic, 136
Complementary subpermanent, 164
Complex matrix, 16
Complex number field, 277
Complex numbers, 7
  matrix representation of, 46
Complex polynomial (*See* Poly-
  nomial)
Condensation chain, 135
Condensation method, 129
Condensed matrix, 135
Conformable matrices, 20
Conformable partitioning, 39
Congruency over $\mathscr{C}$, 239
Congruency over $\mathscr{R}$, 239
Congruent matrices, 231
Conjugate matrices, 34
Conjugate permutations, 161
Conjunctive matrices, 242
Conkwright, N. B., 172
Continuant, 129, 172
Contragredient transformation of
  bilinear form, 234
Correlation matrices, 11
$\mathscr{C}$-orthogonal vectors, 254
$\mathscr{C}$-orthonormal system of vectors,
  254
Covariance matrices, 11
Cramer, G., 156
Cramer's rule, 156
Creation matrices, 11
Cross product of vectors, 55
Cyclic determinant, 125
Cyclomatic matrix, 12


Data matrices, 11
Degree of a $\lambda$ matrix, 185

Degree of a polynomial, 180
Depth layers, 103
Depth rank, 103
Derogatory matrix, 213
Determinant(s), 111
  alternant, 126, 127
  characteristic, 199
  Chio's pivotal condensation pro-
    cess for evaluating, 130
  circulant, 125
  continuant, 129, 172
  cyclic, 125
  elementary properties of, 114
  explicit definition of, 111
  inductive definition of, 121
  order of, 112
  orthogonant, 265
  plus, 163
  postulational definitions of, 165
  rank, 149
  sweep-out process for evaluating,
    166
  terms of, 112
  Vandermonde, 126
Diagonal matrix, 26
Diagonal of a matrix, 15
Dimension of $k$-vector space over
  $F$, 292
Dimension of vector space, 296
Directed incidence matrix, 5
Directed line segments, 8
Direct product of matrices, 107
Direct sum of matrices, 41, 108
Distributive laws for Cayley multiplication, 21, 22
Distributive laws for scalar multiplication, 18
Division algorithm for $\lambda$ matrices,
  186
Division algorithm for polynomi-
  als, 181
Division ring, 289
  matrix realization of, 289
Divisor:
  elementary, 196
  left, 187
  right, 187

Divisor (*Contd.*):
  trivial elementary, 196
Dominance matrices, 11
Dot product of vectors, $19n$
Duncan, W. J., $102n$, 218

ECT matrix, 71
Eigenvalue, 199
Eigenvector, 207
Elementary cogredient operation,
  231
Elementary column operations, 70
Elementary column transformation
  matrix, 71
Elementary conjunctive operation,
  243
Elementary divisor, 196
  trivial, 196
Elementary $\lambda$ matrix, 217
Elementary $\lambda$ operation, 191
Elementary row operations, 61
Elementary row transformation
  matrix, 63
Elements of a matrix, 15
Equality of matrices, 16
Equitable matrix, 274
Equivalence:
  $\lambda$, 191
  of pairs of matrices, 218
Equivalent matrices, 72
ERT matrix, 63
Euler-Knopp matrices, 11
Even permutation, 110

Factor, invariant, 196
Factor theorem for $\lambda$ matrices, 189
Factor theorem for polynomials,
  182
Fibonacci sequence, 172
Field:
  of characteristic 2, 282
  general, or abstract, 282
  multiplicative inverse in, 282
  number (*See* Number field)
  pseudo-, 289
Finck, P. J. E., 114
Frame, J. S., 221

Frazer, R. A., 102$n$, 218
Frobenius, G., 96$n$, 149, 195, 215, 221
Frobenius inequality, 96

$\gamma$ matrices, 11
Generalized Hadamard theorem, 177
Generalized Lagrange identity, 175
Gibbs, J. W., 10
Glaisher, J. W. L., 125$n$
Golomb, S. W., 274
Gram, J. P., 270
Gram matrix, 270, 271
Grassmann, H. G., 10, 304
Greatest common divisor, 183
Greenspan, D., 221
Group, 286
    Abelian, 286
    commutative, 286
    matrix realization of, 287
    non-Abelian, 286
    noncommutative, 286

Hadamard, J., 273
Hadamard matrix, 273
Hadamard's theorem, 177
Hamilton-Cayley equation, 202
Hamilton-Cayley polynomial, 202
Hamilton-Cayley theorem, 200
Hamilton, W. R., 10, 46, 200, 304
Hausdorff matrices, 11
Heide, J. D., 172
Heisenberg, W., 10
Hermite, C., 35$n$, 129
Hermitely congruent matrices, 242
Hermitian form, 46
Hermitian matrix, 35
    index of, 244
Hessian matrix, 11
Hidden root, 199
$h$-line, 161
Hohn, F. E., 220

Idempotent matrix, 27
Identity matrix, 25
Identity permutation, 162

Image under a mapping, 297
Improper $\lambda$ matrix, 185
Improperly orthogonal matrix, 247
Incidence matrices, 12, 164
Indefinite matrix, 269
Indefinite quadratic form, 269
Independence of vectors, 176, 178
Index, 210$n$
    of a Hermitian matrix, 244
    of a nilpotent matrix, 27
    of a quadratic form, 241
    of a real symmetric matrix, 240
Indicant of a matrix, 210
Infinite matrices, 11
Inner product of matrices, 158
Inner product of vectors, 19
Input-output matrix, 12
Integral number ring, 277
Invariant factor, 196
    trivial, 196
Inverse golden rule, 64
Inverse matrix, 65
Inverse permutation, 162
Inversion in a permutation, 110
Inversion of a matrix, 99, 154
    by Frame's recursion formula, 221
    by Hamilton-Cayley equation, 220
    by method of partitioning, 100
Involutoric matrix, 28
Isomorphism theorem, 301

Jacobian, 123$n$
Jacobian matrix, 11
Jacobi, C. G. J., 123$n$, 239$n$
Jacobi's identity, 55
Jacobi's theorem, 123
Jordan algebra, 54, 305
    matrix realization of, 305
Jordan, C., 272
Jordan, P., 54
Jordan product of matrices, 54

Kemeny, J. G., 48$n$
Kojima matrices, 11
Kronecker, L., 195, 235

Kronecker product, 107
*k*-stage route, 49
*k*th adjoint, 156
*k*-vector space over *F*, 291
  basis for, 292
  dimension of, 292
  spanned by a system of vectors, 292
  subspace, 291

λ equivalence, 191
λ matrix, 184
  degree of, 185
  elementary, 217
  improper, 185
  leading coefficient of, 185
  nonsingular, 185
  proper, 185
  rank of, 191
  singular, 185
Laplace expansion theorem, 137
  for permanents, 164
Laplace, P. S., 136
Latent root, 199
Latent vector, 207
Leading coefficient of a λ matrix, 185
Leading coefficient of a polynomial, 180
Leading principal minors of a matrix, 264
Left division, 187
Left divisor, 187
Left functional value, 187
Left inverse, 98
Left quotient, 187
Left remainder, 187
Length layers, 103
Length rank, 103
Le Roy matrices, 11
Lie algebra, 54, 305
  matrix realization of, 305
Lie, M. S., 54
Lie product of matrices, 54
Lindelöf matrices, 11
Linear algebra, 304
Linear combination of vectors, 76

Linear dependence of vectors, 174
Linear independence of vectors, 174
Linearly dependent vectors, 76
Linearly independent vectors, 76
Linear space, 294
Linear transformation(s), 9, 43, 298
  associated matrices of, 300
  of bilinear form, 234
  product of, 300
  sum of, 300
(L,L) system, 287
Loop, 288
  associative, 288
  matrix realization of, 289
  nonassociative, 289
Lotkin, M., 102*n*
Lower diagram of a permutation, 160
(L,R) system, 287
  matrix realization of, 287

MacDuffee, C. C., 165
Mapping, 297
  image under, 297
Matric-polynomial representation, 185
Matrix (matrices):
  addition of (*See* Addition of matrices)
  adjoint of, 153
  adjugate of, 153
  *α*, 11
  annihilation, 11
  anticommutative, 23
  Argand, 11
  associated (of a conic), 104
  associated (of a conicoid), 104
  associated with a linear transformation, 300
  augmented, 86
  Bessel function, 11
  *β*, 11
  Borel, 11
  box (*See* Box matrix)
  𝒞, 16

Matrix (matrices) (*Contd.*):
  canonical, 74
  Cayley product of, 20
  characteristic, 199
  cojoint product of, 159
  column-equivalent, 70
  column rank of, 77
  column space of, 292
  communication, 11
  companion, 263
  complex, 16
  condensed, 135
  conformable, 20
  conformable partitioning of, 39
  congruent, 231
  congruent over $\mathscr{C}$, 239
  congruent over $\mathscr{R}$, 239
  conjugate, 34
  conjunctive, 242
  correlation, 11
  covariance, 11
  creation, 11
  cyclomatic, 12
  data, 11
  derogatory, 213
  diagonal, 26
  diagonal of, 15
  directed incidence, 5
  direct product of, 107
  direct sum of, 41, 108
  dominance, 11
  ECT, 71
  elementary column operations on, 70
  elementary column transformation, 71
  elementary row operations on, 61
  elementary row transformation, 63
  elements of, 15
  equality of, 16
  equitable, 274
  equivalent, 72
  ERT, 63
  Euler-Knopp, 11
  $\gamma$, 11

Matrix (matrices) (*Contd.*):
  Gram, 270, 271
  Hadamard, 273
  Hausdorff, 11
  Hermitely congruent, 242
  Hermitian, 35
  Hessian, 11
  idempotent, 27
  identity, 25
  improperly orthogonal, 247
  incidence, 12, 164
  indefinite, 269
  index of Hermitian, 244
  index of nilpotent, 27
  index of real symmetric, 240
  indicant of, 210
  infinite, 11
  inner product of, 158
  input-output, 12
  inverse, 65, 98
  inversion, 99, 100, 154, 220, 221
  involutoric, 28
  Jacobian, 11
  Jordan product of, 54
  Kojima, 11
  Kronecker product of, 107
  $\lambda$ (*See* $\lambda$ matrix)
  left inverse, 98
  Le Roy, 11
  Lie product of, 54
  Lindelöf, 11
  mechanics, 11
  Mittag-Leffler, 11
  modal, 208
  models-parts, 6
  models-subassemblies, 6
  modified triangular form, 61
  multiplication of (*See* Multiplication of matrices)
  negative definite, 269
  negative semidefinite, 269
  nilpotent, 27
  nonsingular, 65
  nonsingular on the left, 98
  nonsingular on the right, 98
  Nörlund, 11

Matrix (matrices) (*Contd.*):
  normal, 259
  nullity of, 84
  order of, 15
  orthogonal, 246
  orthogonally congruent, 249
  orthogonally similar, 249
  over $S$, 278
  partitioned, 37
  Pauli spin, 23
  payoff, 12
  permutation, 163
  positive definite, 269
  positive semidefinite, 269
  properly orthogonal, 247
  quasi-scalar, 26
  $\mathcal{R}$,16
  Raff, 11
  rank of, 81
  real, 16
  realization:
    of division ring, 289
    of group, 287
    of Jordan algebra, 305
    of Lie algebra, 305
    of (L,R) system, 287
    of nonassociative loop, 289
    of (R,L) system, 288
  reduced echelon form, 61
  regular, 264
  right inverse, 98
  roots of, 219
  row-equivalent, 61
  row rank of, 77
  row space of, 292
  $S$, 15
  scalar, 26
  scalar multiplication of (*See* Scalar multiplication of matrices)
  scalar product of, 175, 177
  scattering, 11
  signature of, 241
  similar, 223
  singular, 65
  skew-Hermitian, 35
  skew-symmetric, 31

Matrix (matrices) (*Contd.*):
  spin, 11
  spur of, 18
  square, 15
  square roots of, 56
  stochastic, 11
  strain, 11
  stress, 11
  subassemblies-parts, 6
  submatrix of, 37
  sum of, 17
  symmetric, 31
  tensor product of, 107
  Toeplitz, 11
  trace of, 18
  tranjugate of, 34
  transpose of, 31
  triangular, 123
  undirected incidence, 5
  unitarily conjunctive, 256
  unitarily similar, 256
  unitary, 254
  upper, 28
  upper triangular, 28
  zero, 17
Minimum function, 211
Minor, 120, 136
  $m$th, 136
  principal, 151
  zeroth, 136
Mittag-Leffler matrices, 11
Modal matrix, 208
Models-parts array, 6
Models-subassemblies array, 6
Modified triangular form, 61
Monic polynomial, 180
$m$-plane, 292
$m$th minor, 136
Multiplication of matrices:
  Cayley product, 20
  cojoint product, 159
  direct product, 107
  inner product, 158
  Jordan product, 54
  Kronecker product, 107
  Lie product, 54
  scalar product, 175, 177

Multiplication of matrices
    (*Contd.*):
    tensor product, 107
    *m*-volume, 176

Negative definite matrix, 269
Negative definite quadratic form, 269
Negative semidefinite matrix, 269
Negative semidefinite quadratic form, 269
Neumann, J. von, 291
Nilpotent matrix, 27
    index of, 27
Nonassociative algebra, 54, 304
Noncommutative algebra, 304
Nonsingular collineation, 105
Nonsingular $\lambda$ matrix, 185
Nonsingular on the left, 98
Nonsingular matrix, 65
Nonsingular on the right, 98
Nörlund matrices, 11
Normal matrix, 259
Normal vector, 247
Nullity of a matrix, 84
Number field, 276
    algebraically closed, 278
    closed under conjugation, 278
    complex, 277
    generated by $a_1, \ldots, a_n$, 280
    rational, 277
    real, 277
Number ring, 276
    closed under conjugation, 278
    of Gaussian integers, 277
    integral, 277
    trivial, 277
    with unity, 277

Odd permutation, 110
One-sided inverses, 98
Order of an algebra, 305
Order of a determinant, 112
Order of a matrix, 15
Orthogonally congruent matrices, 249
Orthogonally similar matrices, 249

Orthogonal matrix, 246
    improperly, 247
    properly, 247
Orthogonal transformation, 253
Orthogonal vectors, 247
Orthogonant, 265
Orthonormal system of vectors, 248
Outer product of vectors, 55

Parker, F. D., 51*n*, 274, 274*n*
Partitioned matrices, 37
Pauli spin matrices, 23
Payoff matrix, 12
Perlis, S., 218
Permanent, 163
Permutation(s), 109
    adjacent transposition, 110
    conjugate, 161
    even, 110
    geometric study of, 160
    identity, 162
    inverse of, 162
    inversion in, 110
    lower diagram of, 160
    matrices, 163
    odd, 110
    product of, 162
    self-conjugate, 161
    transposition, 110
    upper diagram of, 160
Pfaffian, 169
    order of, 169
Pfaff, J. F., 169
Plus determinant, 163
Polynomial(s), 180
    degree of, 180
    of degree minus infinity, 180
    of degree zero, 180
    division algorithm for, 181
    equality of, 180
    greatest common divisor of, 183
    Hamilton-Cayley, 202
    leading coefficient of, 180
    monic, 180
    reciprocal, 267
    remainder theorem for, 182
    virtual degree of, 180

Polynomial(s) (*Contd.*):
  virtual leading coefficient of, 180
  zero, 180
Positive definite matrix, 269
Positive definite quadratic form, 269
Positive semidefinite matrix, 269
Positive semidefinite quadratic form, 269
Postmultiplication, 20
Premultiplication, 20
Principal minor, 151
Principle of duality of determinant theory, 115
Product of linear transformations, 300
Product of matrices (*See* Multiplication of matrices)
Product of permutations, 162
Proper divisor of zero, 30
Proper $\lambda$ matrix, 185
Properly orthogonal matrix, 247
Proper root, 199
Proper subfield, 278
Proper subring, 277
Proper vector, 207
Pseudo-field, 289

Quadratic form(s), 45
  congruent, 46
  indefinite, 269
  index of, 241
  negative definite, 269
  negative semidefinite, 269
  positive definite, 269
  positive semidefinite, 269
  rank of, 241
Quasi-scalar matrix, 26
Quotient, 182
  left, 187
  right, 187

Raff matrices, 11
Rank:
  column, 77
  depth, 103
  determinant, 149

Rank (*Contd.*):
  of a $\lambda$ matrix, 191
  length, 103
  of a matrix, 81
  of a quadratic form, 241
  row, 77
  width, 103
Rational number field, 277
Real matrix, 16
Real number field, 277
Reciprocal polynomial, 267
Reduced echelon form, 61
Regular matrix, 264
Regular ring, 291
Remage, R., 102$n$
Remainder, 182
  left, 187
  right, 187
Remainder theorem for $\lambda$ matrices, 188
Remainder theorem for polynomials, 182
Right division, 187
Right divisor, 187
Right functional value, 187
Right inverse, 98
Right quotient, 187
Right remainder, 187
Ring:
  Abelian, 281
  additive inverse in, 281
  commutative, 281
  division, 289
  of Gaussian integers, 277
  general, or abstract, 281
  left unity element of, 282
  number (*See* Number ring)
  regular, 291
  right unity element of, 282
  with unity, 282
  unity element of, 282
  zero element of, 281
(R,L) system, 287
  matrix realization of, 288
$\mathscr{R}$ matrix, 16
Roots of matrices, 219
Rotations in 3-space, 264

Row-equivalent matrices, 61
Row rank of a matrix, 77
Row space, 291
Row space of a matrix, 292
Row vector, 15
Royal, J. W., 175
(R,R) system, 287

Sarrus, P. F., 114
Sarrus rule, 113
Scalar, 15
Scalar matrix, 26
Scalar multiplication, 17
 associative law, 18
 distributive laws, 18
Scalar product, 17
Scalar product of complex vectors, 177
Scalar product of matrices, 175, 177
Scalar product of vectors, $19n$
Scattering matrix, 11
Schur's theorem, 259
Schwartz, J. T., $56n$
Scott, R. F., $125n$
Secular value, 199
Self-conjugate permutation, 161
Sfield, 289
Signature of a matrix, 241
Similar matrices, 223
Singular $\lambda$ matrix, 185
Singular matrix, 65
Skew-Hermitian matrix, 35
Skew-symmetric matrix, 31
Skew-symmetric matrix of a vector, 55
$S$ matrix, 15
Smith, H. J. S., 191, 195
Smith normal form, 194
Snell, J. L., $48n$
Spin matrices, 11
Spottiswoode, W., $125n$
Spur, 18
Square matrix, 15
Square roots of matrices, 56
Stochastic matrices, 11
Stoll, R. R., 165

Strain matrices, 11
Stress matrices, 11
Subassemblies-parts array, 6
Subfield, 278
 proper, 278
Submatrix, 37
Subpermanent, 164
 complementary, 164
Subring, 277
 proper, 277
Sum of linear transformations, 300
Sum of matrices, 17
Sweep-out process, 166
Sylvester, J. J., 10, $84n$, $123n$, $129n$, 178, 195, 215, $239n$, 272
Sylvester's dialytic method of elimination, 178
Sylvester's law of inertia, 239
Sylvester's law of nullity, 84
Symmetric matrix, 31
System of distinct representatives, 164
System(s) of linear equations, 86
 augmented matrix of, 86
 consistent, 87
 equivalent, 87
 extension of solutions, 96
 homogeneous, 89
  trivial solution of, 89
 inconsistent, 87
 linearly independent solutions of, 92
 matrix of, 86
 nonhomogeneous, 89
 solution of, 86, 169

Tensor product, 107
Theorem(s):
 Autonne's, 271
 Cauchy's, 156
 Cramer's rule, 157
 division algorithm for $\lambda$ matrices, 186
 division algorithm for polynomials, 181
 factor, for $\lambda$ matrices, 189
 factor, for polynomials, 182

Theorem(s) (*Contd.*) :
  generalized Hadamard, 177
  generalized Lagrange identity, 175
  Hadamard's, 177
  Hamilton-Cayley, 200
  isomorphism, 301
  Jacobi's, 123
  Laplace expansion, 137
  Laplace's, for permanents, 164
  remainder, for $\lambda$ matrices, 188
  remainder, for polynomials, 182
  Schur's, 259
  Sylvester's law of inertia, 239
  Sylvester's law of nullity, 84
  Toeplitz's, 263
Thompson, G. L., 48$n$
Toeplitz matrices, 11
Toeplitz's theorem, 261
Trace, 18
Tranjugate matrix, 34
Transposed matrix, 31
Transposition, 110
  adjacent, 110
Triangular matrix, 123
Trivial elementary divisor, 196
Trivial invariant factor, 196
Trivial number ring, 277

Undirected incidence matrix, 5
Unitarily conjunctive matrices, 256
Unitarily similar matrices, 256
Unitary matrix, 254
Unit vectors, 79
Upper diagram of a permutation, 160
Upper matrix, 28
Upper triangular matrix, 28

Vandermonde, A. T., 126$n$, 127
Vandermonde determinant, 126
Vector(s) :
  characteristic, 207
  column, 15
  $\mathscr{C}$-orthogonal, 254

Vector(s) (*Contd.*) :
  $\mathscr{C}$-orthonormal system of, 254
  cross product of, 55
  dot product of, 19$n$, 55
  independence of, 176, 178
  inner product of, 19, 55
  latent, 207
  linear combination of, 76
  linear dependence of, 174
  linear independence of, 174
  linearly dependent, 76
  linearly independent, 76
  normal, 247
  of a skew-symmetric matrix, 55
  orthogonal, 247
  orthonormal system of, 248
  outer product of, 55
  product, 8, 55
  proper, 207
  row, 15
  scalar product of, 19$n$, 55
  space, 294
    basis of, 296
    dimension of, 296
    spanned by a set of vectors, 296
  unit, 79
Virtual degree of a polynomial, 180
Virtual leading coefficient of a polynomial, 180

Wallace, A. D., 159
Webster, N., 282$n$
Wedderburn, J. H. M., 10
Weierstrass, K., 165, 195
Weisner, L., 278
Wellstein, J., 58
Width layers, 103
Width rank, 103

Zero formulas, 122
Zero matrix, 17
Zero polynomial, 180
Zeroth minor, 136

A CATALOGUE OF SELECTED DOVER BOOKS
IN ALL FIELDS OF INTEREST

# A CATALOGUE OF SELECTED DOVER

# BOOKS IN ALL FIELDS OF INTEREST

CELESTIAL OBJECTS FOR COMMON TELESCOPES, T. W. Webb. The most used book in amateur astronomy: inestimable aid for locating and identifying nearly 4,000 celestial objects. Edited, updated by Margaret W. Mayall. 77 illustrations. Total of 645pp. 5⅜ x 8½.
20917-2, 20918-0 Pa., Two-vol. set $8.00

HISTORICAL STUDIES IN THE LANGUAGE OF CHEMISTRY, M. P. Crosland. The important part language has played in the development of chemistry from the symbolism of alchemy to the adoption of systematic nomenclature in 1892. ". . . wholeheartedly recommended,"—Science. 15 illustrations. 416pp. of text. 5⅝ x 8¼. 63702-6 Pa. $6.00

BURNHAM'S CELESTIAL HANDBOOK, Robert Burnham, Jr. Thorough, readable guide to the stars beyond our solar system. Exhaustive treatment, fully illustrated. Breakdown is alphabetical by constellation: Andromeda to Cetus in Vol. 1; Chamaeleon to Orion in Vol. 2; and Pavo to Vulpecula in Vol. 3. Hundreds of illustrations. Total of about 2000pp. 6⅛ x 9¼.
23567-X, 23568-8, 23673-0 Pa., Three-vol. set $26.85

THEORY OF WING SECTIONS: INCLUDING A SUMMARY OF AIR-FOIL DATA, Ira H. Abbott and A. E. von Doenhoff. Concise compilation of subatomic aerodynamic characteristics of modern NASA wing sections, plus description of theory. 350pp. of tables. 693pp. 5⅜ x 8½.
60586-8 Pa. $6.50

DE RE METALLICA, Georgius Agricola. Translated by Herbert C. Hoover and Lou H. Hoover. The famous Hoover translation of greatest treatise on technological chemistry, engineering, geology, mining of early modern times (1556). All 289 original woodcuts. 638pp. 6¾ x 11.
60006-8 Clothbd. $17.50

THE ORIGIN OF CONTINENTS AND OCEANS, Alfred Wegener. One of the most influential, most controversial books in science, the classic statement for continental drift. Full 1966 translation of Wegener's final (1929) version. 64 illustrations. 246pp. 5⅜ x 8½.    61708-4 Pa. $3.00

THE PRINCIPLES OF PSYCHOLOGY, William James. Famous long course complete, unabridged. Stream of thought, time perception, memory, experimental methods; great work decades ahead of its time. Still valid, useful; read in many classes. 94 figures. Total of 1391pp. 5⅜ x 8½.
20381-6, 20382-4 Pa., Two-vol. set $13.00

YUCATAN BEFORE AND AFTER THE CONQUEST, Diego de Landa. First English translation of basic book in Maya studies, the only significant account of Yucatan written in the early post-Conquest era. Translated by distinguished Maya scholar William Gates. Appendices, introduction, 4 maps and over 120 illustrations added by translator. 162pp. 5⅜ x 8½.
23622-6 Pa. $3.00

THE MALAY ARCHIPELAGO, Alfred R. Wallace. Spirited travel account by one of founders of modern biology. Touches on zoology, botany, ethnography, geography, and geology. 62 illustrations, maps. 515pp. 5⅜ x 8½.
20187-2 Pa. $6.95

THE DISCOVERY OF THE TOMB OF TUTANKHAMEN, Howard Carter, A. C. Mace. Accompany Carter in the thrill of discovery, as ruined passage suddenly reveals unique, untouched, fabulously rich tomb. Fascinating account, with 106 illustrations. New introduction by J. M. White. Total of 382pp. 5⅜ x 8½. (Available in U.S. only)　23500-9 Pa. $4.00

THE WORLD'S GREATEST SPEECHES, edited by Lewis Copeland and Lawrence W. Lamm. Vast collection of 278 speeches from Greeks up to present. Powerful and effective models; unique look at history. Revised to 1970. Indices. 842pp. 5⅜ x 8½.　20468-5 Pa. $6.95

THE 100 GREATEST ADVERTISEMENTS, Julian Watkins. The priceless ingredient; His master's voice; 99 44/100% pure; over 100 others. How they were written, their impact, etc. Remarkable record. 130 illustrations. 233pp. 7⅞ x 10 3/5.　20540-1 Pa. $5.00

CRUICKSHANK PRINTS FOR HAND COLORING, George Cruickshank. 18 illustrations, one side of a page, on fine-quality paper suitable for watercolors. Caricatures of people in society (c. 1820) full of trenchant wit. Very large format. 32pp. 11 x 16.　23684-6 Pa. $4.50

THIRTY-TWO COLOR POSTCARDS OF TWENTIETH-CENTURY AMERICAN ART, Whitney Museum of American Art. Reproduced in full color in postcard form are 31 art works and one shot of the museum. Calder, Hopper, Rauschenberg, others. Detachable. 16pp. 8¼ x 11.
23629-3 Pa. $2.50

MUSIC OF THE SPHERES: THE MATERIAL UNIVERSE FROM ATOM TO QUASAR SIMPLY EXPLAINED, Guy Murchie. Planets, stars, geology, atoms, radiation, relativity, quantum theory, light, antimatter, similar topics. 319 figures. 664pp. 5⅜ x 8½.
21809-0, 21810-4 Pa., Two-vol. set $10.00

EINSTEIN'S THEORY OF RELATIVITY, Max Born. Finest semi-technical account; covers Einstein, Lorentz, Minkowski, and others, with much detail, much explanation of ideas and math not readily available elsewhere on this level. For student, non-specialist. 376pp. 5⅜ x 8½.
60769-0 Pa. $4.00

ART FORMS IN NATURE, Ernst Haeckel. Multitude of strangely beautiful natural forms: Radiolaria, Foraminifera, jellyfishes, fungi, turtles, bats, etc. All 100 plates of the 19th-century evolutionist's *Kunstformen der Natur* (1904). 100pp. 9⅜ x 12¼. 22987-4 Pa. $4.50

CHILDREN: A PICTORIAL ARCHIVE FROM NINETEENTH-CENTURY SOURCES, edited by Carol Belanger Grafton. 242 rare, copyright-free wood engravings for artists and designers. Widest such selection available. All illustrations in line. 119pp. 8⅜ x 11¼. 23694-3 Pa. $3.50

WOMEN: A PICTORIAL ARCHIVE FROM NINETEENTH-CENTURY SOURCES, edited by Jim Harter. 391 copyright-free wood engravings for artists and designers selected from rare periodicals. Most extensive such collection available. All illustrations in line. 128pp. 9 x 12. 23703-6 Pa. $4.00

ARABIC ART IN COLOR, Prisse d'Avennes. From the greatest ornamentalists of all time—50 plates in color, rarely seen outside the Near East, rich in suggestion and stimulus. Includes 4 plates on covers. 46pp. 9⅜ x 12¼. 23658-7 Pa. $6.00

AUTHENTIC ALGERIAN CARPET DESIGNS AND MOTIFS, edited by June Beveridge. Algerian carpets are world famous. Dozens of geometrical motifs are charted on grids, color-coded, for weavers, needleworkers, craftsmen, designers. 53 illustrations plus 4 in color. 48pp. 8¼ x 11. (Available in U.S. only) 23650-1 Pa. $1.75

DICTIONARY OF AMERICAN PORTRAITS, edited by Hayward and Blanche Cirker. 4000 important Americans, earliest times to 1905, mostly in clear line. Politicians, writers, soldiers, scientists, inventors, industrialists, Indians, Blacks, women, outlaws, etc. Identificatory information. 756pp. 9¼ x 12¾. 21823-6 Clothbd. $40.00

HOW THE OTHER HALF LIVES, Jacob A. Riis. Journalistic record of filth, degradation, upward drive in New York immigrant slums, shops, around 1900. New edition includes 100 original Riis photos, monuments of early photography. 233pp. 10 x 7⅞. 22012-5 Pa. $6.00

NEW YORK IN THE THIRTIES, Berenice Abbott. Noted photographer's fascinating study of city shows new buildings that have become famous and old sights that have disappeared forever. Insightful commentary. 97 photographs. 97pp. 11⅜ x 10. 22967-X Pa. $4.50

MEN AT WORK, Lewis W. Hine. Famous photographic studies of construction workers, railroad men, factory workers and coal miners. New supplement of 18 photos on Empire State building construction. New introduction by Jonathan L. Doherty. Total of 69 photos. 63pp. 8 x 10¾. 23475-4 Pa. $3.00

# CATALOGUE OF DOVER BOOKS

GEOMETRY, RELATIVITY AND THE FOURTH DIMENSION, Rudolf Rucker. Exposition of fourth dimension, means of visualization, concepts of relativity as Flatland characters continue adventures. Popular, easily followed yet accurate, profound. 141 illustrations. 133pp. 5⅜ x 8½.
23400-2 Pa. $2.75

THE ORIGIN OF LIFE, A. I. Oparin. Modern classic in biochemistry, the first rigorous examination of possible evolution of life from nitrocarbon compounds. Non-technical, easily followed. Total of 295pp. 5⅜ x 8½.
60213-3 Pa. $4.00

THE CURVES OF LIFE, Theodore A. Cook. Examination of shells, leaves, horns, human body, art, etc., in *"the* classic reference on how the golden ratio applies to spirals and helices in nature . . . . "—Martin Gardner. 426 illustrations. Total of 512pp. 5⅜ x 8½.
23701-X Pa. $5.95

PLANETS, STARS AND GALAXIES, A. E. Fanning. Comprehensive introductory survey: the sun, solar system, stars, galaxies, universe, cosmology; quasars, radio stars, etc. 24pp. of photographs. 189pp. 5⅜ x 8½. (Available in U.S. only)
21680-2 Pa. $3.00

THE THIRTEEN BOOKS OF EUCLID'S ELEMENTS, translated with introduction and commentary by Sir Thomas L. Heath. Definitive edition. Textual and linguistic notes, mathematical analysis, 2500 years of critical commentary. Do not confuse with abridged school editions. Total of 1414pp. 5⅜ x 8½. 60088-2, 60089-0, 60090-4 Pa., Three-vol. set $18.00

DIALOGUES CONCERNING TWO NEW SCIENCES, Galileo Galilei. Encompassing 30 years of experiment and thought, these dialogues deal with geometric demonstrations of fracture of solid bodies, cohesion, leverage, speed of light and sound, pendulums, falling bodies, accelerated motion, etc. 300pp. 5⅜ x 8½.
60099-8 Pa. $4.00

*Prices subject to change without notice.*

Available at your book dealer or write for free catalogue to Dept. GI, Dover Publications, Inc., 180 Varick St., N.Y., N.Y. 10014. Dover publishes more than 175 books each year on science, elementary and advanced mathematics, biology, music, art, literary history, social sciences and other areas.